AHRC Research Centre for Textile Conservation and Textile Studies

SECOND ANNUAL CONFERENCE

The Future of the 20th Century: Collecting, Interpreting and Conserving Modern Materials

POSTPRINTS

AHRC Research Centre for Textile Conservation and Textile Studies

SECOND ANNUAL CONFERENCE 26–28 July 2005

The Future of the 20th Century: Collecting, Interpreting and Conserving Modern Materials

POSTPRINTS

Edited by Cordelia Rogerson and Paul Garside

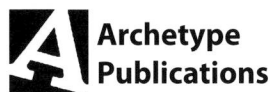

First published 2006 by Archetype Publications Ltd.

Archetype Publications Ltd.
6 Fitzroy Square
London W1T 5HJ

www.archetype.co.uk

Tel: 44(207) 380 0800
Fax: 44(207) 380 0500

© Copyright is held jointly among the authors and Archetype Publications

ISBN 1-904982-17-4

British Library Cataloguing in Publication Data
A catalogue record for this book is available from the British Library.

The views and practices expressed by individual authors are not necessarily those of the editors or the publisher.

All rights reserved. No part of this publication may be reproduced, stored in a retrieval system, or transmitted, in any form or by any means, electronic, mechanical, photocopying, recording or otherwise, without the prior permission of the publisher.

Scentsory Design, Ventolin, Lycra, Nomex, LEGO, Tinytalk, Melinex, Correx, Tyvek, Plastazote, Ethafoam, Ageless, Pyrex, Cryovac BDF-200, Vilene, Lycra and Dorlastan are registered trademarks

Printed on acid-free paper

Typeset by Kate Williams, Swansea
Printed and bound by Gutenberg Press Limited, Malta

Contents

Foreword — vii
Cordelia Rogerson and Paul Garside

Creating and interpreting objects

Scentsory Design: the emotional living tissue — 3
Jenny Tillotson

Can an artist create permanence from transience? The Schmuck Quickies of Yuka Oyama become durable — 11
Cordelia Rogerson and James Beighton

Interpreting the woven devoré textile — 18
Andie Robertson

What makes a textile modern? The recycling of clothing in the Punjabi shoddy trade — 24
Lucy Norris

Collecting modern textile materials

In pursuit of forgotten fibres? The development, disappearance and rediscovery of regenerated protein fibres — 33
Mary M. Brooks

'A bomb in the collection': researching and exhibiting early 20th-century fashion — 41
Alexandra Palmer

Early elastic threads and fibres in clothing — 48
Laura Petzold

Material challenges

Identifying modern materials: taking it to the collection — 55
Paul Garside and Paul Wyeth

Man-made fibres from polypropylene to works of art 61
Thea van Oosten, Ineke Joosten and Luc Megens

Probing the microstructure of protein and polyamide fibres 67
Paul Garside and Mary M. Brooks

Investigating cellulose nitrate degradation caused by fungal attack 72
Margarida Silva

Polyurethane foam: investigating the physical and chemical consequences of degradation 77
Paul Garside and Doon Lovett

Sticky oilskins and stiffened rubber: new challenges for textile conservation 84
Irene Skals and Yvonne R. Shashoua

The effect of acid dyes on the photodegradation of knitted nylon conservation support net 92
M.K. Sinha, R.M. Christie and R. Shamey

Freezing the present to preserve the future 100
Yvonne R. Shashoua

The pits of despair? A preliminary study of the occurrence and deterioration of rubber dress shields 107
Anna Hodson

Conservation applications: object studies

A global challenge: the search for conservation solutions for Eero Aarnio's Globe/Ball chair 117
Joelle Wickens

A study of sequins on a Cantonese opera stage curtain 122
Angela Cheung

Wet look in 1960s furniture design: degradation of polyurethane-coated textile carrier substrates 128
Tim Bechthold

Storage issues for contemporary textile art: a solution for one example 134
Rosemary Baker

Television puppets from the 1960s and 1970s: creation, materials and conservation 137
Rebecca Smith

The treatment of the light-damaged nylon component of a flight suit used 139
during the test flights of Concorde c.1968
Anna Hodson

Modern textile materials in practice at the State Hermitage Museum 142
Elena Mikolaychuk

Foreword

Modern materials, whether as art or everyday objects, are the basis of the contemporary material world. Accordingly objects encountered within museums and collections increasingly represent a broad spectrum of materials whose preservation may be without precedent. The conference considers the conservation of modern materials in the textile field as a subject in its own right with topics ranging from familiar textile types, such as costume, to more unusual applications in suitcases, wall hangings, furniture and theatre scenery; other papers prompt the reader to reconsider what makes a textile modern. Delegates in attendance represented 18 countries and hence indicate a truly international interest in this diverse and significant discipline.

Written accounts of the vast majority of the oral and poster presentations at the conference are presented here. The papers are structured to transport the reader from the creation and interpretation of artworks into the realm of understanding their deterioration pathways, to arrive at a number of considerations for their treatment. Presenting humanities-based papers alongside those with a more scientific content, and the words of artists together with those of curators and conservators, is a deliberate strategy to represent the wide spectrum of subjects involved in the preservation of modern textile materials.

While hailing the conservation profession as a multidisciplinary practice is not original, the following papers truly demonstrate the different perspectives and approaches that may be taken when considering the treatment of artefacts.

Each paper was peer reviewed and the contribution made by these anonymous individuals is gratefully acknowledged. Credit is also due to the authors who were subjected to the arduous editing process and for their good nature in accepting and discussing any amendments that were suggested.

The efficient running of the conference is testament of the attention to detail and foresight of the organising committee including Chris Bennett, Maria Hayward and Nell Hoare. Indispensable financial support was provided by the Arts and Humanities Research Council, the British Academy and Willards Developments Ltd. Thanks are also due to Mike Halliwell for his technical support, and to the student helpers in attendance throughout the proceedings: Konstantinos Hatziantoniou, Wendy Hickson, Bernice Morris, Rachel Rhodes and Branwen Roberts. Finally, the session chairs, Mary Brooks, Dinah Eastop, Yvonne Shashoua, Anita Quye and Lynda Hillyer are thanked for their input and commentary and, importantly, for ensuring the speakers were thorough but succinct.

Cordelia Rogerson, Paul Garside
March 2006

Creating and interpreting objects

Scentsory Design: the emotional living tissue

Jenny Tillotson

ABSTRACT This paper explores Scentsory Design: responsive fabrics that go beyond passive microencapsulated techniques through the inclusion of microfluidic scent delivery systems that sense and respond to psychological and environmental changes, in order to enhance human wellbeing, avoid skin allergies and prevent insect-borne diseases. Fashion is about displaying personal identity information. Scentsory Design is about creating a scent bubble around the user, enhancing the visual message of fashion with medical, sensory and psychological wellbeing. By adding aroma to the fashion domain, it is possible to create smart fabrics with radical, active properties, which support current colour and shape-changing electro-textile research. In this paper, the development of a collection of responsive brooches is described that dispense fragrances, triggered by sensors that react to an individual's body state and the environment. The paper concludes by proposing fabric as an emotional living tissue that resembles human skin, offering social and therapeutic value in a desirable context, i.e. clothes that reduce anxiety and depression, prevent mosquito bites and replace the use of alcohol as a binder.

Keywords: scent, multisensory, microfluidics, wellbeing, electro-textiles, smart second skin, aroma-chology

Introduction

This study describes the invention of a new method of aroma delivery, funded by an Innovation Award from the Arts and Humanities Research Council (AHRC). In collaboration with analytical chemists Professor Andreas Manz (pioneer of lab-on-a-chip) and Dr Gareth Jenkins from the Institute of Analytical Sciences in Germany, small microfluidic devices were implanted into responsive jewellery. The study also investigates the extent to which microfluidics embedded in an emotional living tissue can deliver fragrances inspired by the quote: 'Fashion is the recognition that nature has endowed us with one skin too few, that a fully sentient being should wear its nervous system externally' (Ballard 1997). The purpose of the fabrics is to reduce the percentage of alcohol in perfume and to improve quality of life, not only to benefit human wellbeing through olfaction stimulation of the autonomic nervous system, but as a novel communication system to send an aroma message that could be informative, protective, seductive or healing.

Furthermore, the paper serves to introduce a wider audience of curators, cultural historians and conservators to the artistic intent and development process of innovative products that use unconventional construction techniques and combinations of materials. Not only does the creative process reveal the multidisciplinary collaborations necessary to develop the work but also its evolution mirrors the equally multidisciplinary approach that will be needed to preserve the items for the future. Such interactive objects present curatorial and conservation problems hitherto unknown when they are displayed, since the objects are unique in conception and construction. For example, smell is a sensation not normally exercised in visitors by museum artefacts and adds a dimension to exhibits not often encountered by curators and conservators. In all probability, elements of Scentsory Design research will continue to be exhibited and preserved in collections. Documenting the development process will, therefore, elucidate why and how Scentsory Design originated and functions, so that it can be interpreted appropriately now and for the future.

Background

Evolving from the PhD work on smart second skin at the Royal College of Art (RCA), London, in 1997 and BA fashion communication work on multisensorial surfaces at Central Saint Martins College of Art and Design, London, in 1991, this research straddles the science/art boundaries. It bridges the disciplines of nanotechnology, analytical chemistry, perfumery, electrical engineering, aroma-chology, fashion, textiles and neuroscience. The PhD described a multisensorial approach to biomedical designs, recognising that all senses interact. A smart second skin fabric is a membrane of micro-tubes fused together with microfluidics, electronics and yarns embedded in clothing, to create a scent delivery system that adds function to fashion. At the RCA, the research was demonstrated by an interactive installation that pulsed coloured fluid around a transparent sculpture to illustrate colodours: colour therapeutic scent delivery for different emotions and moods. The spectrum of scents gave the impression the sculpture was creating an olfactory experience and was a direct reference to the aroma rainbow emitted from the Scent Organ in the novel *Brave New World* (Huxley 1932): 'The Scent Organ was playing a delightfully refreshing herbal capriccio – ripping arpeggios of thyme and lavender, rosemary, basil, myrtle, tarragon, a series of daring modulations through the spice keys into ambergris and a slow return through sandalwood, camphor, cedar, and new mown grass.'

Smart second skin and touch

In this context, the membrane is analogous to the body and human skin thereby facilitating interaction between the two membranes, using the blood signals and bodily fluids of the human system. Clothing becomes an almost living organism, as an internal pump represents the heart of the fabric and the tubing as the nervous and respiratory system as represented by the smart second skin dress, funded by an AHRC Small Award in the Creative and Performing Arts in 2002 (Fig. 1).

As a conceptual piece, the dress interacts with human emotions whereby the aroma dimension is an integral part of the user's sensory experience. It is constructed from two layers of white organza silk with medical polyvinyl chloride (PVC) tubes embedded in between containing coloured liquid that demonstrate a selection of different fragrances.

Smart second skin paves the way to an expanded life, making the most of what the senses have to offer. As a 'living tissue', the purpose is to increase creativity, expressions and visions, spark little reminders, expand colour, texture, sounds and taste, and push the boundaries of the senses of which the user was unaware. It transforms negative mood states into good sensations, releasing scents to help sleep, boost confidence, relax, energise, arouse, increase self-esteem, expand the imagination, bring people out of their shells, define self-image and stimulate the user's sense of wonder.

The smart second skin dress was installed as an interactive, touchable-olfactory display piece, as part of the Touch Me exhibition at the Victoria and Albert Museum (London) during the summer of 2005. The exhibition presented a selection of playful, innovative products by international new media artists and textile designers that explored technology and interfaces to collect data, change shape or colour or send out specific messages. The smart second skin dress, alongside the other exhibits, took museum visitors beyond their conventional position, forcing them to consider senses beyond the boundaries of typical exhibits. The curators Lauren Parker and Hugh Aldersey-Williams wanted the audience to question the sense of 'touch' in a world that is surrounded by new technologies. They were overwhelmed by the sophistication of crudely designed contemporary gadgets that appear to be obsessed with miniaturisation over usability, while the user's experience is often an alienating one that avoids social contact (for example, the new chip and PIN credit card system that no longer requires the shop assistant to look at the customer). The exhibition also explored the pleasures, uncomfortable reactions, sensations, delights and tactility of modern materials which inform the user's understanding of the world, for example, ice jewellery that fits the body by Naomi Filmer, pull-up vases by Davidson and Gitta Gschwendtner, braille clothing by textile designer Shelley Fox and hug seats by Yoshi Saito (Rattray 2005).

Our smell system

While vision is unquestionably our most important sense, when it comes to smell and gathering an emotional response, scent is a more powerful trigger since 75% of the emotions that we generate on a daily basis are affected by smell (Lindstrom 2005). Odours drive our emotions, warn us of danger, influence our body chemistry and steep us in luxury. We begin our life with smell as we form a bond with our mothers. Since smell signals have a direct access to the emotional centres of the brain, the emotional shading of our lives is influenced by the smells around us.

There are specific areas in the brain where smell memories are received and stored. Smell information travels from the olfactory bulb to centres of the brain that handle strong emotions such as aggression, fear and sexual arousal. This centre also plays a significant role in selecting and transmitting information between our short- and long-term memories, evoking memories from the past, and can be of great benefit to the elderly (especially those with Alzheimer's disease). Smells arouse emotions of sadness, loss, love, disgust, longing and passion, buried deep in our subconscious. Only a few molecules from an odour are required to convey a message to the brain, creating a smell image – from a flower, a memory or place, a person or time, an olfactive evocation, or alternatively an aggression alarm or warning signal of danger lurking from where the sense of smell appears principally to have evolved.

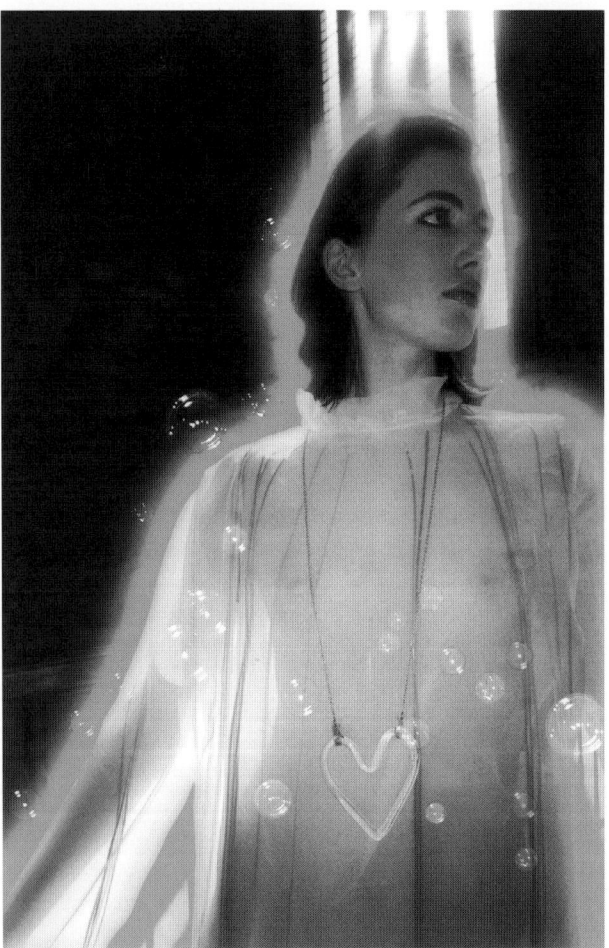

Figure 1 Smart second skin dress (Plate 1 in the colour plate section).

History of perfumery

There are fashions in perfumes as there are in clothes. Each era has its favourite scents that evoke the charms of each succeeding age. Aromatherapy is a therapeutic treatment that combats psychological problems such as stress, anxiety and depression, by enhancing wellbeing, promoting good health and complementing other forms of therapeutic treatments. Records have shown that aromatherapy oils were used as far back as 3000 BC. The overall holistic approach claims to be more efficient when used for the benefit of the 'whole person', both physically and emotionally. Each essential oil claims to have unique therapeutic properties that are antiseptic, anti-infectious or antispasmodic. Others claim to ease pain and digestion, stimulate circulation, heal skin disorders, increase physical energy levels, facilitate feelings of relaxation, benefit problems relating to stamina, reduce depression, nausea, insomnia, radiation burns and enhance the immune function (Buckle 1999).

In 1989, the Sense of Smell Institute developed a new partnership of perfumery and science called aroma-chology, whereby perfume companies employed sensory psychologists to work alongside perfumers. Its purpose was to elicit various feelings and emotions using headspace technology, making it possible to analyze and synthetically reproduce odours given off by almost any element, recreating the desired smell in a fragrance and allowing a new palette to take shape. Unlike aromatherapy (which has no science to back it up), aroma-chology is not concerned with the therapeutic effects on mental or physical conditions, but with the temporary effects of fragrance on feelings and emotions through stimulation of olfactory pathways in the brain. It measures the effects of blends of odorants and single natural and synthetic odour materials, through electrical brain activity, physiological parameters such as heart rate and skin conductance, cognitive functions and voluntary and involuntary behaviour (Jellinek 1999).

Odour delivery systems

The kernel problem in any olfaction project is the classic issue of delivery. To research an emotional living tissue, it has been crucial to delve into nature and divide odour delivery into non-biological and biological systems. Scentsory Design fabrics emulate aspects of biological events in clothes of a radical design, in which one of the primary functions of the fabrics is to act as a global, sophisticated odour communication system. Non-biological systems are exemplified by the passive action of non-interactive fragrance delivery techniques on skin, clothing and in the home. These include conventional perfume bottles, diffusion room-freshener devices, fabric conditioners and microencapsulation methods. Microencapsulation is the process whereby tiny particles are surrounded by a coating made of small capsules with useful properties. Once the coating is broken, droplets of fragrances in the capsules are released.

Biological systems are concerned with body odour in mammals, which are both active and interactive. Thus, an animal using all its senses will sense biologically relevant odours, which leads to activation by pumping the odour glands. A biological system is clearly human as it is garment oriented in its novel futuristic aspects. Secondly, it introduces the notion of pulsing a fragrance around a fabric and imitating biological references to odour glands. Since our sense of smell is so idiosyncratic, however, it would be impossible to attempt to change people's moods using one single fragrance that would have the same effect on everyone. If a scent is to be feasible and have the desired effect it must be used at the right time and in short bursts.

Lab-on-a-chip

Microfluidics is a new technology which involves the design and production of devices that deal with extremely small volumes of fluids. These devices can combine electrical and mechanical components down to a characteristic length scale of 1 μm. Microfluidics is the generic technology of manipulating fluids on a chip, including the integration of pumps, valves, mixers and reaction chambers that enable the fabrication of microreactors and lab-on-a-chip devices (Brunnschweiler *et al.* 2000).

Transporting smell is a more complex process than simply concentrating on the mechanics and electronics of pulsating delivery systems. This research is dealing with microfluidics that are then miniaturised and embedded into fabrics. Difficulties, however, could arise with chemical issues relating to all the discussed areas, including evaluation, chromatography (a method of scientific analysis and calculation in chemistry and perfumery), and the compatibility of tubing, microfluidic materials, threshold timing, fabric and fibre compatibility. Other problems to consider are fabrics that leak perfume when they are crumpled or sewn with a machine needle, or how to preserve new materials over long periods of time.

The primary ethical concerns relate to the potential for odour pollution and the pulsing of chemicals onto a localised area. The advantage for this research, however, is that it allows for the targeted delivery of minute droplets of scent that is more efficient and economic in use, focusing on intimate and personal use rather than generalised and higher volume use. The fragrance industry is taking odour pollution seriously and addressing fragrance-sensitivity issues. As there is an increase of asthma in children, which could be connected to odour-injected pathways, it is important to emphasise that scent delivery with microfluidic technology will be minimised and controlled accordingly.

Other research that explores chemical warfare has identified the principles for the defence mechanism in bombardier beetles that squirt predators with a high-pressure jet of boiling liquid in a rapid-fire action (Eisner and Aneshansley 1999). Along with natural fluidic references to the human body, it is the innovative delivery system in the bombardier beetle that offers further inspiration for the fabric development in this paper. Microfluidics, to some extent, replicates the firing chambers that bombardier beetles facilitate to mix their deadly poison and pulse at immense speed from their tail pipes (Fig. 2).

The fluidic structure is also similar to the inkjet printing process that pumps ink using atomiser techniques. All insects have an acute sense of smell, each with their own unique scent. Their robot-like response to odours and delivery system are necessary for the delicate mechanisms behind the design work for Scentsory Design fabrics, and confirmation that nature has solved the problem.

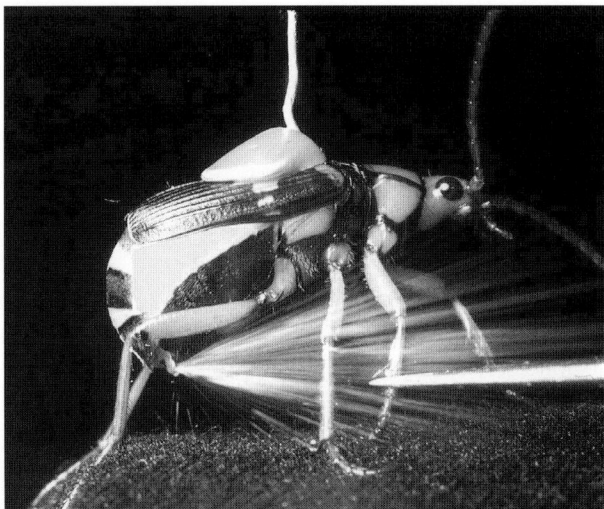

Figure 2 Bombardier beetle (photo: courtesy of Professor Thomas Eisner, Cornell University).

The living tissue

In order to research the purpose of a Scentsory Design fabric, it is necessary to compare the fabrics with the dynamic properties of human skin, which is a tough, waterproof, continuous living tissue and the largest organ in the body. The skin is not merely a thin boundary protecting the inner person from the outer world, but a multilayered organ called the dermis which has its own nervous system and blood supply. New cells are constantly pushed to the surface, changing function and shape on their long journey. Not only is skin an excretory organ and defence barrier, holding the internal organs together, it is a prime source of research relevant to Scentsory Design fabrics because beneath the surface of the skin, complicated areas are found such as muscles, hair shafts, sebaceous and sweat glands and nerve endings that respond to pressure and stimuli. Secondly, a third of the body's blood is pumped from the heart to the skin. Thirdly, the skin is the major interaction point with the surrounding world, receiving sensory messages from the external environment that are then passed to the brain.

Skin is perforated with approximately two million sweat pores, distributed unevenly around the body, e.g. palms of the hands, forehead, nose, armpits, groin and soles of the feet. Sweat glands lie deep in the dermis and spiral through layers of horny cells and out of a tiny pore. Not only do these glands predominantly produce salty fluids, but also pheromones from the modified scent sweat glands, aiding sexual attraction. Pheromones or social odours are chemical messages produced by one member of a species that influence the physiology, hormone levels and behaviour of another member of the same species. Much of the research in Scentsory Design fabrics will include work with human sex pheromones (Thornhill and Gagestad 2002). Thus, there are a number of likely social applications for the emotional living tissue such as sending scent messages to a potential partner (Fig. 3).

Comparing the similarities between layers of human skin and an emotional living tissue, it is essential to distinguish Scentsory Design fabrics as the surrounding shell or smart second skin that is protecting the smart interface. It contains a skeleton intelligence of an array of sensors and tubing deep inside that is similar to the body's capillaries and internal nervous network system. Once the initial trigger occurs, the smart second skin has the capability to read the body's physical and mental state, i.e. detect stress and respond accordingly by pulsing fragrance to the nose receptors.

Digital fragrance

The aim of Scentsory Design is to use a variety of scents as creative tools to improve mental wellbeing, by embedding them in responsive hi-tech clothing that offers social and therapeutic value in a desirable fashion context. A future shaped by new technologies and materials presents exciting possibilities for the future of the fashion industry. Further examples of research at the intersection of electrical engineering, biotechnology, chemistry and materials science include spray-on dresses, 'growable' suits, evening gowns that change colour according to the mood of the user, programmable jackets, coats that display downloadable designs and self-cleaning shirts (Lee 2005).

Although electro-textiles such as smart conductive fibres in wearable circuits are being developed to process environmental and biomedical data for the military to protect soldiers from the enemy and the elements, these textiles are too brittle to wear. There is little evidence of similar electronic textile research to be found on a micro scale, however, which seeks to remedy the limitations of current work on scent-output devices suitable for custom control applications. Recent digital fragrancing research includes Pinoke by Aromajet and Trisenx

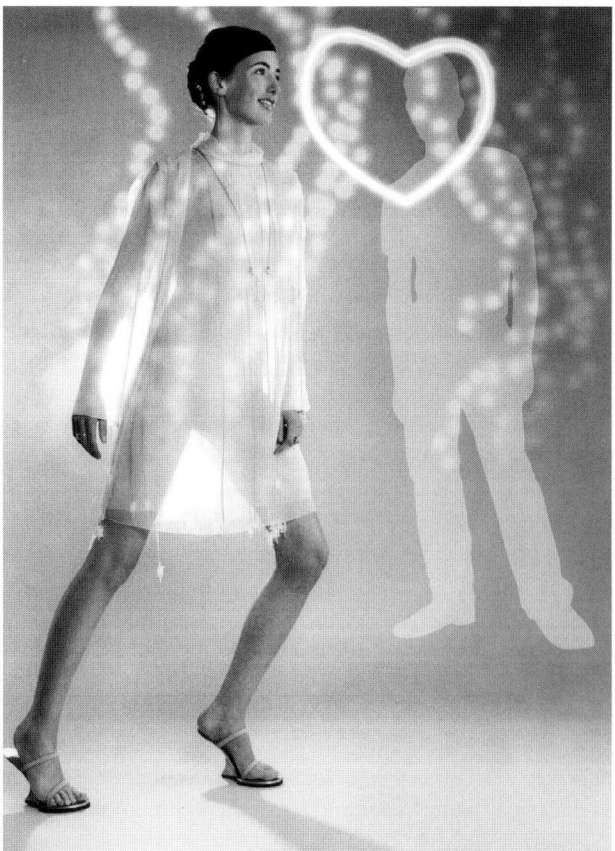

Figure 3 Human sex pheromones.

(computer games), Kaori web service by K Opti, Scent Mail by Telewest (web messaging systems) and ScentStories by Procter & Gamble (scent-emitting CD player), but these technologies are too big to embed in fashion items. The textile industry has benefited from microencapsulated scented fabrics since the 1970s and more recently applications have included moisturising, deodorising, vitamin, insect repellent, anti-cellulite and anti-stress fibres. These standard techniques, however, are not active because they are unable to detect stress and respond to other feelings from which the user could be protected (e.g. fear or sadness).

The basis for Scentsory Design is supported by research that has demonstrated how olfactory substances are capable of increasing an individual's wellbeing through changes in electrical brain activity, evidence of how scent chemicals have the power to evoke emotion (Vernet-Murray et al. 1999). As a result, it is anticipated that while the properties of the emotional living tissue will benefit everyone, it will be of special value to those susceptible to anxiety and depression. Clothes could stimulate the adrenal cortex and boost therapeutic qualities by combining the confidence enhancing and social acceptability of fashion design with the positive psychological benefits of manipulating moods. Recent research proves that the benefits of fragrance include the balancing of the nervous system, improving concentration, promoting a positive mood, reducing rising blood pressure caused by stressful events, reducing heart rate, muscle stiffness, fear and the stress of unpleasant medical procedures, e.g. MRI scans (Warrenburg et al. 2003).

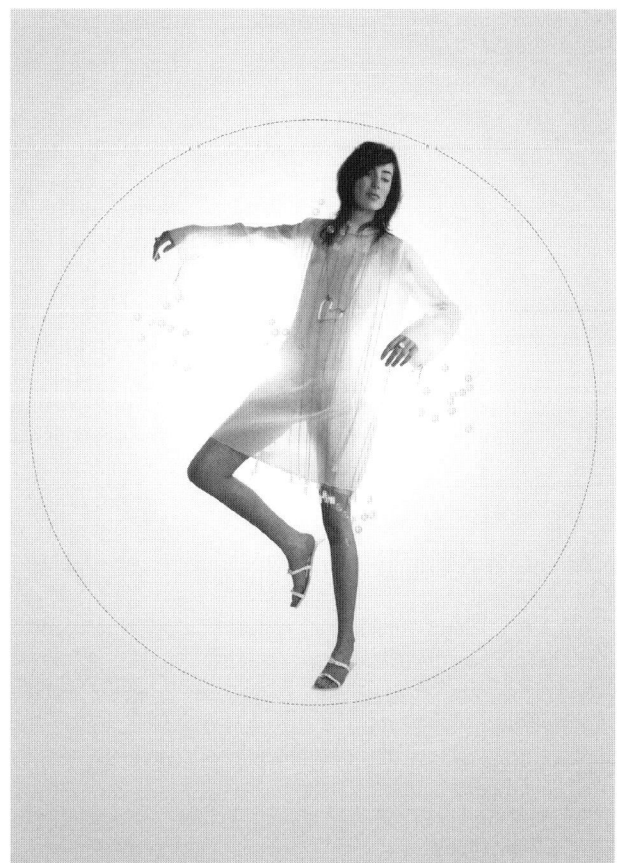

Figure 4 Scent Bubble.

Experiments

A brooch was designed by fusing microfluidic components, dispenser nozzles and reservoirs that dispense airborne nano-litre-sized droplets of scent into the air, to form a scent bubble around the user, so that a coded scent message is delivered to certain areas of the body (Fig. 4).

The aroma mix uses a higher percentage of fragrance concentrate in ethanol since the brooch does not require alcohol, unlike mass-market eau de toilette that contains 99.7% pure ethyl alcohol (Pybus and Sell 1999). Alcohol has been used in perfumery since the 18th century as a neutral solvent in preparing fragrances. It is added to the concentration as a vehicle for the oil, modifying its intensity and making it easily applicable to skin. Our skin, however, was never designed as a vehicle to hold perfumes; alcohol-based perfumes tend to dry it out, causing severe rashes and skin disorders. In this instance the technology developed in this paper delivers scent that is inherently lighter and less likely to irritate sensitive skin.

The technology created for this study has also provided a new way to send a scented message over a wireless network. Scent Whisper links a remote sensor in a spider brooch with a fragrance-dispensing unit in a bombardier beetle brooch, to create a jewellery set that constitutes the wireless web. A secret message is 'scent by a wireless web' by the user who whispers into the spider; this message is then transmitted to the beetle worn by a partner. The spider's sensor, which is implanted in its abdomen, records the humidity of the partner's breath and the beetle releases a scent onto a localised area (Fig. 5).

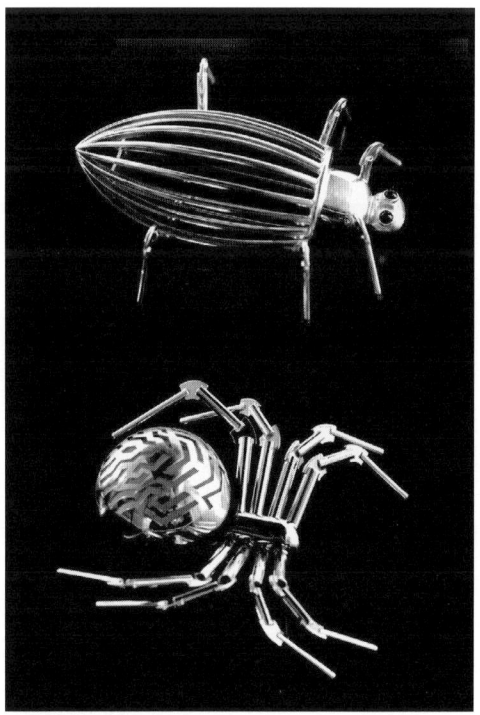

Figure 5 Scent Whisper.

Applications

There are a number of applications where scent could be delivered using the technology described in this paper.

Responsive environments, upholstery or wallpaper

- As an aromatic system for the sensory impaired.
- To ambiently communicate abstract information (such as the status of the stock market) by releasing certain scents when changes in information occur, for example if the market goes up or down as suggested by the Dollars & Scents ambient stock market display research (Kaye 2001).
- To assist odour annoyance in environmentally sensitive public spaces, should a user enter a perfume-free zone. The device would halt perfume release as the user approaches a space that communicates information about a person who is allergic to certain components in perfume (Fig. 6)

Medical healthcare

- As an insect repellent that directs a spray towards a localised area of the body.
- As an alternative to an audio ring tone for mobile phones.
- A drug delivery dispenser that sprays Ventolin molecules for the prevention of asthma.
- To suppress or enhance appetite through odour (Hirsch *et al.* 2003).

Fabric research

This research will advance the future development and functionality of Scentsory Design fabrics so that they are capable of responding to biological conditions triggered by the body's physical signals. The fabrics will house individual therapeutic scent reservoirs that allow refilling at the source, connected to a micro-bore tube capillary flow system. Current micro-bore tubing is finer than a human hair and is so miniaturised that it rivals nature's own capillaries.

The fabrics will be protected in water-resistant polymers and embedded in a membrane without causing intermittent waste. This allows the exact amount to be delivered in response to the user's state from an embedded scent recipe palette, eliminating problems related to odour time span and creating an economical system whereby little scent is wasted. The fabric research could offer three examples depending on the requirements of the user:

1 Spraying scent directly onto the skin to contribute towards psychological wellbeing: we are entering a new age of perfumery that will have a radical impact on mental health. Smells engender emotional responses and the raw materials used to create many of them have mood-enhancing effects. Once sprayed directly onto human skin, fragrances can enhance the personality and identity of the user. Each scent smells differently from one person to another and every scent affects a person in a different way. Recently, the World Health Organization (WHO) Global Burden of Disease Survey estimated that by the year 2020, depression will be the illness of the age second only to heart disease. If fabrics developed for an 'emotional fashion' collection can reduce the need for traditional antidepressant routes to treatment, including the unpleasant side effects such as headaches, insomnia, sweating and agitation, then this research will have considerable social value. Sleep disorders are extremely common in the United States: 33%

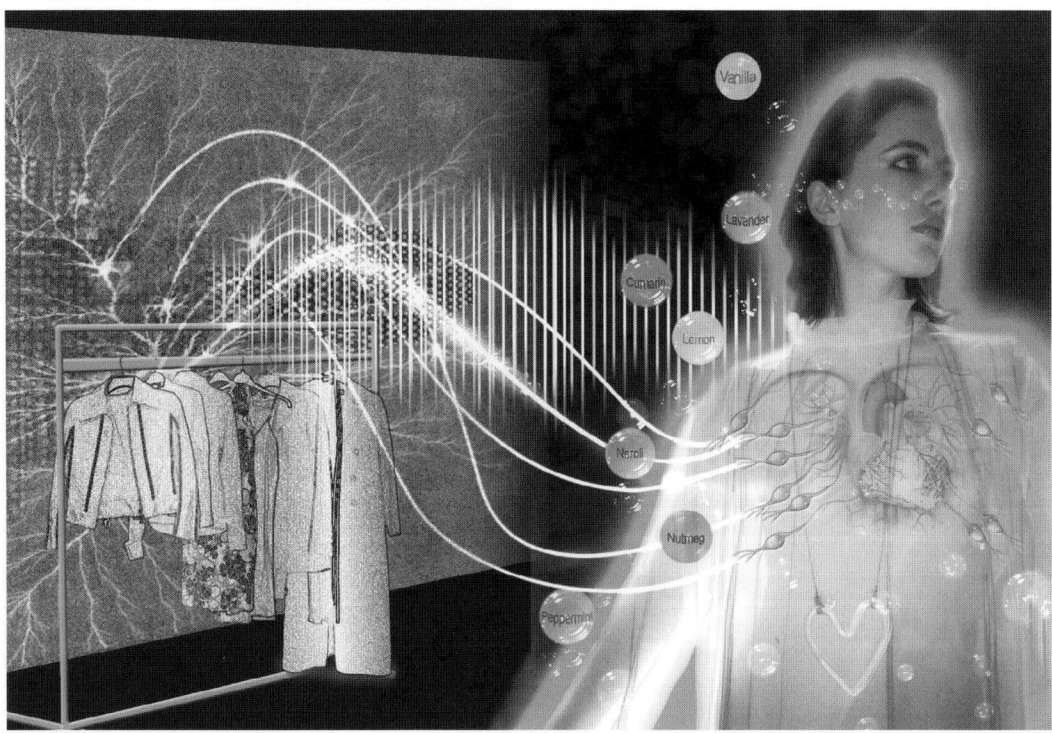

Figure 6 Wireless perfume-free zone.

of the adult population experience bouts of insomnia while 9–12% experience chronic insomnia (Ford and Kamerow 1989). Fabrics that pulse minute droplets of scent throughout the night with properties to encourage sleep could be invaluable.

2   Pulsing scent through fabric surfaces and away from the body: studies at the American Academy of Dermatology suggest that up to 10% of the American population experience a reaction to cosmetics containing alcohol when applied directly onto the skin. The fabric proposed in this research endeavours to replace astringents that burn sensitive skin. The clothing itself will act as a new medium for the fragrance industry to sell perfume. The key and novel advantage of the delivery system developed for the jewellery is that the fragrance obviates the need for skin contact by solvents. The direct spraying of perfume means no additional chemicals are needed either for evaporation (e.g. alcohol) or for propellant (as used in deodorant sprays). The delivery will be in response to the user's real-time needs, i.e. activated by a variety of body sensors (increased heart rate, galvanic skin response, temperature) or sound (dance music, insect noises, high decibels), communication from other users (symbiotic response), direct user request etc.

3   Spraying scent away from the body to create an active mobile barrier: it has been predicted that by 2010 half of the world's population (nearly 3.5 billion people) will be living in areas where malaria is transmitted and it will be at least a decade before a vaccine for the disease will be readily available (Greenwood *et al.* 2005). Consequently there is a need to expand existing repellent control methods. WHO currently estimates that approximately 300 million people worldwide are affected by malaria, with more than 120 million clinical cases and between 1 and 1.5 million deaths each year (WHO 2005). Scentsory Design research cannot include the treatment of malaria, but instead the disease could be prevented by repelling the mosquitoes that carry the parasite. Female mosquitoes are attracted to human body odour and carbon dioxide in breath. Ingredients that repel mosquitoes do so by repelling them from the host (negative hedonics), by distributing their ability to find the host or by distributing their ability to feed on human flesh (Warren 1998). A wearable screen that repels mosquitoes could significantly reduce malaria and other airborne insect diseases by accurately targeting the release of minimal amounts of repellent, thereby removing the need to apply copious amounts of greasy sprays directly on the skin.

Conclusion

The jewellery results are a snapshot of scent-output devices in clothing, as fashion products and technology will evolve in the future. Further steps include creating the colour-changing colodours effect by combining controlled scent delivery with thermochromic ink technologies developed by creative laboratories around the globe, for example, by International Fashion Machines (a design and research company based in Boston, Massachusetts).

Our sense of smell is now recognised as more important for our daily life, especially following the recent scientific breakthrough in olfactory reception and the awarded Nobel Prize for Medicine and Physiology in 2004 (Buck and Axel 1991). As a result, a wider spectrum of multisensory research will be available to artists and designers. It is clear from the evidence presented herein that this research will lead to further development and design implications in fashion, textiles, interior design, healthcare and other systems that use smell to convey information.

Scent, with all its complexities, will become a user-friendlier medium, as new media art continues to be a rapidly changing and dynamic field of creative practice, straddling conventional categories and crossing disciplinary boundaries, and challenging our assumptions about art. Scentsory Design creates interfaces with the art world, museum culture, the media, publishing and academia, allowing people to learn coded scent messages, whether for the purposes of wellbeing, communication or simply having fun.

It is within the realm of technology that fashion will find its innovative future. As more is discovered about olfaction science and electro-textiles develop into wearable, seamlessly lightweight fabrics that are less brittle and easy to power and preserve, there will be room for many beneficial applications in this new and exciting field of Scentsory Design.

Acknowledgements

The author would like to thank the AHRC for funding research and development work mentioned in this paper, colleagues at the University of the Arts London, Professor Manz and Dr Jenkins at the Institute for Analytical Sciences in Germany, Professor Eisner and Jofish Kaye at Cornell University, Adeline Andre, Wendy Latham and Guy Hills.

References

Ballard, J.G. 1997. *Artforum* (as cited in the art journal, quoted by Camilla Nikerson) December: 75.

Brunnschweiler, A, Koch, M and Evans, E. (2000) *Microfluidic Technology and Applications*. Microtechnologies and Microsystems Series. Baldock: Research Studies Press.

Buck, L and Axel, R. (1991) 'A novel multigene family may encode odourant receptors: a molecular basis for odour recognition', *Cell* 65: 175–87.

Buckle, J. (1999) 'New roles for essential oils in healthcare', *Aroma-Chology Review* VII(4): 1.

Eisner, T. and Aneshansley, D.J. (1999) 'Spray aiming in the bombardier beetle', *Proceedings of National Academy of Science* 96: 9705–9.

Ford, D.E and Kamerow, D.B. (1989) 'Epidemiologic study of sleep disturbances and psychiatric disorders', *Journal of the American Medical Association* 262: 1479–84.

Greenwood, B., Bojank, K., Whitty, C. *et al.* (2005) 'Malaria', *The Lancet* 365(9469): 1487–98.

Hirsch, A.R. *et al.* (2003) 'The effects of odour on weight perception', *Chemical Senses* 28: 210.

Huxley, A. (1932) *Brave New World*, ch. 11. Harmondsworth: Penguin Books.

Jellinek, J.S. (1994) 'Aroma-chology: a status review', *Cosmetics and Toiletries Magazine* 109: 83–101.

Kaye, J. (2001) *Symbolic Olfactory Display*. Master's thesis, Massachusetts Institute of Technology Media Laboratory, Boston. MA.

Lee, S. (2005) *Fashioning the Future: Tomorrow's Wardrobe*, 17. London: Thames & Hudson.

Lindstrom, M. (2005) *Brand Sense: How to Build Powerful Brands through Touch, Taste, Smell, Sight and Sound*. London: Kogan Page.

Pybus, H. and Sell, S. (1999) *The Chemistry of Fragrances* (RSC Paperbacks), appendix 1. Cambridge: Royal Society of Chemistry.

Rattray, F. (2005) 'Smell, sight, sound, touch, taste: a sense liberated', *V&A Magazine* June 2005.

Thornhill, R. and Gagestad, S.W. (2002) 'Human sex pheromones', *Aroma-Chology Review* X(2): 6–7.

Vernet-Murray E., Alaoui-Ismaili, O., Dittmar, A., Delhomme, G. and Chanel, J. (1999) 'Basic emotions induced by odourants', *Journal of the Autonomic Nervous System* 25 September: 176–83.

Warren, C. (1998) 'Human and insect olfaction make news at AChemS', *Aroma-Chology Review* VII(2): 3–4, 7.

Warrenburg, S., Christensen, C., Wilson, W. *et al.* (2003) *Fragrance Research: Measuring the Emotional Power of Fragrances*, ESOMAR Seminar, Amsterdam, 16–18 March. Amsterdam: ESOMAR.

WHO (2005) 'Burden of mental and behavioural disorders: depressive disorders', in *The World Health Report*, ch. 2. Geneva: World Health Organization.

The author

Jenny Tillotson is a Senior Research Fellow at Central Saint Martins (CSM) University of the Arts, London, Fellow of the Institute of Nanotechnology and Associate of the British Society of Perfumers. Prior to her academic work she was a fashion stylist and Sensory Designer for Charmed Technology. She has exhibited widely including the Victoria and Albert Museum, Tate Modern, e-Culture Fair, Cheltenham Science Festival and many other international events.

Address

Jenny Tillotson, Central Saint Martins, University of the Arts London, The Innovation Centre, Southampton Row, London WC1B 4AP, UK (j.tillotson@csm.arts.ac.uk).

Can an artist create permanence from transience? The Schmuck Quickies of Yuka Oyama become durable

Cordelia Rogerson and James Beighton

ABSTRACT Japanese-born jewellery artist Yuka Oyama is best known for her Schmuck Quickies performances (literal translation 'Quick Jewellery'), in which audience members become participants as she creates spontaneous pieces of jewellery for each of them. The objects created through this performance are, of their nature, ephemeral, and fall outside the context of art-historical study from the moment that the performance ends. During a recent residency at the Middlesbrough Institute of Modern Art, Oyama used the experiences and materials recovered from the Schmuck Quickies performances to develop a series of studio pieces, made as a limited edition series. This development has shifted the context through which Oyama's work can be interpreted and places them within the reach of the private and institutional collector. The study unveils how an artist has considered the legacy of her work in the public domain and has modified her techniques and conceptual framework in an endeavour to create more durable artworks for the future. As conservation also now becomes an issue, curatorial and conservation expertise must negotiate a practice to evaluate the best course of preservation for the works.

Keywords: jewellery, performance art, participation, transience, permanence, recycling

Introduction

Japanese-born jewellery artist Yuka Oyama (b. 1974) is best known for her Schmuck Quickies ('Quick Jewellery') performances in which audience members become participants as she creates spontaneous pieces of jewellery for each of them. The jewellery is deliberately temporary and transient, lasting only for the duration of the performance, because neither the materials nor the construction techniques, when combined, possess longevity and durability. Since 2002, she has undertaken a number of Schmuck Quickies performances in Japan, Germany and the UK.[1] As a result of a residency during the latter part of 2004 at Middlesbrough Institute of Modern Art (mima) in the UK, Oyama changed her working practice. The artist moved from the immediacy of creating jewellery directly onto the body or the wearer's garment to more process-intensive batch production pieces created in the studio that necessarily and intentionally have increased durability and longevity. By tracing the new development, the role of transience in her work will be analyzed to assess how it has altered in response. The study unveils how an artist has considered the legacy of her work in the public domain and has modified her techniques and conceptual framework in an endeavour to create more durable artworks for the future. In this instance, rather than longevity solely being the concern for a curator or conservator, the artist has engaged with the idea in response to her personal expectations and aspirations for her artwork and growth as an artist.

Moreover, the analysis of Oyama's work marks a collaboration between the two professions of curatorship (James Beighton) and conservation (Cordelia Rogerson) because it highlights similar areas of awareness and cognition encountered by each when contemporary art made from modern materials is studied. Any analysis of Oyama's work inevitably involves examining the ideas and materials within her work in tandem, since these two components are entirely enmeshed. In order to curate and consider the appropriate preservation needs of her jewellery, the curator and the conservator need to equally comprehend these concepts to address their own area of specialism. The information presented within this paper is based upon the participation and observation of both authors in Schmuck Quickies performances, oral testimony of the artist and through curation (by JB) of Oyama's residency.[2]

The concept of performance art

Artworks in which performance is the dominant mode of expression, and where the actions of an individual at a particular time and place constitute the artwork, is an ever-developing creative and artistic practice that has evolved since the latter half of the 20th century. In contrast to, and sometimes in reaction against, traditional genres within the visual arts, such as painting and sculpture, where posterity is an assumed motivation of the artist in the production of the object, impermanence is central to performance art. The concept is literally embodied and demonstrated by the French artist Orlan (b. 1947). Since 1990 Orlan has been undergoing a series of plastic surgical operations, transforming her face based on seven classical models of beauty (Dusinberre 2004; Wilson *et al.* 1996). Her art resides in the 'performance' of the surgery itself, and a presentation of her resulting face in its bruised and changing state. Her concepts are multilayered, embracing psychoanalysis and feminism, but it is possible to read her performance as a flirtation with the idea of conserving the human face in its youthful 'beauty'. Her performances point to the impossibility of eternal youth, which media celebrity so desires.

Oyama's jewellery operates within the realm of performance art. Like Orlan, her artwork is non-permanent, but plays with ideas of permanence. Whereas Orlan uses her own body as art, Oyama works on the bodies of others. As with much contemporary artist-jewellery, her work is concerned with how the jewellery object interacts with the body as a site for presentation and she has taken this concern to its logical conclusion by working directly onto the body. There is an almost missionary zeal underlying the Schmuck Quickies performances. She intends for her work to impart the courage of self-expression upon the wearer, against the uniformity that society encourages. In her artist statement she remarks 'as I sat in my studio making jewellery and staring at views onto the street, where everyone was looking grey, correct and clean, I felt an acute need to shake everything up' (Oyama 2004).

She goes on to explain how she sees her works as the antithesis of global standardisation, focusing instead on the individuality and personality of the wearer, to which she aims to give expression. She states:

> Through my work, I seek to demonstrate the individuality and locality that resides in each person and culture, instead of passively consuming given standardized products and ideas. Wearable objects are used as media to express my ideas, since I am interested in the inseparable relationships between the wearable objects and human beings. The presence of a person enlivens expressions in the objects, while the objects that are worn elevate the emotional integrities of the wearers, as well as tell stories about them. My work recaptures the direct contact and personal interaction that once existed between the dressmaker and the wearer. I make body adornments that grow out of the personalities and interests of the wearers (Oyama, 2004).

The performance unfolds upon a stage set out like a hairdresser's salon. A participant comes forward from the spectators and sits on a folding chair in front of a mirror. Oyama then starts a conversation by asking 'what kind of jewellery can I make for you?' (Fig. 1).

Schmuck Quickies: a collaboration between maker and wearer

Through the ongoing, casual conversations with the participant, Oyama shapes jewellery pieces that are an interpretation of the wearer's character as revealed through the conversations. Frequently the jewellery is physically united with the clothes of the wearer and it becomes an extension of the individual; an outward, unconventional and intimate expression of their persona, created through spontaneous collaboration between

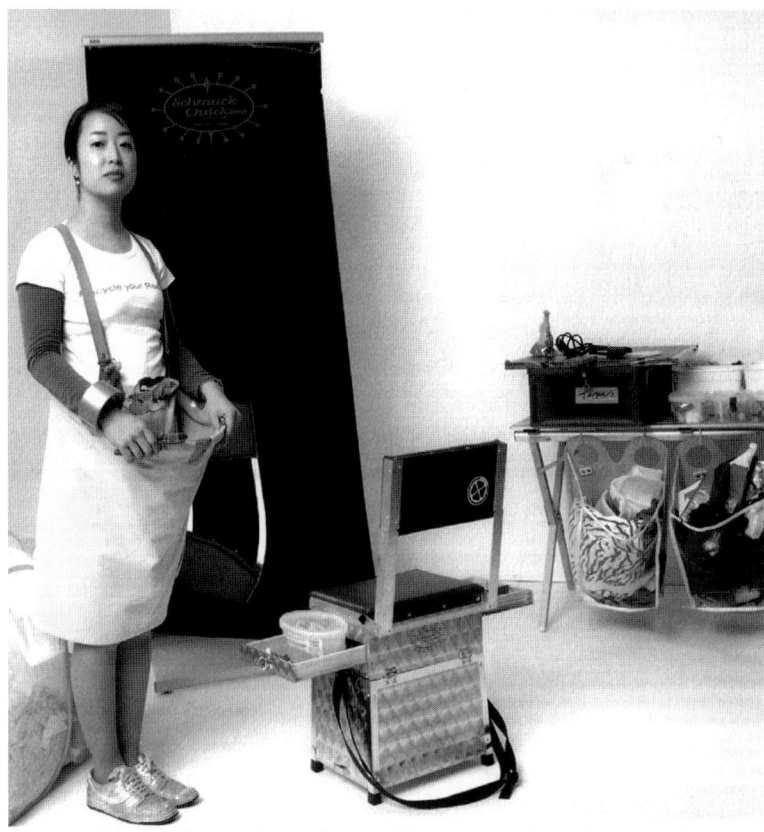

Figure 1 What jewellery can I make for you? (Plate 2 in the colour plate section.)

the artist and subject. The jewellery acts as a momentary visual representation of that person's individuality; their inner self is temporarily exposed to the outside world. Oyama explains how she observes change in the participants as the jewellery begins to take shape upon them, 'there is a transformation or shift in the emotional integrity of the participant. It comes from the radiation of the person. They become more radiant.'[3]

Because of the intimate union between the jewellery and the garment upon which it is constructed, Oyama's work is often destroyed within hours of the performance ending, as the act of undressing frequently involves dismantling the jewellery as well. Even where the object is constructed in such a way that it can survive being taken off the body, Oyama is acutely aware that, from her perspective, the piece ceases to exist from the moment that the wearer leaves the stage. She will never see the object again, and can never capture the collaborative spark that brought the object into existence in the first place. For this reason, an essential element of the performance involves documentary portraits being taken of each participant wearing their creation. These portraits serve the dual purpose of giving a permanent record and presenting an anthropological survey of attitudes towards performance art, jewellery and self-expression across different cultures.

Schmuck Quickies: the materials

All materials used in Schmuck Quickies are recycled and the kind of materials employed is rooted in the locality of the performance, and gathered in the area. When performing at the O-Bon festival in Niigata, Japan, for example, Oyama used recycled kimono fabrics in her jewellery, since these were manufactured regionally (Beighton *et al.* 2005). When creating Schmuck Quickies during her residency, the materials were gathered from industrial recycling units within Middlesbrough. The materials reflect the locality by presenting a less obvious product of Middlesbrough's industrial economy, namely clean waste resulting from over-ordered components, discarded packaging etc. The range of potential components at Oyama's disposal during her residency in Middlesbrough was comprised predominantly of plastics, foams, adhesive tapes and textile materials (Fig. 2). The construction techniques that naturally suited both the materials and circumstances of the performances included wrapping, adhering, stitching and pinning (Fig. 3).

During her performances, the materials selected for each jewellery item were matched to the individual participants. She gained an impression of their personality during the performance and, in collaboration with them, drew out the strongest facets with her work. The pieces are rooted to the individual's clothing and therefore become a part of the individual's repertoire of self-expression. The specific material nature of the components and techniques used in the jewellery is therefore essential in reflecting both Middlesbrough as a town and the impact of this upon the wearer's psyche. This sociological aspect of Oyama's work cannot be ignored by the curator, thereby elevating the materiality of the jewellery to a position of primary importance.

Furthermore, Oyama's interest in recycled materials is derived from the premise that much can be revealed about

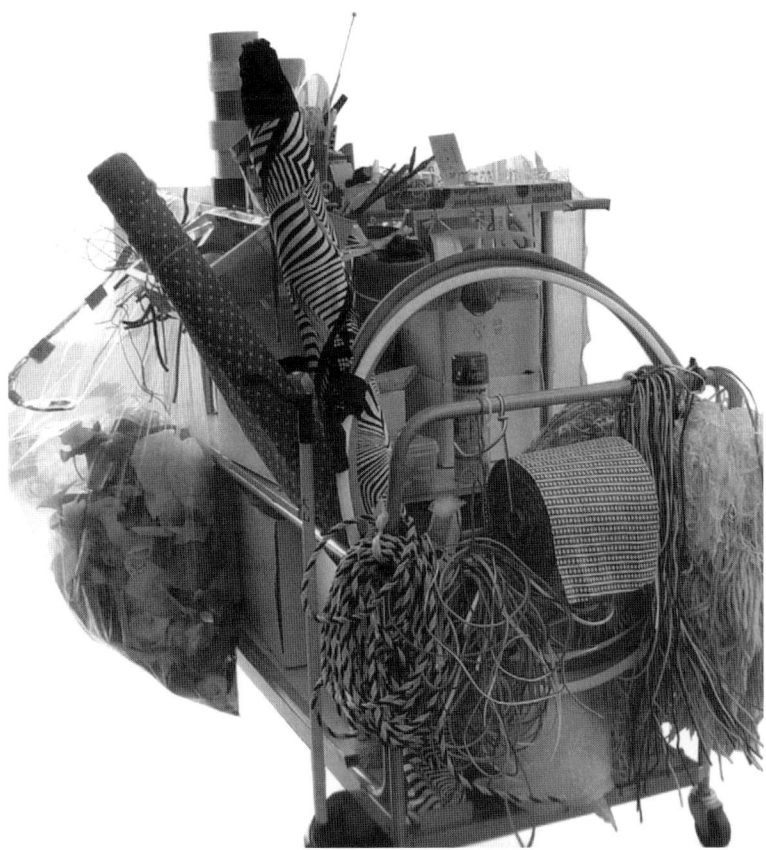

Figure 2 The cart of materials that forms the basis of Oyama's Schmuck Quickies (Plate 3 in the colour plate section).

a person's character by what they throw away. The artist recognises that the material detritus of human existence reveals a poignant account of human interaction with our environment and the values we place upon material goods. In her words she is 'reassessing what is available in real life'.[4] By sensing artistic potential in the materials that are deemed to be refuse by others she is altering the context in which they are understood, and thereby extending the normal and expected lifespan of the components, albeit in the short term. Both the material and the human wearer are reinterpreted by Oyama's performances. Materials are temporarily transformed and are refreshed and invigorated. Not only is their status enhanced by her recontextualisation but they become capable of expressing the character of the locality and the personality of those who live in the environs of Middlesbrough (Fig. 4).

A discrepancy exists, however, between the idea of greater longevity entrenched in the use of recycled materials for Schmuck Quickies and the expected short lifespan of the jewellery artworks. The work created is destroyed as soon as it is taken off. The transient nature of performance art is in opposition to Oyama's intent of demonstrating the creative potential of refuse thereby providing an alternative to throwing it away. She explains the dilemma by stating: 'The experience remains forever in your head. It is the experience that matters.' By articulating this standpoint it becomes clear that the memory of the performance, and the resonance it has for the human subject, is the only element of longevity within her work.[5] Even Oyama admitted later that she lamented that her physical art was destroyed so readily and rapidly.

On a more practical conservation level the plastics, foams and adhesive tapes used for Schmuck Quickies in Middlesbrough and the spontaneous construction techniques are ideally suited to temporary short-lived artwork. The materials used for Schmuck Quickies degrade and alter rapidly in conservation terms, but because the art is destroyed within a few hours, no consideration for preservation and conservation is necessary. For Schmuck Quickies, conservation simply is not an issue since the objects are not available for collecting or wearing in the long term.

The development of After Schmuck Quickies: the ideas

During her artist residency at mima, Oyama nurtured and acted upon an aspiration to move Schmuck Quickies through to a new body of work where the finished jewellery and construction techniques have greater longevity. She longed for her, now absent, process of developing and building work and desired that her work remained wearable for longer and have the capacity to be sold and acquired for collections. Pieces would be more slowly considered and ultimately be more durable, but significantly would draw their inspiration from her previously made spontaneous transient pieces in performances undertaken in Middlesbrough. She named these pieces After Schmuck Quickies.

At first the desire to move back to the studio seems surprising. The immediacy of Oyama's performance and the way

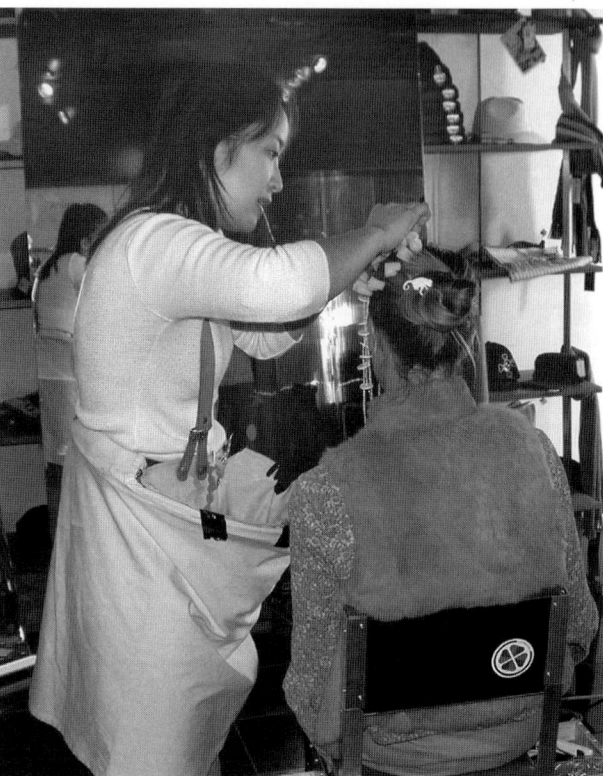

Figure 3 Schmuck Quickies performance in progress, Psyche, Middlesbrough, 26 October 2005 (Plate 4 in the colour plate section).

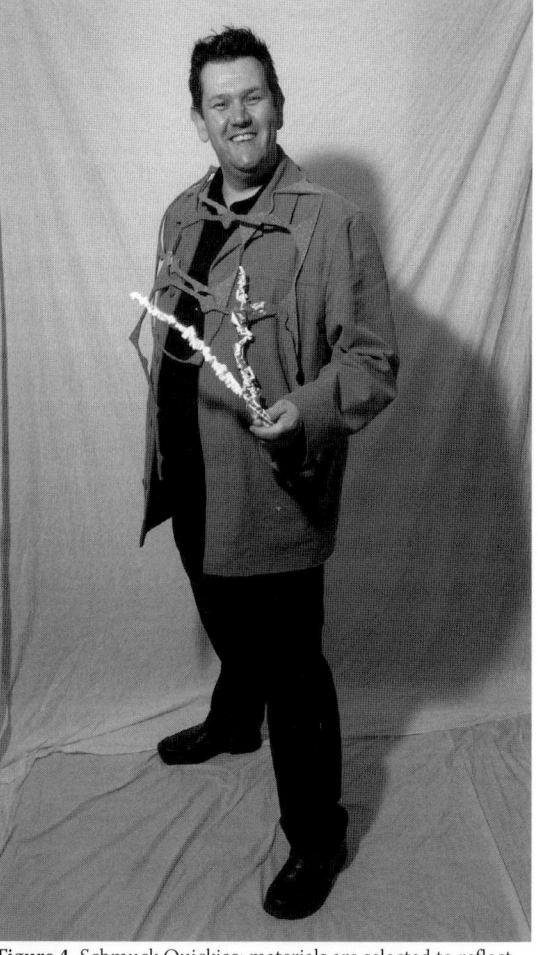

Figure 4 Schmuck Quickies: materials are selected to reflect the location and character of the wearer (Plate 5 in the colour plate section).

that it plays with the boundaries between fine art and craft practice are incredibly attractive factors for the curator. As already intimated, curating performance art does not involve considerations for preservation and storage associated with a permanent collection. Schmuck Quickies provides an opportunity for direct audience involvement with jewellery, which in itself can act as a valuable interpretative tool where contemporary jewellery is a part of a gallery's artistic programme. In contrast, there is an acute danger that studio-based practice could seem detached from the audience, with the work losing its edge. Further consideration of Oyama's trajectory, however, reveals this to be an understandable move, and represents a growing awareness on the part of the artist of what she is actually achieving and the place of her work within the jewellery discipline.

Oyama has communicated that her long training as an artist and in metalsmithing and jewellery-making techniques has left her with a feeling of detachment, which she considers counterproductive at this time in her life. Working in isolation within an artist's studio caused her thinking to become too focused and introspective, limited in its exposure to other areas of life. Schmuck Quickies set out to address this imbalance: 'in working with other people I can accept more ideas'. Oyama now feels able to bring her experiences back to the studio.[6]

The move back to the studio results from a changing understanding of impermanence in her work and the role that both the artist and wearer can play in this. There are three factors to consider in this shift. First, there is the role that the participants have played in shaping Oyama's ideas. As documentation of her Schmuck Quickies performances in Japan and elsewhere has revealed, some participants expressed a desire for Oyama to slow down and create pieces that they could wear again in commemoration of the occasion of the performance. In Schmuck Quickies, the artist acted like a historical master of a Japanese tea ceremony, nominating objects as holders of beauty and directing the ceremony in which they are to be used. After Schmuck Quickies places the wearer in a more prominent position: Oyama takes responsibility for nominating the objects, but through giving the wearer control over the way it is worn, the ceremonial aspect of adorning is left with the wearer.

Secondly, there is a certain pragmatism, as Oyama creates objects that can be sold through galleries and collected by museums; again this is partly a response to a participant's comment that she will never make a living from Schmuck Quickies. Rather than interacting with one individual only, the new work can interact with several since the batch production process allows more than one piece of each design to be constructed.

Thirdly, the newer work aspires to reconcile the conflict between the desire to extend the life of recycled materials through changing the context in which they are appreciated, and the fact of their destruction once dismantled at the end of the evening. In utilising traditional jewellery-making techniques and recognisable fastening elements such as brooch pins, hairslides and clips, which are applied to throwaway materials, Oyama succeeds in heightening the sense of the object as a jewel and therefore something to be treasured rather than discarded. This transformation of the mundane has a certain resonance with Japanese tradition, the culture of *Wabi Sabi* wherein beauty is discovered in the everyday (a mass-produced rice bowl or a stem of bamboo, for example), but more importantly it re-situates Oyama's work within a Western art-historical context. The nomination of ready-mades as works of art has a history that goes back, most famously, to Marcel Duchamp's *Fountain* of 1917, and is a concept that is echoed in some of the most important works from the new jewellery movement active during the late 1970s and 1980s (Schwarz 1997). The pioneering jewellers of the time paved the way for the use of any material for jewellery that has expressive and artistic potential and set a precedent for Oyama's use of material. In 1967, Gijs Bakker presented a neck ornament and bracelet constructed from aluminium stove pipes which he had anodised (Joris 2000). During the 1980s, Otto Künzli exhibited brooches made from, among other materials, drawing pins, postcards and wallpaper (Turner 1996). More recently, Sigurd Bronger has exploited a technique which parallels Oyama's. He constructs elaborate, precisely engineered jewellery findings made of gold and brass. These mounts are used as settings – not for precious gems but for everyday artefacts such as plastic erasers, balloons and shoe heels.[7] Bronger's art achieves a detached sense of irony through its exacting workmanship in contrast to that of Oyama who deliberately seeks to present her work as impulsive and spur of the moment.

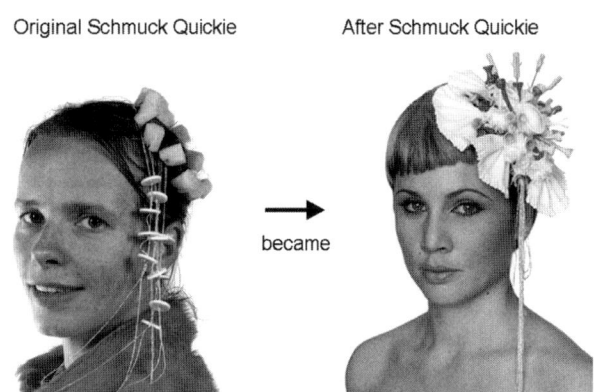

Figure 5 Schmuck Quickies and After Schmuck Quickies (derived from the former Schmuck Quickies) (Plate 6 in the colour plate section).

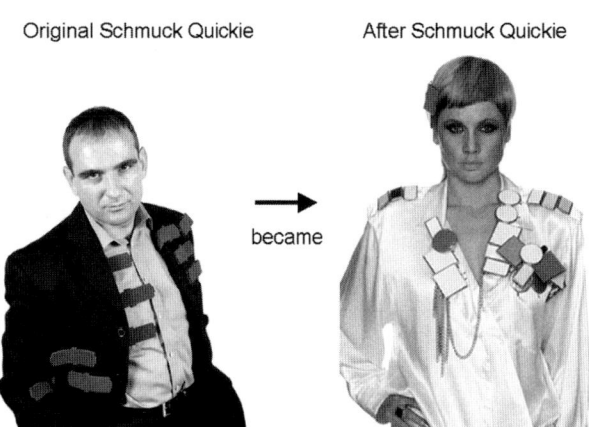

Figure 6 Schmuck Quickies and After Schmuck Quickies (derived from the former Schmuck Quickies) (Plate 7 in the colour plate section).

After Schmuck Quickies: the materials

What Oyama did not anticipate during the development of After Schmuck Quickies is that the concept of transience that underpinned her performances, but which did not sit entirely comfortably with her recycling of materials, would still exist in her new work. It was her materials of choice that dictated this.

The materials found and located in Middlesbrough, which are so representative of its modern-day industry and thus provided a distinctive flavour for Schmuck Quickies created there, were also essential in Oyama's new work. Without the plastics and foams that are central in Middlesbrough's industrial surplus and waste, the work would lose the spirit of locality and hence some of the pivotal meaning to the art. After Schmuck Quickies, therefore, uses many similar materials to the earlier Schmuck Quickies. Adhesive tape plays a less prominent role but foams, rubber and a variety of plastic components alongside textiles are present. The major differences between the earlier and later work are the construction techniques employed and how the jewellery pieces function to adorn a body.

Since the pieces have been created more slowly, in an artist's studio rather than during a performance, Oyama has employed more precise and detailed procedures. Stitched elements have been more delicately undertaken while adhered elements are less significant and obvious. The pieces appear more finished, more controlled and are generally smaller than her earlier work. While the workmanship of Oyama intentionally does not meet the sophisticated level of Bronger (discussed above), After Schmuck Quickies presents work that is far more structured and contrived by comparison with her former pieces (Figs 5 and 6).

Moreover, the jewellery is adorned on the body in a more conventional jewellery manner. Schmuck Quickies were inextricably secured to the clothing and body of the wearer: After Schmuck Quickies are removable stand-alone items. They may be worn then removed by the wearer on numerous occasions; the timing is dictated by the wearer not the artist, unlike the performance pieces.

Inevitably what the new pieces lack is the actual spontaneity of creation despite Oyama's efforts to retain this visually with her construction techniques. Yet the materials components of her work determine that the jewellery is still somewhat ephemeral and fragile in nature when considered in conservation terms, and compared with the majority of contemporary artist jewellery being created in the late 20th and early 21st century.[8] Foams, rubber and adhesives in all probability will degrade and deteriorate swiftly. The concept of transience, therefore, remains due to the very specific material requirements of the work. Undoubtedly, though, Oyama's newer artworks will last significantly longer than her original performance pieces. In this respect, her aims have been successful. She has furthered the lifespan of the recycled materials considerably, providing an extended opportunity to extract and demonstrate the artistic potential of society's waste, which was the element that was absent in her performance pieces. Her work reveals the fallacies of our current throwaway culture; objects we readily discard, without due thought, may still have potential for constructive and serviceable use. The model of rubbish theory, devised by Michael Thompson, which explicates how goods can move from the status of useful to rubbish to that of desirable commodity once more, provides a helpful foundation for analyzing Oyama's practice (Thompson 1979). Her artistic creativity is the impetus for reinstating the worth and utility of materials that have already been discarded. The wearer is forced to consider the origins of the materials and reinterpret their importance as art. Moreover, she illustrates how we can relate seemingly worthless objects intimately to our surrounding environment and our own identities. The contemporary industrial identity of Middlesbrough and its inhabitants are encapsulated within the art. In the symbolic transition from 'throwaway' to 'treasured', Oyama has created an artwork that challenges our perception of waste. The act of repeatedly wearing and exhibiting After Schmuck Quickies provides a more extensive platform for engaging with the debate she presents.

Oyama may have fulfilled her own artistic potential with After Schmuck Quickies but in doing so she has inadvertently transferred any considerations of transience and longevity in the jewellery from herself to the new custodians of the work. Preservation and conservation were not relevant concepts to her initial Schmuck Quickies output since they were destroyed almost instantly, but her later work has implications for preservation in which she is inexperienced. While conceptually, longevity is probed by the development of Schmuck Quickies morphing into After Schmuck Quickies, the reality is that museums and collections acquiring Oyama's work will see the limits of longevity made manifest. The work will be destroyed ultimately by unstoppable degradation processes or by the intended function – the action of wearing – rather than deliberately being dismantled. No longer is it forced out of existence by being taken apart or even being thrown away. Arguably, on one level the ultimate demise of Oyama's work provides an additional stratum to her practice. The materials and, therefore, the jewellery, proceed to their own natural conclusion and hence the true and maximum possible lifecycle of the materials is demonstrated. From a resource management and preservation perspective, Oyama's newer work presents conservation challenges without precedent in jewellery collections. The conservation of contemporary artist jewellery and the implications of deterioration on the artistic intent is only just being considered.[9] Ideally, curators and conservators will work in tandem with Oyama, as her work ages, to comprehend how any changes to the material components and structure either enrich or detract from its meaning. Whatever the outcome, the extended presence of After Schmuck Quickies compared with Schmuck Quickies will compel the artist, as well as curators and conservators who come into contact with it, to consider how the cyclic process of production then abandonment of materials, followed by artistic creation and subsequent ultimate destruction of the artworks, correlates on both physical and conceptual levels.

Conclusion

The concept of transience both within the material and the idea that was evident in the original Schmuck Quickies still remains in After Schmuck Quickies. The difference is that the expectation for her later work to last must be realistic and understood in the context of her ideas. Oyama's artwork remains fundamentally temporary rather than unequivocally permanent. Her constructions are unlikely to last longer than a few decades without undergoing visual and physical changes as a result of the inevitable degradation pathways of the component materials. Transience remains a central theme to both her artistic intent and her physical choice of materials. What Oyama has produced is a more emphatic exploration of the recycling and presence of materials in society, the resonance of location and the individual within materials and artworks, and an expression of the temporality of existence.

The residency at Middlesbrough provided an environment for artistic development, the exact direction of which could not necessarily be predicted by the artist or the curator at the outset. Oyama has now produced collectable work as evidenced by the fact that two museums are currently poised to purchase some pieces of After Schmuck Quickies. Oyama is still developing and refining her ideas and the exact nature of her work in the future may well be modified again. What is important is that her artistic intent, conveyed by the ideas and materials of her jewellery, is communicated and understood and accepted by those professionals who are responsible for the collections within which it is held.

Acknowledgements

The authors would like to thank Yuka Oyama for her participation in the performances and residency at Middlesbrough Institute of Modern Art between October and December 2004. Thanks also to mima for facilitating and supporting the work undertaken by the authors and the Royal College of Art, London, for supervising and assisting Cordelia Rogerson's research undertaken as part of her PhD studies.

Notes

1. For some of Oyama's past performances see: www.dearyuka.com.
2. The authors both participated in and observed a Schmuck Quickies performance held at Psyche, Middlesbrough, 26 October 2004. Oyama's residency at mima ran from October 2004 until December 2004.
3. Yuka Oyama, oral testimony recorded by Cordelia Rogerson, 27 October 2004.
4. *Ibid.*
5. *Ibid.*
6. *Ibid.*
7. A collection of *Portable Instruments* by Sigurd Bronger is held in the International Contemporary Jewellery Collection at mima.
8. Cordelia Rogerson's PhD research based at the Royal College of Art/V&A conservation department, and ongoing at the time of writing, considers the vulnerability to deterioration of contemporary artist jewellery created within the last few decades.
9. Cordelia Rogerson's PhD research (see note 8) also includes an assessment of how material deterioration of jewellery influences the interpretation of artistic intent.

References

Beighton, J., Smith, C., Bustard, J. and Worsdale, G. (2005) *Yuka Oyama Schmuck Quickies.* Middlesbrough: Middlesbrough Institute of Modern Art.

Dusinberre, D. (2004) *Orlan Carnal Art.* Paris: Flammarion.

Joris, Y. (ed.) (2000) *Jewels of Mind and Mentality: Dutch Jewelry Design 1950–2000.* Rotterdam: 010 Publishers.

Oyama, Y. (2004) *Schmuck Quickies Artist Statement*, distributed during Schmuck Quickies performance, Psyche, Middlesbrough, 26 October 2004.

Schwarz, A. (1997) *The Complete Works of Marcel Duchamp.* London: Thames & Hudson.

Thompson, M. (1979) *Rubbish Theory: The Reaction and Destruction of Value.* Oxford: Oxford University Press.

Turner, R. (1996) *Jewelry in Europe and America: New Times, New Thinking.* London: Thames & Hudson.

Wilson, S., Onfray, M., Stone, A.R., Serge, F. and Adams, P. (1996) *Orlan: This is my Body ... This is my Software.* London: Black Dog Publishing Limited.

The authors

- Cordelia Rogerson lectures part-time at the Textile Conservation Centre where she has worked variously as a textile conservator and tutor since 1997. Since 2002 she has been a part-time PhD candidate at the RCS/V&A studying the preservation of contemporary artist jewellery.
- James Beighton is Curator of Craft at the Middlesbrough Institute of Modern Art where he specialises in the fields of contemporary artist jewellery and British studio ceramics.

Addresses

- Cordelia Rogerson, Lecturer, Textile Conservation Centre, University of Southampton, Park Avenue, Winchester, Hants SO23 8DL, UK (C.Rogerson@soton.ac.uk).
- James Beighton, Curator of Craft, Middlesbrough Institute of Modern Art, c/o 57 Gilkes Street, Middlesbrough TS1 5EL, UK (James_Beighton@middlesbrough.gov.uk).

Interpreting the woven devoré textile

Andie Robertson

ABSTRACT The devoré technique is unique within the field of textile designing due to its reliance upon fibre destruction as opposed to added embellishment in the creation of fabric patterning. Since the later part of the 19th century, devoré manufacturing has been used to both create and decorate a range of constructed textiles, including plain and pile woven fabrics, lace textiles and latterly, knitted devoré textiles. The development of metallic thread, viscose and cellulose acetate woven devoré textiles during the 1920s and 1930s established the contemporary devoré processes that are in use today.

The woven devoré process continually attracts new designer-makers. Its popularity with consumers is reflected in its continuous presence within leading fashion and interior fabric collections. Accordingly, the breadth of materials used within modern woven fabrics for devoré is ever increasing. Despite its continuing popularity with designers, the history of the devoré technique is neglected and poorly understood, and its presence in textile collections sporadic. This paper aims to track the technical and design history of devoré fabrics as an introduction for curators and conservators to encourage the identification, collection and preservation of this important class of textile.

Keywords: woven devoré, patterning, process development, preservation

Introduction

Generally classed as a textile-finishing technique, the devoré process of textile ornamentation involves burning away single or multiple fibres from a constructed fabric. The process differs from other 'special finishes' because of its reliance upon structural enhancement by fibre destruction rather than decoration through added embellishment (Lyle 1976). The appeal of the devoré technique to designers and manufacturers can be measured by its repeated adaptation and application. Since its development in the middle of the 19th century, the devoré method of fabric procedure has been employed by some of the textile industry's leading designers and engineers. While initially intended for use with fibres and threads that were previously unworkable, the devoré process, over time, was adapted to create and decorate a broad range of constructed textiles, including plain and pile woven fabrics, lace textiles and latterly, knitted devoré textiles. The textiles were used as fashion and furnishing fabrics. During the 1920s and 1930s, the development of metallic thread, viscose and cellulose acetate woven devoré textiles established the contemporary devoré processes that are still in use today (Fig. 1).

It might be assumed, therefore, that the origin and evolution of devoré should be established and widely understood, with the relationship between the various forms of chemically manufactured textiles such as woven and chemically etched lace and embroideries widely acknowledged, yet this is not the case. Both lace and embroidery have at times been highly fashionable and well documented, yet the manufacturing origin of woven devoré is little known and, until recently, received limited consideration or review. To date, the occurrence of devoré textiles in collections is sporadic and they are repeatedly overlooked.

A lack of clarity regarding the manufacturing origin may be responsible for the textile industry and the textile historian's neglect of woven devoré history. The simplicity of the textile's final appearance often belied the complexity and diversity of the fabric's manufacturing process. Moreover, the similarity of woven devoré textiles to woven textiles patterned during fabric construction clearly influenced the lack of recognition of these textiles. It may be that the design and patterning of the woven devoré textile has been perceived by historians as being imitative of established woven fabrics and, therefore, lacking in originality of pattern design or manufacturing.

For the most part, woven devoré textile history has been fashioned from a present-day point of view rather than based on any historical evaluation. Sweeping statements regarding its origins and employment during the 20th century are made with no real foundation. Specifically it has been velvet devoré that has received the most coverage over the last 20 years. The contemporary fashion industry's dependence on, or rather fascination with, velvet and pile woven devoré fabrics constructed of viscose and silk fibres, has seemingly overshadowed the adventurous range of natural and synthetic fibres that have previously been employed in conjunction with various forms of the process (Fig. 2).

Specifically, silk viscose velvet has become so associated with woven devoré, followed closely by silk viscose satin, that it would be easy to assume that this was the limit of devoré application (Fig. 1). Furthermore, limited past references regarding the nature and breadth of the process have done little to aid designers and researchers in discovering the likely manufacturing history of the process. This suggests the manufacturing techniques used in the creation of woven devoré textiles have disappeared from the textile industry's scientific memory.

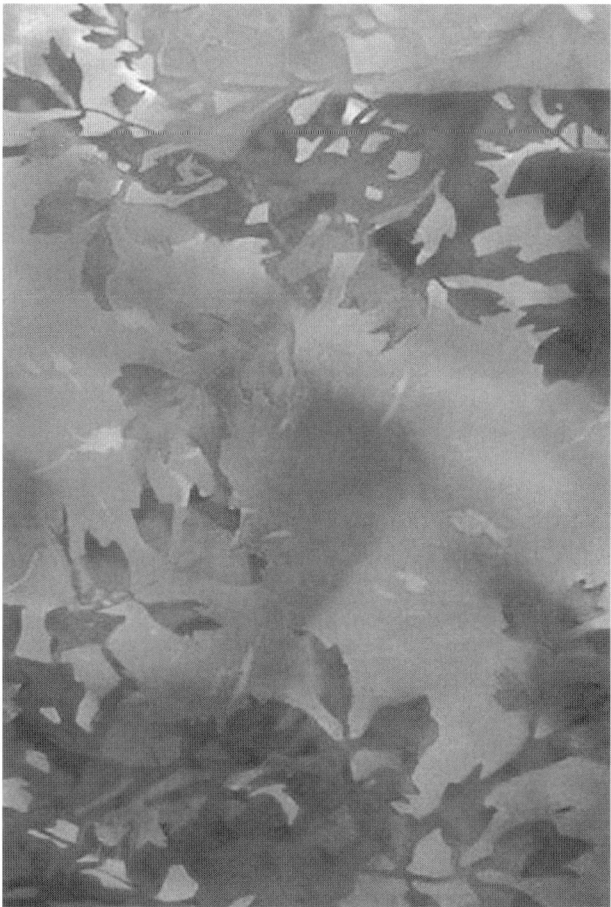

Figure 1 Silk satin viscose devoré print (Pippa Tinning) (Plate 8 in the colour plate section).

This paper aims to track the neglected technical and design history of the devoré process from its origins to the 1920s, the period of most active development. The breadth of fibres and processes used for the devoré technique, as well as the economic factors surrounding its development are considered. The paper will serve as an introduction for curators and conservators to encourage the identification and understanding, collection and preservation of this important class of textile. A study of patents detailing the technical processes that were developed over time and contemporary advertisements, alongside examination of textiles made using the devoré process, have provided much of the information to reveal the history presented here.

Technical origins

Devoré, or burn-out manufacture, is reliant upon the careful arrangement of vulnerable fibres and resistant fibres, which differ chemically from each other or are treated to behave differently, within one textile. The underlying structure of the textile, of which there are numerous combinations, has to survive the removal of part or whole areas of fibre by an application of, or immersion in, a destructive chemical reagent or solvent. By doing so, a decorative pattern is produced on the textile surface. An elementary understanding of the chemical composition of fibres, and in turn, their differential solubility or vulnerability to heat or applied destructive agent is thus needed.

The modern devoré process traces its origins to mid-19th century wool carbonising and those textile industries involved in the processing and cleaning of wool. Cellulose debris trapped in wool fibre was removed by saturating the wool stock in a chemical bath and then heating to ensure chemical disintegration of the contaminating matter, as substantiated by the textile patents of Steiger 1881, Suter 1883, Chaux 1883 and Scheppers 1887.

From an historical viewpoint, the development from wool stock cleaning by carbonising to the creation of yarns and textiles by carbonising appears to have been a logical evolution. Although the carbonising-based devoré process was eventually directed towards the patterning and decoration of constructed textiles, it was initially employed in the destruction of supplementary fibres purposely added to the constructed textile. For instance, a fabric backing used as a temporary scaffold in the embroidering of lace, a supporting scaffold yarn for staple fibres, or a temporary yarn included within a constructed textile during yarn manufacturing or textile construction. When no longer required, the temporary support was in each case removed by carbonising. Moreover, by using such a process the textile engineer could remove anywhere between 5% and 50% of a fabric's original structure or a supplementary yarn. As a result, the scaffold or supporting yarn was repeatedly employed in association with innovative fibres that were, at the outset, difficult to weave because of their delicacy or elasticity.

Examination of the historical patent record for textile design and manufacturing, chemical textile finishing and related chemical processes, including single fibre destruction techniques, shows that the seemingly diverse procedures of devoré or burn-out manufacture are based upon the carbonising concept. Each, however, uses one of three different procedures, these being:

1. Chemical disintegration as used in carbonisation where a textile is saturated in a chemical bath and then heated to ensure fibre removal.
2. The dissolving of a fibre using a suitable solvent.
3. Carbonisation produced by applied means (such as printing) in conjunction with heating.

Supplementary processes adapted to work with these procedures are:

4. Pre-treatment of fibres to either accelerate or resist carbonisation.
5. Printed resists to selectively protect or encourage fibre degradation.
6. Printed solvents to weaken fibre structure and the saponification of fibres to either encourage or allow fibre destruction, all of which can be used in localised fibre removal.

In contrast to the carbonising devoré process, the printing of a chemical to encourage fibre disintegration appears to have been for the most part developed by woven textile manufacturers, velvet finishers and associated printed textile industries. As a technique it was particularly favoured by velvet manufacturers because it was compatible with existing practices of printing and finishing, in much the same way that devoré fabric creation by carbonising seemingly suited those involved in the wool extraction industry.

As a form of fibre destruction, chemical printing could be accomplished by spraying, roller printing, stencil plates and later on screen printing. The process was also frequently used in combination with heating techniques that were already employed by velvet and other woven fabric finishers. Moreover, fibre destruction could be localised to specific areas of a textile which allowed for even the most delicate of patterned effects. For this reason it became a style of devoré manufacturing popular with small-scale manufacturers and the designer crafts-maker, such is its simplicity and reliability of production. Furthermore, larger textile manufacturers recurrently employed the applied chemical method of devoré fabric production for two major reasons: first, the applied method could be completed with great rapidity; secondly, devoré patterning could be altered almost instantaneously to suit design trends in patterning, whether on woven, knitted or stitched textiles.

A 'modern' product using 'modern' processes: early 20th century developments

Decorative fabric manufacturers began to employ standard printing machinery and equivalent patterns in the manufacture of pile devoré textiles towards the end of the 19th century. The textile industry was on the threshold of significant manufacturing change. Hand-machine methods of production were being replaced by more fully automated manufacturing and the development of woven devoré manufacture seemingly coincided with the reduction in hand-finishing practices. New methods of patterning both plain and pile woven fabric were to emerge during this period, with engineers in America and France pioneering decorative chemical etching techniques.

At the turn of the century, woven devoré fabric manufacture, together with embroidery lace textile processes, was part of the shift toward mass-produced commodities supplied to an even greater range of consumers. The devoré textile of the early 20th century was designed for both fashion and the interior. Its structural patterning was styled to reflect the designs of other figured textiles, often imitative of bespoke woven fabrics, with machinery commonly used in the creation of printed textiles adapted for use in devoré manufacturing. Meanwhile, the increasing sophistication of regenerated cellulose fibres began to impact upon woven devoré manufacture, with the treatment of fabrics containing the filament registered between 1910 and 1913. These regenerated cellulose fibre devoré processes were the technical basis of all 20th-century viscose devoré manufacturing, including contemporary procedures.

Accordingly, the woven devoré textile manufactured between 1900 and 1914 was a modern product. In all instances, its production during this period involved a wide range of machine-manufacturing processes, from the fabric's printing to its final washing and brushing. As a consequence, the manufacturing cost of the devoré textile was considerably lower than the labour-intensive, hand-manufactured textiles it was aiming to compete with in the marketplace. The standard of manufacture of the woven devoré textiles during this period is thought to have been variable, in part because roller-printed fabrics had a reputation for being low quality with their applied images and patterns often inaccurately aligned, but also because of the economic fluctuations in material costs experienced in the years prior to the outbreak of the First World War. The devoré textile seemingly occupied a unique position within the fabric marketplace; its clear patterned weave structure ensured it wasn't commercially bracketed with the printed reproduction textiles at the lower end of the market, yet it clearly wasn't intended to be marketed as a sophisticated, hand-woven or bespoke Jacquard woven textile. Overall, its method of devoré manufacture enhanced its visual styling, establishing it within the novelty and fancies category, a status the woven devoré textile – and particularly the pile woven devoré textile – retained for most of the 20th century.

Economic downturn and recovery during the 1920s

Progress in woven devoré textile manufacturing processes, cut short by economic downturn and the outbreak of the First World War, did not really resume until the latter part of the 1920s. The economic burden of the war continued to affect fibre and fabric production well into the middle of the decade. Immediately after cessation, British fabric manufacturers faced high-level demand from consumers for an insufficient supply of textile goods. The situation, however, was quickly followed by a rapid increase in productivity, brought about by the recommencement of civilian fabric manufacturing, which resulted in a reduction of fabric prices (Laver 1961). A broader selection of textile goods now reached a wider spectrum of consumers.

Economic stability in Europe and America, however, gave rise to a market for bold and highly decorative textiles, and consumers an opportunity to wear more novel styled dress textiles. Although only a small number of woven devoré textiles survive from this period, those fabrics that have been discovered are predominantly pile woven in style. Often they are constructed from mixed fibres such as silks, cottons, artificial silks or metallic threads. While plain and pile woven fabrics were equally popular weaves within patent registrations, figured pile woven textiles tended to dominate in magazine reporting and advertisements during the mid to latter part of the decade. The lack of surviving plain woven devoré textiles within textile collections may be the result of the greater popularity of the pile woven textile, or a difficulty in distinguishing devoré woven from traditionally patterned woven textiles. More likely, it may be because of the increased vulnerability of plain woven fabrics to wear and tear.

Despite the absence of woven devoré patents in the early part of the decade, at least 18 inventors registered some 16 woven devoré patents, with over half of the techniques being directed towards the utilisation of new 'artificial silk' fibres such as cellulose acetate. Woven devoré processes thus contributed to the development of novel cellulose acetate fabrics at the time.

The treatment of acetate fibre fabrics with a devoré technique was a meeting of two mutually compatible processes. The manufacturing basis of both cellulose acetate fibre and however, to show that interior woven fabrics such as mohair and cotton pile textiles were definitely patterned by devoré, and lighter weight linens and cottons etched with a gauze weave were created by a woven devoré process.

Development of devoré textile design during the 1920s

The patterning of woven devoré textiles in the 1920s, as revealed by patent specifications of the period, was typically floral, abstract or geometric in design, with reference made to checks and stripes, although hand-painted splashed and irregular drip (spotted) marks were also popular. Moreover, several of the 1920s woven devoré patents refer to the manufacture of brocades or brocaded styled fabrics, supporting the idea that woven devoré textiles were designed to be structurally reflective of current trends in woven textile manufacturing, rather than solely plagiaristic of historical patterning. The brocaded textile, traditionally a heavy fabric woven with gold and silver patterning, was widely employed as a fashion fabric during the latter part of the decade. Brocades were frequently categorised as a figured multicoloured woven fabric, with a Jacquard manufactured pattern in low relief (slightly raised) and formed by floating weft (filling) yarns. Metal brocades, including tinsel brocades and small patterned lamés were hugely popular with fashion designers and magazine editors alike.

The development of novel brocade fabrics was indicative of the changeable and exciting fibre trends of the decade, initiated at the beginning of the century. The inclusion of new fibres such as lightweight metallic threads or artificial silk within brocaded textiles was all the more significant because woven devoré processes also relied upon the mixing of chemically diverse yarn types. Arguably, woven brocades and devoré textiles inadvertently shaped each other's development and style at the time. Natural cottons, silks and wools had been consistently employed in woven devoré textiles, however, they too were now combined with metallic threads, cellulose acetate and other new yarns such as crêpes, viscose rayon and other artificial silks.

The patterning of woven devoré textiles, despite a move towards a distinctive recognisable fabric style, almost certainly drew upon historical influences. During the 1920s, the reproduction of historical decorative arts, such as the shapes, colours and patterns of textiles, was widespread practice. The employment of historically referenced patterns in 1920s woven fabric was not regarded as a plagiaristic activity on the part of the woven textile designer. Imitating former fabric designs clearly stimulated contemporary woven fabric design innovation during the decade; the interpretation and modern adaptation of past imagery or patterning can be evidenced in new or modern textiles of this period.

Extensive analysis of the patterning trends of woven devoré revealed that there were clearly two distinct approaches to a fabric designed by devoré. The first style of devoré patterning removed little fibre, creating an etched effect into the fabric so that the lighter, now visible ground, created the design. The second patterning style differs from the etched style in that it involves the removal of a large amount of fibre. An early example

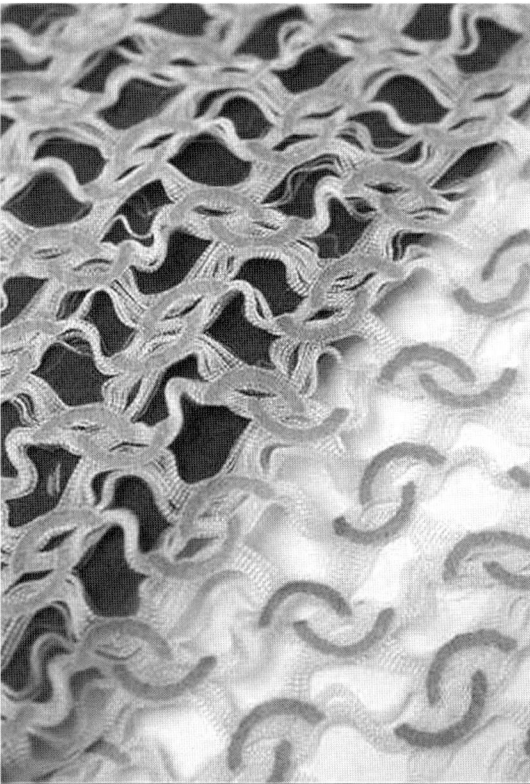

Figure 2 Latex printed silk cotton elasticated fabric (Claire Harbin) (Plate 9 in the colour plate section).

the devoré technique of chemical disintegration (specifically simple dissolution) were technically ideally suited. Since this innovative cellulose acetate fibre was eventually to become regarded by its inventors and their textile customers (both fashion and interior fabric manufacturers) as an economic substitute for cotton and other cellulose fibres, it was employed as a direct substitute for cellulose fibres previously incorporated within woven devoré textiles. Chemical technology associated with cellulose acetate devoré techniques gradually became more complex, possibly as a result of the pioneers of cellulose acetate fibre development being involved in the development of suitable cellulose acetate/woven textile devoré techniques. Despite this, little obvious change in the patterning of the actual devoré textile occurred when cellulose acetate fibre was utilised. Only the lustre of the acetate cellulose fibre would have revealed the fabric's true fibre origin, and even this changed when new delustering treatments were developed to soften the sheen of the cellulose acetate fibre.

In conjunction with cellulose acetate woven devoré development, woven devoré processes that utilised natural fibre combinations were patented by a range of textile manufacturers. The rationale of these technologically straightforward processes was the production of highly decorative and elaborately styled fabrics for use as commercial dress textiles. While both plain woven and pile woven textiles feature within the American and British patent record, the surviving woven devoré textiles discovered within fashion textile and vintage clothing archives suggest that the woven devoré fabrics of the 1920s were principally designed for highly decorative women's apparel. Although interior devoré fabric patents were registered during this period, no actual examples of interior devoré fabrics have as yet been found. There is sound evidence,

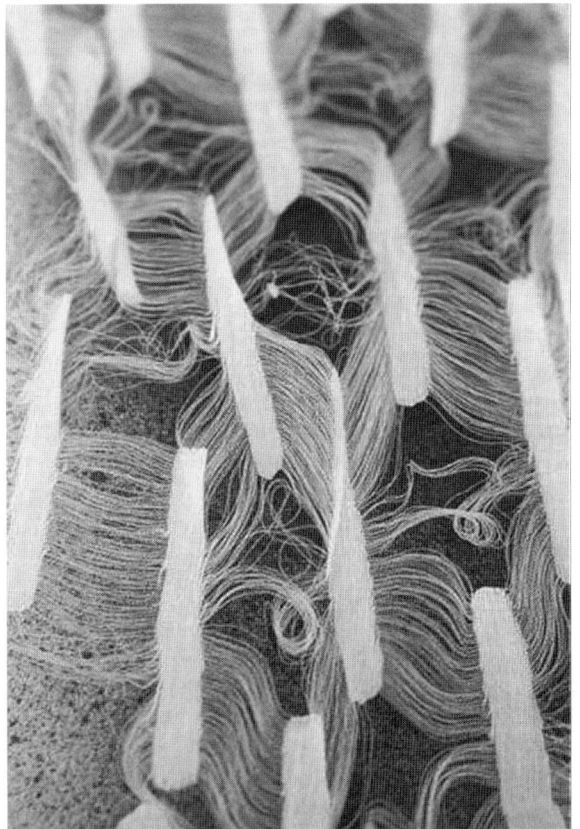

Figure 3 Silk viscose plain weave fabric devoré, printed. Sketchbook swatch (Claire Harbin) (Plate 10 in the colour plate section).

The evolution of fashionable and bold 'modernist' repeat patterns, simple in their composition, colouring, shape and linear outline, were also a popular choice for the design of woven devoré fabrics. The technique was suited to the style because chemically etching geometric shapes from a comparable geometric structure, whether a plain or pile weave, tends to work well. Furthermore, the woven devoré technique allowed the designer to create multidirectional patterning, for instance angular, linear or silhouette shapes diagonally across the fabric, on the bias. Essentially, the composition of a design could transform from minimally etched design in one area to mass fibre removal in another, as dictated by some modernist designs. Such an adaptable and versatile style of devoré designing was employed successfully in conjunction with velvet and lamé pile woven textiles towards the end of the decade. Moreover, in terms of woven devoré fabric patterning, the use of repetitive geometric designs of a small to medium scale was also highly effective and a common feature of many figured textiles of the time.

The widespread use of this style of patterning on woven devoré textiles led to it becoming a design characteristic of the pile woven devoré textile, still popular to this day. Meanwhile, the successful association between small-scale repetitive patterning and pile woven devoré textiles may explain why the term *façonné* has in the past been commonly used to describe pile woven devoré textiles (Anstey and Weston 1997). More usually, the term is applied to figured Jacquard woven fabrics with a pattern of small, scattered motifs, that are generally singly coloured.

of this cut-out style devoré is illustrated by Otto Timme's patent of 1902 whereby a devoré print removes large areas of the pile to reveal the ground, while the remaining pile creates a relief pattern (Timme 1902). In Timme's patent, a three-colour floral design and a devoré paste were roller printed simultaneously onto the fabric and then processed by steaming. Based on the date of registration, the visible ground was likely to have been a plain weave in cotton and wool pile.

As the patterning and structural design of woven fabrics became more trend directed by manufacturers and designers, and reflected in the seasonal forecasts within women's magazines, the design of the woven devoré textile was also likely to have undergone rapid design changes. With the influence of exoticism, motif and visual imagery from China, Japan, ancient Egypt, Persia and Africa permeating the haute couture houses and the fashion designs of the avant-garde, the textile manufacturers purportedly translated these new themes into woven devoré textile patterning. Printed woven devoré processes could be speedily adapted to suit prevailing tastes in fibre trends and patterning. Woven devoré manufacturers could thus modify their fabric designs to exploit the popular designs of the printed and figured woven textile market.

Another trend within textile print design, during the early to middle part of the decade, towards simplification of established complex forms and shapes, particularly florals, is similarly reflected within woven devoré production. At this time, over half the patents registered suggest floral patterning as a suitable design for woven devoré fabrics. Of the woven devoré textiles still in existence, abstract florals or geometric-styled patterning predominate.

Cutting through the confusion: devoré manufactured pile woven textiles

The trend for woven (non-devoré) geometric figured pile fabrics during the mid to latter part of the 1920s may have led consumers to consider the devoré manufactured pile woven textile a member of the 'cut velvet' textile family. Although probably cheaper to purchase than the fully woven fabric, the pile woven devoré textile was not necessarily the low-quality product often supposed by some textile commentators. Cut velvet fabrics, figurative pile designs woven to show areas of pile against plain woven ground work, were simply and economically replicated using the chemical printed and printed resist devoré techniques. The pile textile used in their replication, however, need not have been inferior in quality. In reality, it may have been equal to or superior in yarn quality or manufacture to other figured fabrics of the time.

With the development of new transparent velvets with their transparent ground weave during the mid part of the 1920s, the pile woven devoré process truly had the potential to establish itself as a viable alternative to the woven cut pile fabrics (Linton and Pizzuto 1961). Transparent velvets were noteworthy as a style of fabric, at the time, because they could be worn in many guises throughout the day. Some examples of transparent cut velvets still in existence may in fact have been patterned by devoré processes. There appears to be a general impression among textile historians and archivists that natural silk pile fabrics were the predominant fabrics employed with the technique during the period. Furthermore, the seemingly widely held assumption that a fabric constructed of a silk transparent ground weave and viscose or cellulose acetate, tinsel pile could not have been treated by the devoré process, is not supported by the patent record.

The full impact of the relationship between the 1920s transparent velvet and woven devoré processes perhaps only becomes evident some 60 years later. In the late 1980s, the resumption of the manufacture of transparent velvet devoré, this time using a fabric with silk ground and viscose pile, sparked what can now only be described as the longest running revival of the 1920s styled woven devoré processes. By studying revival fabrics of the 1980s alongside 1920s textiles, it becomes clearer that the original style was reflective not only of the exotic inspiring the dress and textile styling of the 1920s but also the move towards freedom of movement in women's dress. The manufacture of elegant pile fabrics with lighter gauze or ground weave in the 1920s was in response to the 'simplified and unstructured silhouettes of the period' (Fashion Institute of Technology 1991–92). Moreover, devoré textile manufacture, at least in part, still apparently relies heavily upon its past styles and influences for the creation of a contemporary devoré fabric.

Collecting and preservation: the logical conclusion for documenting the history of devoré textiles

Despite the devoré technique being a scantily documented and analyzed technology, the increase in woven devoré manufacturing in the early 1920s considerably influenced woven devoré textile development of the late 20th century. Consequently, the limited collecting of woven devoré textiles is surprising, as woven devoré has repeatedly attracted the interest of the textile industry's leading manufacturers and engineers. If a woven fabric appeared to have been manufactured by devoré, it is questionable whether it would have been considered worthy of collecting and preservation. The exceptions to this rule would be textiles considered to be particularly high quality or of design significance. The woven devoré textile could, however, be as exclusive as any other highly patterned fabric. The association of the devoré process with mass production, middle market fashion and novelty textile manufacturing, has seemingly contributed to the widespread lack of preservation of these textiles. Furthermore, an outcome of the renewed interest in past woven devoré production is the realisation that little is known of the historical development of the process, which the author has aimed to address.

An additional hindrance to the presence of devoré in textile collections is the problem of distinguishing a devoré textile from a traditionally woven textile. For instance, recognising a cut velvet fabric as distinct from a devoré velvet fabric is often only possible by looking closely at the weave configuration. Analyzing the ragged or fluffy quality of fibre ends is often the only way to determine woven devoré manufacturing. Such expertise and experience is currently all too rare among textile specialists, historians and conservators. The creation of a digital record that would gather together information regarding the design and fabric structure, state of preservation and museum or archive location would perhaps be a start in providing the information that is currently absent from textile scholarship.

Although devoré is a destructive process that seemingly contradicts the idea of textile creation, the woven devoré process of fibre removal by etching a design into the structure of a textile appears to cause little long-term damage to the remaining fabric (Fig. 3). In the author's experience, the conservator only has to contend with the normal and often difficult issues surrounding fibre degradation and modern textile preservation. These issues notwithstanding, a comprehensive selection of devoré textiles within textile collections would not appear to present particularly complex or resource-stretching problems for a museum. Instead, the presence of devoré textiles in museums would inform and educate visitors about an important and influential realm of historic textile production that still has resonance in textile manufacture today.

References

Anstey, H. and Weston, T. (1997) *The Anstey Weston Guide to Textile Terms*. Cambridge: Woodhead Publishing.

Chaux, L. (1883) *Art of Manufacturing Woolen Fabrics from Short-Staple Fibre*. US Patent 288,015.

Fashion Institute of Technology (1991–92) *Extravagant Lengths: Velvet, Plush and Velveteen*. Exhibition at the Fashion Institute of Technology, New York, 19 November 1991–11 January 1992.

Laver, J. (1961) *Between the Wars*. London: Vista Books (with quotes from Right Hon. C.F.G. Masterman, 1922, *England after the War: A Study*, London, Hodder and Stoughton).

Linton, G.E. and Pizzuto, J. (1961) *Applied Textiles: Raw Materials to Finished Fabrics*, 88. New York: Duell Sloan and Pearce.

Lyle, D.S. (1976) *Modern Textiles*. New York: John Wiley and Sons Inc.

Scheppers, E. and Scheppers, E. (1887) *Art of Making Fabrics from Coarse Long-Staple Wool or Hair*. US Patent 362,317.

Steiger, J. (1882) *Improvements in the Production of Embroidery*. GB Patent 4143.

Suter, F. (1883) *Process of Producing Open-work Fabrics*. US Patent 280,094.

Timme, O. (1902) *Method of Producing Figured Pile Fabrics*. US Patent 705,977.

The author

Andie Robertson is currently a Senior Lecturer on the BA (Hons) Textiles and Surface Design Course at Buckinghamshire Chilterns University College, coordinating print and dye technology. Her current design interests include devoré textile innovation, a response to her research into the history of the devoré textile, 1880 to the present. Andie is now developing a devoré research website, which will focus on historical and contemporary devoré textiles. She is also creating a new body of devoré textile work to be exhibited in 2007/08.

Address

Dr Andie Robertson, Buckinghamshire Chilterns University College, Queen Alexandra Road, High Wycombe, Bucks HP11 2JZ, UK (arober01@bcuc.ac.uk).

What makes a textile modern? The recycling of clothing in the Punjabi shoddy trade

Lucy Norris

ABSTRACT This paper seeks to broaden the frame of reference of 'modern materials' by arguing that, alongside innovative developments in fibre technology, contemporary techniques of production and culturally specific consumption practices are necessarily implicated in the concept. The industrial recycling of textiles in the modern world involves the processing of a wide variety of both natural woollen and plastic fibres whose provenance may be deeply obscured, as their successful reincarnation relies upon the total destruction of the original fibre source.

Cast-off clothing from the West is sorted, baled and shipped around the world by the container load to recycling centres in the developing world. The garments are then shredded, carded and re-spun. The town of Panipat in north India is one of the largest shoddy recycling centres in the world, creating new 'Indian' products. The reclaimed shoddy thread is either knitted up into sweaters or woven into blankets, shawls and suiting. It is very much a material of the moment, whose past history must be eradicated, yet the end products must appeal to local consumer preferences to be successful. Therefore new images which resonate with both Indian traditional values and those perceived to be of a modern global society are highly esteemed.

Keywords: recycling, wool, shoddy manufacture, Indian iconography, ethnography, materiality

Introduction

This paper will demonstrate the value that ethnographic research can bring to the study of materials. It suggests that longer-term, qualitative research in the field, the dominant methodology of ethnographers, offers a unique perspective that complements those obtained through conservation practice, the study of museum collections and historical research on archives and supplementary material. Ethnographic research not only provides a useful contribution to the search for new methodologies, policies and approaches for the consideration of modern materials in particular, but can expose the subtle transformations that take place in their creation that other disciplinary approaches may inadvertently miss.

In the study of material culture by ethnographic means, one of the most successful methodologies is that of following the lifecycle of objects, espoused by Appadurai (1986), where objects pass in and out of circulation, at one point a commodity, before being singularised through incorporation into the social milieu of a person (Kopytoff 1986). The project referred to below forms part of such a study, following the recycling of cast-off Western clothing as it is processed in India. This research into contemporary recycled textiles from India problematises the concept of 'modern materials' through the study of their manufacture, design, marketing and consumption. It stresses the usefulness of maintaining a broad frame of reference when attempting to consider what is 'modern' within the field of textiles. While innovative developments in fibre technology may be thought to constitute modern materials, the study of the conditions of modernity, such as contemporary techniques of production and the division of labour, the design, marketing and branding of goods and culturally specific consumption practices are necessarily implicated. The examples presented here briefly help to both define the category of the 'modern' while revealing the essentially blurred nature of its boundaries through the constant creation, destruction and recycling of the object.

The industrial recycling of textiles in the modern world involves the processing of a wide variety of both natural woollen and plastic fibres whose provenance may be deeply obscured. This invisibility fulfils the expectations of Michael Thompson's rubbish theory, where transient goods have to decline in value to a point where they are consigned to the category of rubbish before they can be 'reclaimed' and invested with new value as durable items (Thompson 1979). For such a transformation to be effected, an element of duplicity has to take place, a sleight of hand whereby the object can be represented to us once more as a desirable thing. When cast out as rubbish it is interesting to note that objects are essentially 'worthless', they have been taken out of circulation; objects in museums are equally said to be 'priceless' while out of the market. Rubbish is usually dealt with behind the scenes of everyday life, just as museum collections are kept hidden in stores or segregated from the viewer in the context of a museum display. This secrecy is especially important in India, where the reuse of materials such as clothing, so closely associated with the body, presents a potential problem for consumption due to cultural beliefs in pollution and attitudes to hygiene. For this reason, their successful reincarnation relies upon the total destruction of the original fibre source and their re-presentation as new, culturally appropriate objects for sale.

Tracing the cycles of production, consumption and disposal of used clothing, from the domestic sphere to that of fibres traded on the global market, allows for a re-examination of a multiplicity of concepts often conceived in opposition to each other, principally those of 'old' and 'new', 'clean' and 'dirty', 'pure and impure', 'fake' and 'authentic' and 'traditional' and 'modern'. These recycled textiles are extraordinarily complex composite objects that provide a peculiarly pertinent example of the difficulties involved in defining the 'modern'.

Casting out

As many of us will know, clothes which may have once constituted a vital part of our self-image can come to be seen as no longer useful – an embarrassment even, a surplus of waste that could be put to better use. No longer desired, these discards are peeled away from the body, as a snake sheds its skin, through routine iconoclastic practices of detachment and riddance. Indeed, it is the effacing of former selves that is the true goal of periodic wardrobe clearances, just as its counterpart – hoarding unused clothing – can be associated with retaining former selves (Banim and Guy 2001). Strategies of handing on to family and friends, and domestic recycling practices such as patchwork and quilting often deliberately preserve the bodily traces and sentimental reminders of former owners and act as repositories of memory (Stallybrass 1993). Many clothes, however, appear to be simply thrown out into the unknown.

In the West, this often means giving them to charity organisations or commercial collecting companies (Palmer and Clark 2004; Gregson et al. 2000). Used clothing, originating in developed countries which cannot be resold locally, is then bought by international recycling businesses who resell it as a commodity on the global market, where its final destination is often the developing world (Fig. 1).

There it can be regarded as a useful resource, for example in an impoverished economy such as in various sub-Saharan African countries, whose inhabitants are unable to afford new clothing. Hansen's work on Zambia has highlighted the complex structure through which local middlemen buy up old clothes (*Salaula*), which are refashioned by tailors and become highly desirable clothing for the upwardly mobile (Hansen 2000). But they can also be viewed as a threat to indigenous industries, resulting in various levels of protectionist import restrictions or outright bans (for example, in India).

During fieldwork in Delhi and the Punjab, the author encountered huge piles of cast-off imported clothing, mountains of trash awaiting transformation into treasure. There is a flourishing black market trade in selling used clothing on the streets directly to consumers in India, and this is facilitated by cross-border smuggling and official corruption. The research that is the focus of this paper, however, deals with the little known legal trade in imported woollens. Container loads of old jumpers and coats are shipped into massive warehouses in the Punjab in northern India, where crucially they remain invisible to all except those who work with them. These constitute the raw material from which new, 'Indian' products will be manufactured using the shoddy process. The outline of this trade and the technology of the process have been described elsewhere in more detail (Norris 2005), so only a brief overview is provided below.

The recycling of imported clothing

The shoddy trade originated in the UK at the turn of the 19th century with the development of the 'pulling process'. Woollen garments are shredded in order to reclaim the fibres; this is then turned into shoddy yarn, which can be used for knitting and weaving. Pure virgin new wool is expensive to buy in the

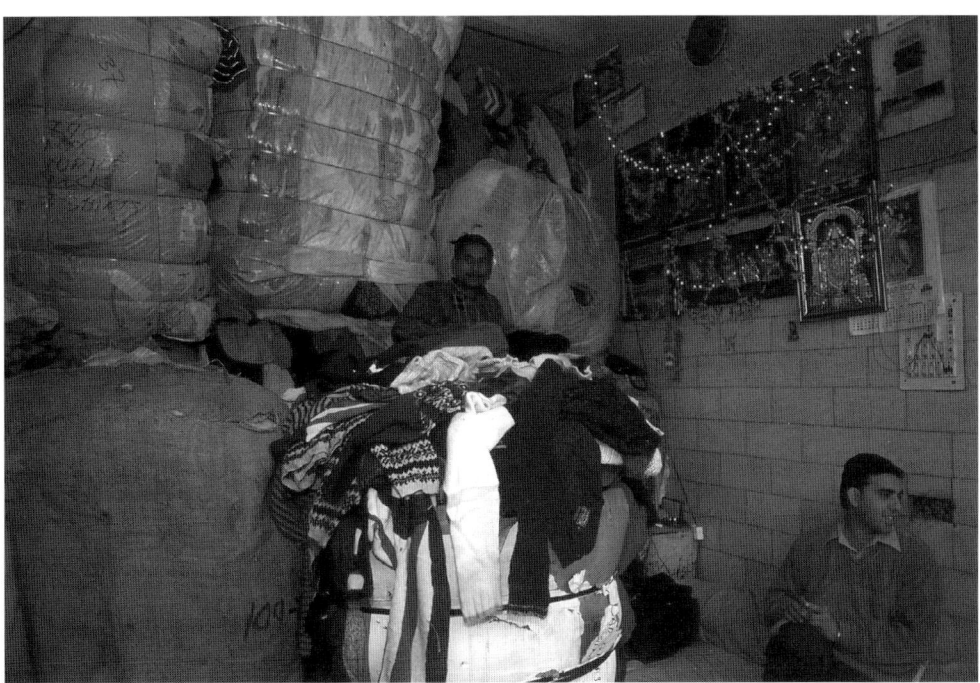

Figure 1 A bale of cast-off clothes sold for export (photo: Tim Mitchell) (Plate 11 in the colour plate section).

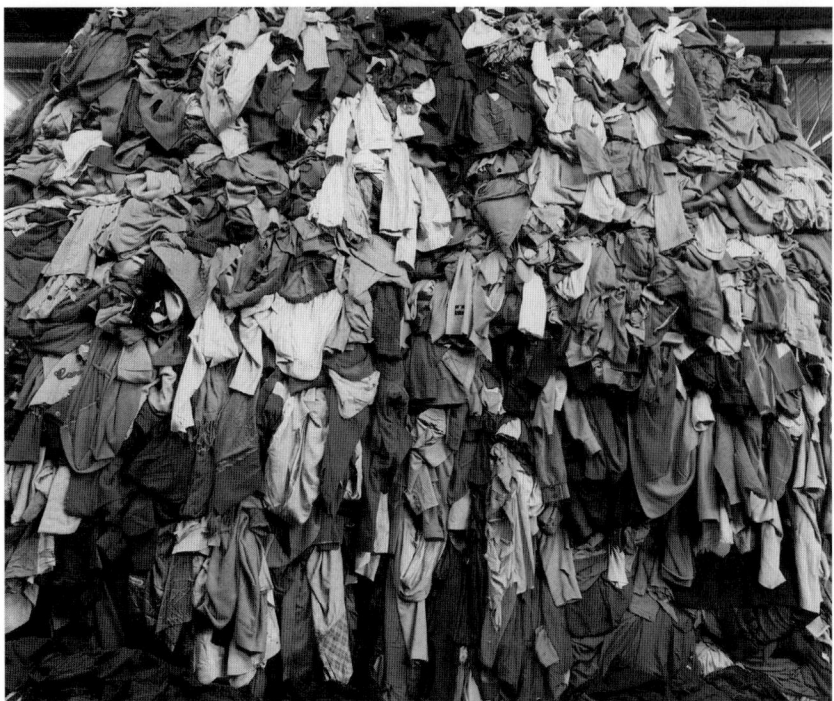

Figure 2 Piles of 'mutilated hosiery' sorted by colour and waiting to be processed (photo: Tim Mitchell) (Plate 12 in the colour plate section).

international market – much of it comes from Australia and New Zealand. Many developing countries such as India lack the natural resources to produce their own wool in sufficient quantities and cannot afford to import either the new wool or invest in the latest spinning and weaving technologies it requires in order to manufacture good quality garments. A synthetic hosiery business exists in India, based in Ludhiana, but the cost of the fibre is linked to the international price of oil and at times can become prohibitively expensive for the poorer sections of society. The availability of cheap, lower quality shoddy wool therefore supplies the lower end of the indigenous market with affordable clothing and blankets.

In India, the import of woollen rags is allowed to fuel the shoddy recycling industry now located in the Punjab. Imported rags are a permitted commodity that attracts a tariff of only 40% whereas used clothing is currently banned altogether (as of December 2004). In order to try to control the smuggling of illegal imports of wearable garments, the Indian government insists that all used clothing is slashed at source by large machines known as mutilators, creating a product generically known as 'mutilated hosiery' (Fig. 2).

In Delhi, many of the used clothing importers deal in both illegal garments and legal mutilated hosiery, diverting the former into the black market and the latter to the recycling mills in the Punjab, but it is difficult to estimate numbers. Many traders are local operators who buy and sell small quantities from middlemen depending on the vagaries of cash flow and market demand. Several successful dealers have, however, managed to establish global networks and amass large fortunes in the trade, and Indian owners are now replacing traditional Jewish firms in the United States.

A few hours drive north of Delhi, Panipat is known to the educated elite and tourists as the site of three major battles in the 16th and 17th centuries which helped to establish the Moghul empire, but it now markets itself on its website as the major centre for the recycling of mutilated hosiery in India – if not the world – with up to 300 mills processing shoddy from old clothing. There is a similar industry in Amritsar, near the Pakistan border, though they claim that the quality is higher there due to the softness of the local water.

The process of manufacturing shoddy yarn is to a large extent determined by the actual garments available, and the availability of the technology for their transformation through capital investment in machinery (itself recycled from Italy and Poland) and raw materials. By this stage, clothing quality is measured wholly in terms of fibre length, strength and colour rather than style or form of the garments.

There are three main grades of 'woollen' clothing imported into India:

1 'Commercial all wool' (CAW): 70–80% wool hosiery, i.e. jumpers, scarves and hats.
2 'Acrylic loose knits' (ALK): 100% synthetic hosiery, i.e. jumpers.
3 'Original woollen rag' (OWR): 70–80% wool cloth, i.e. old coats.

When buying clothing in the international marketplace, the reputation of the source companies in the West is very important in judging the probable fibre quality and colour of the bales' contents. Reputable Western companies are known to employ careful sorters and graders: they have a first and a second grading and do not mix fibres. The names of firms such as Allied (Houston), International (Texas) and HB (Hebden Bridge, UK) act as brands that guarantee quality; for example, HB is known for the best quality English wool coats and jumpers.

Buyers favour new and 'bright' shades rather than dull; 'bright' means intense, such as royal blue, while 'dull' can refer to 'English' shades, pastels as well as greys, browns and olive greens. The purity of colour in the used clothing is extremely important; the fibres are not re-dyed, and the purer and brighter the colours obtainable in the raw material, the more valuable the end product. Buyers cannot choose colours and are reliant on what they find in the bales. If certain shades are fashionable in India, such as camel was in 1999, the recyclers cannot provide them in any quantity; their poor clientele have to take what comes. Yet the huge scale of the trade results in a wide array of colours mixed in bales, so bulk buying allows for a degree of control over the design of end products. The OWR from northern hemisphere winter coats, however, is overwhelmingly comprised of dark greys, browns, greens and blacks, in checks, tweeds and so on, with only a few groups of bright shades. Manufacturers with old equipment can make a profit by using these materials, but the end products are unsurprisingly sludgy colours and receive the lowest market rates.

This mutilated hosiery is sorted further into a range of 'colour families', e.g. peacock, *rani* (hot pink), *mehendi* (henna), American beauty (a bright red shade), saffron and checked. Larger colour families are next sorted into subgroups of light pastel, dark and bright shades. The 'fancy' groups – multi-coloured checks and tweeds – are occasionally re-dyed black once sorted.

The garments must then be turned into pieces of cloth, ready for the mechanised shoddy process. This is carried out solely by women workers, squatting outside the factory buildings, often accompanied by young children and babies. They strip the clothing of all their non-fibre components, such as labels, buttons, zips, press studs, leather patches and trimmings, shoulder pads and linings. Buttons are burnt off and discarded, while the mainly synthetic coat and jacket linings are removed and bought by local rubbish dealers and carted away on overflowing trucks. The remaining cloth is then cut into pieces about 50 cm square by the same women, who sit on traditional wooden vegetable cutters with curved steel blades.

Although most of the by-products can be reused, the garment labels themselves are not recycled at all. Once advertising information vital to the value of the original garment, this is now worthless, or may even detract from the re-evaluation process. Labels are complex documents: in addition to advertising well-known brands such as Gap, Levi's, Calvin Klein etc, they provide the material evidence for the history of the 20th-century transnational trade in new clothing, charting the temporal and spatial network of fibre sources, designs, manufacture and retail companies across the globe which make up a garment. In aggregate, a pile of labels on the floor reveals a staggering mosaic of global transactions, which must be eliminated during the transformation from rubbish into desirable new products.

The clothes are then broken down into a tangled mess of fibres and soaked overnight in a solution to which is added 'bettering' oil (a type of diesel), to enable them to pass through the machinery easily and prevent the build-up of static electricity. The fibres are then fed through huge mechanised processes to tease, card and spin them into new thread which is wound onto spools made from old newspapers.

The purer the wool content in used clothing and the longer the fibre staple, the better the quality; a stronger, finer thread can be produced by recycling. The oldest recycling machines can only produce poor quality, thick, heavy wool products with a short staple, hence returning a lower profit on the finished goods, while newer machinery can produce a finer yarn. Woollen yarn is classified into 'counts': the higher the count, the finer the yarn. The quality of the end product therefore depends upon both the raw material and the technology available.

These spools of coloured yarn are then either woven or knitted into new products. Different production methods and end products are centred in different areas, although not exclusively. Panipat manufacturers can achieve only an 8½ to 10 count thread as their machines are the oldest; in Ludhiana they make 10, 12 and 14 count threads; and in Amritsar 12 to 13 count. Panipat is known mainly for its cheap shoddy blankets and shawls made from OWR, Ludhiana is a centre for synthetic hosiery and knits products from recycled ALK as well as using new materials such as cotton, while Amritsar weaves better quality woollen cloth from CAW to be made up into clothing for the Indian middle classes and export garment trade.

The suiting from Amritsar capitalises on the straplines of modernity to create an up-to-date fabric with an international feel – this is achieved by copying well-known designs such as tartans and checks in modern colours and weaving phrases such as 'Millennium 2000' and 'Star Plus' into the selvedge (Norris 2005). Below, however, another recycled product made from this shoddy yarn is focused on, i.e. low-quality blankets from Panipat. In contrast to the suiting from Amritsar, Panipat manufacturers appear to utilise traditional design motifs to 'Indianise' their products.

Indianisation: the vase of flowers motif

Panipat mills produce a range of very low-quality sludgy coloured blankets and shawls, many of which are used as army blankets or bought by relief agencies. Men's shawls and suiting are also woven from old coats of indeterminate colour. By using yarn with a higher percentage of acrylic, however, brightly coloured shades can be woven alongside darker ones, in bold stripes, checks and repeating motifs using computer-aided design (CAD). Woven on jacquard looms in long lengths, they can then be 'fluffed' up through steam treatment, creating thicker products that are then cut into blanket sizes; their edges are bound with satin fabric and a new product label is sewn on. These are exported to Russia and other developing countries or sold in India, each fetching between 100 and 150 rupees. Some shopkeepers order in bulk and distribute them via wandering sellers, who make perhaps 10 rupees from each one. They are also sold through Sunday street markets such as that in Delhi opposite the Red Fort, for several hundred rupees.

These blankets have no clues as to the origin of their fibres and the labels proclaim them to be Indian products with names of the gods often emblazoned in eye-catching logos. Indian buyers from the middle classes would consider it unthinkable to purchase cloth made from such inferior materials and claim to not knowingly buy second-hand cloth in the market for fear of ritual pollution and hygienic contamination (although it is

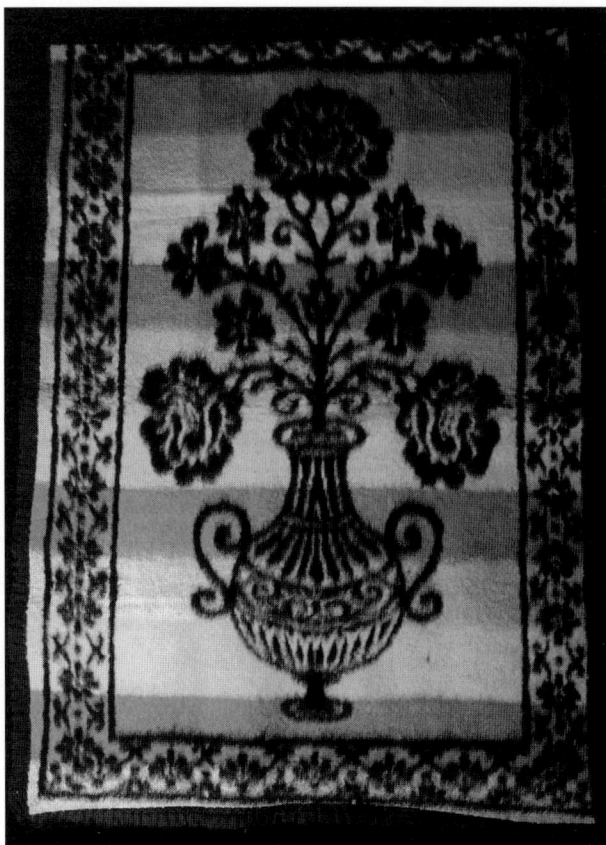

Figure 3 A baby's blanket with vase of flowers design made from recycled jumpers (photo: Lucy Norris) (Plate 13 in the colour plate section).

certain that some people do surreptitiously buy up second-hand clothes). But unknowingly, customers in India and around the world are buying products made from very old and dirty used clothing. All of these blankets examined in India by the author were thick and gritty to the touch, with a greasy texture and unpleasant smell. At no point, from being ejected from the Westerner's wardrobe, via travelling across the globe, being shredded and rewoven, are these garments ever washed; the human dirt and environmental pollution encountered along the way literally cling to their very fibres. They still possess a certain greasiness and whiff of mechanical oil about them, but such smells could be attributed to the laundering process which uses petroleum derivatives to clean stains from saris.

The successful recycling of such material to a wider discerning market requires manufacturers to skilfully navigate such customers' value systems through various strategies of image transformation and marketing. These blankets have new labels and logos attached which appeal to the Indian market and are wrapped in plastic proudly emblazoned with 'Product of India'. A smaller baby's blanket made from a fluffed acrylic-wool mix (now in the Pitt-Rivers Museum, Oxford), bought for about 250 rupees, claims to be an 'Indian' product. It features a vase of flowers set against a striped background with a floral border (Fig. 3). It is one of a whole genre of blankets that feature versions of either the 'vase of flowers' design, latticed outlines or niches filled with flower motifs and flowering trees.

What is striking about these fluorescent blankets in highly contrasting colours offset against black edging and backgrounds is the familiarity of the designs. These motifs are based upon an extremely old repertoire: the brimming vase of flowers motif was known in early Indian culture by the Sanskrit term *pūrna kalasha* or *pūrna kumbha* (Kramrisch 1946). It is a symbol of abundance and thus a boon, found in Buddhist temples from at least the 1st century AD (Knox 1992). The motif is also found in early Islamic art from Pakistan and Iran, where flowers and trees are symbols of paradise on earth, so it can also be said to be Indo-Persian in origin. Certainly during the fruitful hybridity of Moghul art in the 16th and 17th centuries, the design known as *pūrna ghata* was incorporated into columns, friezes, ceramics and textiles, and was also influenced by the increased contact with foreign cultures, especially European baroque (Koch 1982, 1991). Indeed Koch argues that it was used as a three-dimensional element after woodcuts by Lucas Cranach the Elder were circulated in 1531 (Koch 1991). During the reign of the Moghul emperor, Shah Jehan, in Delhi, naturalistic flowery plant motifs found in architecture were also influenced by European herbals; artistic expressions of a paradise on earth existing at the imperial palaces sat alongside Islamic injunctions against the depiction of animated beings. Their use in relief work and composite inlays was taken from Florentine techniques. The design has also been found on an 18th-century tile from Multan, Pakistan (Yaldiz *et al.* 2000).

There is a clear correlation, however, with their continued use over a long period of time in textiles as well as solid architectural form. Numerous examples abound, which will be familiar to textile historians, and they are linked with variations of the tree of life motif, rooted in hillocks, Chinese-influenced clouds or bare-rooted. The vase of flowers is connected to the development of the *boteh* found on Kashmiri shawls from the 17th and 18th centuries, and which developed into the ubiquitous 'paisley' motif (Ames 1986; Irwin 1955, 1973). Indeed Ames suggests that at the height of the shawl's popularity, when worn by Josephine during the French Napoleonic empire, the influence of Greek classical urns was also essential to the cultural aesthetic, and that this was transmitted back into Kashmiri design after the Sikhs took over Kashmir in 1819. Similar designs find a place in 18th-century Kashmiri *millefleur* carpets, examples of which are discussed by Walker (1997) and are still popular today (Gans-Ruedin 1984). Painted textiles from 1700 onwards, such as the sought-after *palempores* and chintzes imported into Europe from further south in the Indian subcontinent, also featured versions of the vase of flowers; these were heavily influenced by the circulation of European prints and Chinese ceramic and textile designs, and thought to be later adaptations of the tree of life (Barnes *et al.* 2002; Irwin and Brett 1970; Irwin and Hall 1971).

It is clear that a hybridity of design influences linked to local and national shifts in power and the opening up of new trading opportunities found expression in a variety of material forms, both ephemeral and portable such as textiles, and in massive stone palaces and temples. The overall design repertoire, however, is still identified strongly with the Indian subcontinent and a local consumption base that supersedes either a Hindu or Muslim design aesthetic. Recycling this image once more in the 21st century has reclaimed the anonymous fibres woven into the blankets as Indian products, attaching the perceptual qualities of a familiar image popularly understood to be both indigenous and a lucky symbol onto the product.

Conclusion

The conditions of modernity result in the existence of a surplus of goods in the developed world that can become rubbish, yet be recycled once more on the other side of the globe within a very different cultural framework. Stretched commodity chains embedded within complex social structures enable end products to be culturally reincorporated due to their obscure origins – new meanings are interwoven with the fibres as they are re-formed into artefacts once more. Entrepreneurs rely on invisibility and secrecy to produce a valued new product from the heaps of unwanted stuff which they collect surreptitiously. The investigation into the murkier workings of the used clothing trade reveals a complex pattern of continuously ravelling and unravelling relationships ever ready to respond to challenges in the market to create modern hybrid products. As their constituent fibres pass through reincarnations from virgin wool jumpers sold by global brands to dirty old blankets remade and repackaged as Indian products, at what point can these textiles be described as 'modern'? Or are they merely moments of composition in ever-evolving processes of creation and destruction, where the origins of culturally complex objects are deliberately concealed and can only be discovered by researching the networks through which they pass?

References

Ames, F. (1986) *The Kashmir Shawl and its Indo-French Influence.* Woodbridge: Antique Collector's Club.

Appadurai, A. (ed.) (1986) *The Social Life of Things: Commodities in Cultural Perspective.* Cambridge: Cambridge University Press.

Banim, M. and Guy, A. (2001) 'Discontinued selves: why do women keep clothes they no longer wear?' in *Through the Wardrobe: Women's Relationships with their Clothes*, A. Guy, E. Green and M. Banim (eds), 203–20. Oxford and New York: Berg.

Barnes, R., Cohen, S. and Crill, R. (2002) *Trade, Temple and Court: Indian Textiles from the Tapi Collection.* Mumbai: India Book House Pvt Ltd.

Gans-Ruedin, E. (1984) *Der Indisch Teppich.* Herford: Bussesche Verlagschandlung.

Gregson, N., Brooks, K. and Crewe, L. (2000) 'Narratives of consumption and the body in the space of the charity/shop', in *Commercial Cultures: Economies, Practices, Spaces*, P. Jackson *et al.* (eds), 101–21. Oxford: Berg.

Hansen, K.T. (2000) *Salaula: The World of Secondhand Clothing and Zambia.* Chicago and London: University of Chicago Press.

Irwin, J. (1955) *Shawls.* London: Victoria and Albert Museum.

Irwin, J. (1973) *The Kashmir Shawl.* London: HMSO.

Irwin, J. and Brett, K.B. (1970) *The Origins of Chintz.* London: HMSO.

Irwin, J. and Hall, M. (1971) *Historic Textiles of India at the Calico Museum.* Vol 1. *Indian Painted and Printed Fabrics.* Ahmedabad: Calico Museum of Textiles.

Knox, R. (1992) *Amaravati: Buddhist Sculpture from the Great Stupa.* London: British Museum Press.

Koch, E. (1982) 'The Baluster Column: a European motif in Mughal architecture and its meaning', *Journal of the Warburg and Courtauld Institutes* 45: 251–62.

Koch, E. (1991) *Mughal Architecture: An Outline of its History and Development (1526–1858).* Munich: Prestel.

Kopytoff, I. (1986) 'The cultural biography of things: commoditization as process', in Appadurai 1986, 64–91.

Kramrisch, S. (1946) *The Hindu Temple.* Calcutta: University of Calcutta Press.

Norris, L. (2005) 'Cloth that lies: the secrets of recycling in India', in *Clothing as Material Culture*, S. Küchler and D. Miller (eds), 83–106. Oxford and New York: Berg.

Palmer, A. and Clark, H. (eds) (2004) *Old Clothes, New Looks.* Oxford and New York: Berg.

Stallybrass, P. (1993) 'Worn worlds: clothes, mourning and the life of things', *Yale Review* 81(2): 35–50.

Thompson, M. (1979) *Rubbish Theory: The Creation and Destruction of Value.* Oxford: Oxford University Press.

Walker, D. (1997) *Flowers Underfoot: Indian Carpets of the Moghul Era.* New York: Metropolitan Museum of Art.

Yaldiz, M., Gadebusch, R.D., Hickmann, R., Weis, F. and Ghose, R. (2000) *Magische Götterwelten. Werke aus dem Museum für Indische Kunst, Berlin.* Berlin: Staatliche Museen zu Berlin – Preußischer Kulturbesitz, Museum für Indisch Kunst.

The author

Lucy Norris is currently a Research Fellow in the Department of Anthropology, University College London, where she also teaches the material culture of South Asia. Her research interests include material culture studies, theories of value, recycling and global provisioning, the anthropology of cloth and clothing and ethnographic methods. Her PhD research focused upon the recycling of both Indian and Western cloth in India, upon which this paper is based.

Address

Lucy Norris, 62 Mervan Road, Brixton, London SW 2 1DU, UK (Lucy.norris@ucl.ac.uk).

Collecting modern textile materials

In pursuit of forgotten fibres? The development, disappearance and rediscovery of regenerated protein fibres

Mary M. Brooks

ABSTRACT As pressure mounted in mid-20th century Europe and America, planners were increasingly anxious about the availability of wool for military purposes. Research into substitutes focused on transforming milk, soya and maize protein into wool-like fibres; other experiments were undertaken into the fibre-forming potential of egg white, feathers, fish and slaughterhouse waste. Manufacturers and politicians shared a common interest in developing these fibres, named azlons, for the war effort. Companies such as Atlantic Research Associates (ARA) in America and Imperial Chemical Industries (ICI) in England were directly promoting their new milk-based fibres to female consumers, making their purchase a patriotic act. Nowadays, regenerated protein fibres are being marketed as high-quality ecological and biodegradable fibres. This paper presents a brief survey of the development of azlon fibres together with a review of technical aspects of their production. Few examples of mid-20th century azlon textiles and garments have yet been identified in museum collections and accounts of their development no longer appear in standard textile histories. Issues relating to this 'disappearance' from cultural memory are discussed. Thompson's rubbish theory is used to explore attitudes to the collecting of man-made materials. The implications of this for conservation practice in both identifying and treating these fibres is reviewed.

Keywords: regenerated protein fibres, man-made fibres, collecting practices, conservation practices, rubbish theory

Introduction

Fibres made from milk, peanuts, soya beans, feathers, whale flesh, shellfish and even egg white sound like the stuff of fiction – evoking the 1951 Ealing comedy *The Man in the White Suit* starring Alec Guinness as the naive inventor, bewildered by the implications of his invention of an apparently indestructible fabric. In fact, these are innovative fibres linked to a range of political, economic and ecological issues. These regenerated protein fibres, classified as azlons, were first mass produced in a period of threatened deprivation prior to the Second World War when European and American politicians and manufacturers feared for wool supplies.[1] Some, such as the milk and peanut fibres, were produced commercially and extensively marketed. For a variety of reasons, these substitute wool fibres failed to become established and most ceased to be manufactured in the 1950s. Azlons are now undergoing a revival thanks to innovations in fibre technology and the demand for fibres perceived to have a reduced ecological footprint. Textiles and clothing made from soya beans, milk and chitin (shellfish) proteins are being promoted as luxurious high-quality fibres with health-giving properties.

This paper reviews the range of azlon fibres and their history, manufacture and marketing in terms of the pressures which led to their development and disappearance. Since the 1950s, the wartime fibres appear to have disappeared rapidly from cultural memory and tend not to be identified in museum dress collections or discussed in standard textile histories. Using a purely object-based research strategy is therefore difficult. An interdisciplinary approach is necessary, enabling documentary evidence drawn from patents, trade journals, technical literature, manufacturers' data and fashion magazines to be used to support the technical analysis when samples are available for this to be undertaken. The intellectual and physical 'disappearance' of the earlier azlon fibres raises issues relating to curatorial collecting practices and for conservation strategies in their identification and preservation.

Sources for azlon fibres

Almost all soluble proteins may be turned into fibres, although whether or not they are useful fibres is another matter. Some knowledge of the function of proteins is necessary to understand the problems encountered by researchers and manufacturers working to create regenerated protein fibres. Proteins form a vital part of all living things and each has a unique function. They therefore have extremely diverse structures which may be completely fibrous or globular. A regenerated protein fibre is essentially a non-fibrous protein (like milk) which has been reconfigured so that it takes up a fibrous form (like wool). In order to form a fibre, large molecules are required which are capable of crystallising, achieve a reasonable degree of orientation and have a high degree of polarity to give intermolecular cohesion.

Producing a regenerated protein fibre

Researchers, during the early to mid-20th century, acknowledged that transforming a non-fibrous protein into a stable fibrous protein was complex. Wormell, a leading British researcher in this field, noted that unlike 'cellulosic fibres [which] are regenerated immediately on coagulation ... protein filaments have to be cross-linked if fibrous products are to be obtained' (Wormell 1954: 19). The basic manufacturing process involved separating out the protein from the source material, solubilising it so it could be extruded and then treating it so that the fibres thus formed could be stabilised. Five main production stages can be identified:

1. Separation: removing fats and oils from the source protein to form a solid curd.
2. Solubilisation: dissolving the resulting washed and dried curd to form a 'spinning' solution with a high solids content. This was allowed to mature with careful control of oxidation and bacterial activity.
3. 'Wet spinning': forcing this ripened solution through spinnerets into a coagulation bath, resulting in the formation of fibres. Contemporaneous documents refer to this as 'wet spinning' although it is actually an extrusion process. The coagulation bath was usually a salt and acid bath such as sodium, aluminium or magnesium sulphate and sulphuric acid. The osmotic pressure created by the salt caused the diameter of the newly extruded filaments to shrink, thus strengthening them and minimising their tendency to clump together.
4. Insolubilising: stretching and hardening these fibres, often using formaldehyde under acid conditions. The aim of this insolubilisation bath was to enable the formation of a complex network between the protein chains with sufficient cross-links to improve the wet strength of the resulting fibre, but not so many as to create an over-rigid structure (Wormell 1954: 90). Lack of strength in the hot baths required for processing was a persistent problem. A delicate balance of acidity level and temperature was required to improve resistance to boiling water without damaging the fibre's physical appearance or properties. Bobbins or reels were used to collect the filaments which were then pulled through a bath over two glass pulleys (Ramseyer 1941: 12). One of these pulleys revolved faster than the other so the filaments were tensioned or stretched. This process aimed to improve the parallel orientation of the molecules to the fibre length, resulting in better wet and dry strength. Acetylation, sometimes using acetic anhydride at temperatures of 80 °C or above, could be used to improve colour, handle and dyeing performance (Traill 1951: 268).
5. Controlled washing and drying of the rope-like 'tow' was followed by cutting it into staple lengths. Further processing could be undertaken using standard textile machinery.

Although some aspects of the production methods have remained unchanged, several technical innovations have enabled the production of the new azlon fibres. Biochemistry has allowed the modification of the structure of source proteins while fibre strength and modulus is being improved by incorporating a synthetic component, often polyvinyl alcohol (PVA). Bi-component fibre structures are being explored to combine the textural benefits of regenerated proteins with the strength and durability of PVA (Zhang et al. 1999).

Azlon fibres in a cultural and political context

What was behind the great burst of research into regenerated protein fibres in the late 1930s and throughout the 1940s? One important motive was intellectual curiosity coupled with the search for commercial benefits. Researchers were hoping to develop viable fibres from protein sources, comparable to the regenerated cellulosic fibres, which would be like wool or silk but cheaper when used as single fibres or as bulking agents in blended fabrics. Some experiments, such as research into egg-white fibres, never seem to have progressed beyond the laboratory. Those that did have commercial potential became intimately linked with mid-20th-century politics. A 1942 article makes explicit links between wool, described as 'almost as precious as ammunition', military provisions and the need for substitute fibres: 'Every ounce [of wool] ... is needed for our men in the front lines .... warm clothes may become as essential as either food or guns ... Fibers from milk casein, from soybeans and many other sources are making their appearance' (O'Brian 1942: 512–14).

Huge hopes were pinned on such fibres by governments in Europe and America seeking to find substitutes as wool imports were threatened by the ever-increasing spread of the war. As Lundgren and O'Connell, two American researchers working in the US Department of Agriculture Western Regional Laboratories on developing azlons, observed: 'interest in the formation of artificial fibers from proteins has been stimulated by the war emergency' (Lundgren and O'Connell 1944: 370).

An editorial in *Harper's Bazaar* (August 1942: 85), an American fashion magazine aimed at the more affluent female consumer, shows how these concerns were used to market the new fibres, making them a patriotic purchase:

> From the first day forward we are milklings. In one form or another milk is woven tight and fast into the fabric of our lives. Milk straight, separated into cream, churned into butter, ripened into cheese ... Now we will wear milk – dress in new milk-fed clothes based on discoveries that are rocking the fabric industry and taking the sting out of wool shortages ... Out of milk casein, via years of research, guesswork, test tubes, and microscopes: Aralac, a spinning fiber that looks in its raw state not unlike wool; that shares with wool the secret of warmth and resiliency.'

Quirky puns and inspiring images might be used to promote the new milk fibre but the underlying point is serious – such innovative fibres were needed to supply the nation with a replacement for wool in a time of national crisis.

Azlon fibres

Proteins from both animal and vegetable sources were explored for their fibre-forming potential (Table 1).

Table 1 Vegetable and animal sources research as sources for azlon fibres.

Vegetable	Corn (maize) *Zea mays*. L.	
	Peanuts (ground nuts) *Arachis hypogae* L.	
	Soya beans *Glycine max*. L.	
	Sunflower seeds *Helianthus*	
Animal	Egg white	
	Feathers	
	Fish, whale flesh, crab and shrimp shells (chitin)	
	Gelatine from slaughterhouse waste (horns, hooves, animal muscle and hide)	
	Milk	

Animal protein fibres

Adam Millar patented a process for making fibres from proteids in egg, blood, casein (milk) and vegetable albumen in 1898 (BP 6700; US 625,345). He patented a silk-substitute fibre, Vanduara, made from gelatine fibre. The German gelatine fibre Marena, used mainly for brush bristles, was made from the mid-1920s to the late 1950s (Koch 1972: 32). In 1941, German researchers were exploring the potential of tendons and horsemeat as fibre-forming materials and produced Carnofil, used for both surgical and textile purposes (Jaumann 1936). An American research laboratory, the Arthur D. Little Laboratories, also explored this technology simply to demonstrate the value of research. In order to disprove the proverb 'you can't make a silk purse out of a pig's ear'[2] they extracted gelatine from pigs' ears and formed fibres using a method similar to Millar's (Arthur D. Little Laboratories 1921).

Feather fibres

Lundgren and O'Connell developed a process for making fibres from chicken feathers. Contemporaries reported that army blankets made with these fibres became unusable because of the smell (*American Magazine* 1945).

Whale, fish and shellfish fibres

In 1941, the Japanese were exploring the possibility of making a wool substitute from whale or shark protein hardened with formaldehyde, while the Germans produced a 'fish rayon' – Wikilan. Chitin, derived from shrimp or crab shells, is now being produced commercially for fibres which are said to have health benefits.

Milk fibres

Dr Frederich Todtenhaupt seems to have been the first to market a milk fibre commercially. He registered patents for a 'casein silk' fibre in both Germany and America (1904 DRP 170,051; 1906 US 836,788). Technical problems meant it was not successful and Todtenhaupt's company, Deutsche Kunstseidenfabrik, failed. Sadly, he committed suicide in 1919. No further work seems to have been carried out until the opposing forces of surplus (excess milk by-products) and lack (threatened wool imports) encouraged renewed research.

Antonio Ferretti, an Italian researcher, produced a viable milk fibre in the 1930s, patenting 'serin-wool' and 'rayon-wool' (1935 GB 483,731; 1942 USA 2,297,397). He introduced sodium chloride and formaldehyde baths to improve insolubilisation, resulting in fibres with improved wet strength. Snia Viscosa, an Italian textile manufacturer, produced Lanital and Merinova which were used for men's suitings, dress fabrics, knitwear and underwear. Their names evoke the natural fibre: *lana* is Italian for wool while *ovis* is Latin for sheep. By 1937, Lanital was being promoted in America as a fashion fabric by the well-connected socialite Princess Caetanis (Dirks 2000). Mussolini, who was seeking to make Italy economically self-sufficient for political and military reasons, urged Italians to use substitute products: 'Carcadé [hibiscus tea] replaces tea, lignite replaces coal, Lanital replaces wool' (Mana 1998). Ferretti's method was widely licensed to other European manufacturers and continued to be made in Poland until the 1960s.

Courtaulds, an English textile firm, developed a similar milk-based fibre called Fibrolane. In America, manufacturers of regenerated cellulosic fibres, such as the Celanese Corporation, and of synthetic fibres, such as DuPont, also patented regenerated protein fibre processes. Other companies with expertise in casein, such as the dairy company Bordens (who made casein-based adhesives) and the paint firm Gliddens, were also interested in milk fibres but do not seem to have produced any viable fibres. The first American patent for a casein-based fibre was a 1938 public service patent awarded to Stephen Gould and Earl Whittier, both chemists at the US Department of Agriculture's Bureau of Dairy Industry (USA 214, 274).

The first commercial fibre was produced by Atlantic Research Associates (ARA), the research arm of the National Dairy Corporation formed by Thomas McInnerney in 1923. McInnerney was seeking a use for dairy industry by-products and appointed as the company's president Francis Clarke Attwood, a research chemist with a special interest in casein. After four years of research supported by the US Department of Agriculture, Attwood patented a casein fibre derived from milk (1942 USA 287,928). Aralac was named by pairing ARA's initials with *lac*, the Latin for milk. ARA's logo linked the dairy, chemistry and fashion (Fig. 1). The fibre was marketed to civilians in order to release wool and cotton for military use (Dirks 2000: 82). California textile converters developed

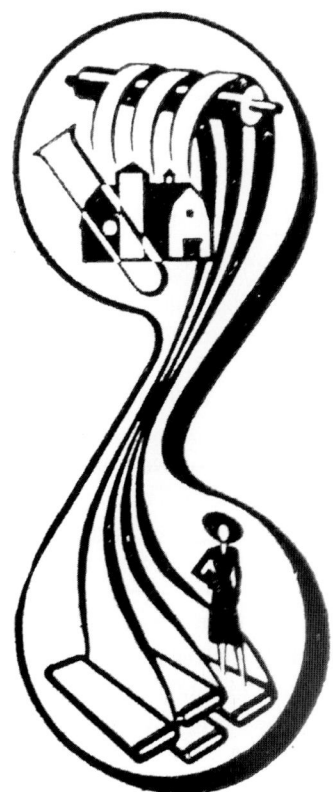

Figure 1 Aralac logo.

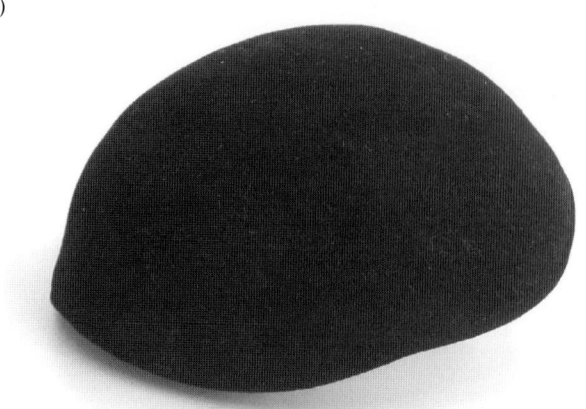

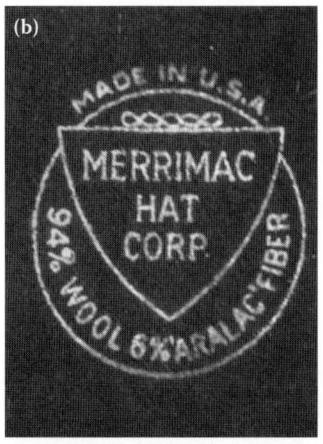

Figure 2 (a) Merrimac hat and (b) maker's stamp, 1940s (author's collection).

versions of Aralac-based fabrics specifically for the West Coast consumer. Hollywood film stars, such as Jane Wyman and Dorothy Lamour, promoted the fabulously named Aralac fibre Vitamin Coat. Despite this apparent success, National Dairy sold their Aralac plant in 1948.

Documentary evidence indicates that Aralac was used in blends for blankets, interlinings and possibly car upholstery (Dirks 2000: 83). Aralac was extensively used in hat felts as a substitute for rabbit hair, unavailable during the war. The well-known hat manufacturer, Merrimac, noted the exact blend of Aralac and wool in this 1940s hat (Fig. 2 a and b). Milk is again being used for fibres in combination with PVA. The Italian company Fabricha is selling milk fibre scarves (Fig. 3) and T-shirts, promoted as a silk-like, environmentally friendly fibre.

Vegetable protein fibres

Soya bean fibres

The Japanese developed a soya bean fibre, Silkwool, in 1938 but this was not a technical success (Araki *et al.* 1946). A number of American firms with expertise in the use of soya for paints or animal feed filed patents but none seems to have produced viable fibres. Both ARA and Ferretti were interested in soya bean protein as an alternative to milk casein but Henry Ford emerges as the unexpected champion of soya bean fibres (Brooks 2005). Soya beans formed an essential part of his vision of 'farm chemurgy'[4] – industrial production based in the countryside. He set up a laboratory, directed by Robert Boyer, to explore the industrial potential of soya beans for cars

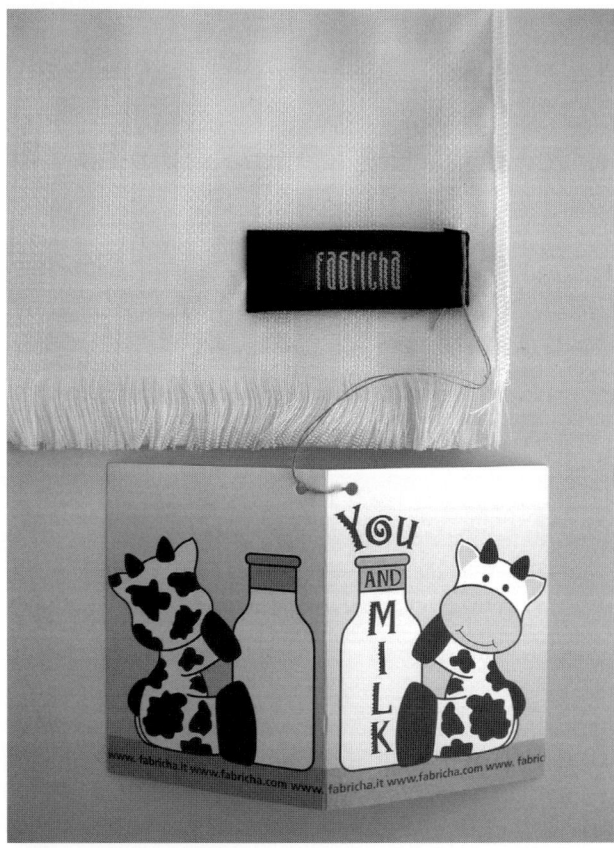

Figure 3 Fabricha milk and PVA scarf (author's collection) (Plate 14 in the colour plate section).[3]

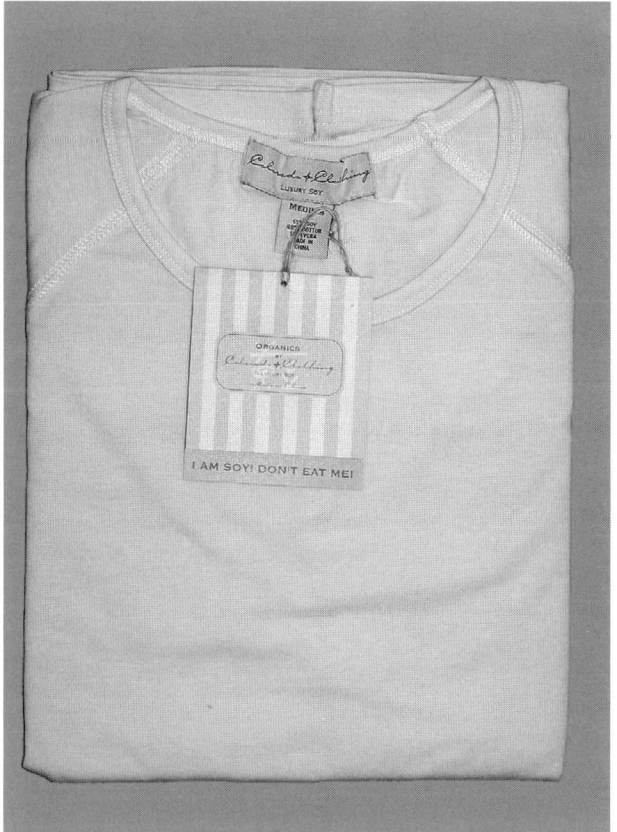

Figure 4 Colorado Clothing Company T-shirt; soya, cotton and Lycra (author's collection) (Plate 15 in the colour plate section).

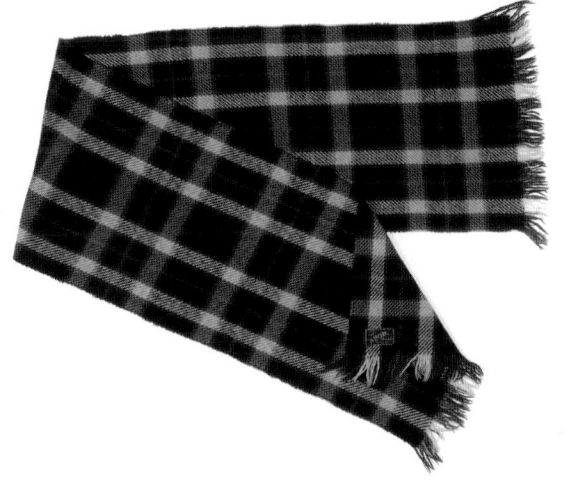

Figure 5 Ardil scarf (Karen Finch Reference Collection; reproduced with permission of the Textile Conservation Centre) (Plate 16 in the colour plate section).

– as plastics, upholstery fabrics and seat wadding. Always a brilliant publicist, Ford had a suit made from a soya bean/wool fabric. Unsuccessful in persuading the armed forces to use this fibre, he sold the process to the Drackett Company, who had expertise in soya curd production. Boyer moved to Drackett with Ford's support and announced improvements to the fibre in 1945. Although Drackett did produce soya bean fibre for hat felts, and possibly some yarn, production ceased in 1949.

Research is now being carried out into soya bean fibres in China and America and the fibre is available commercially (Fig. 4). Chinese versions usually combine soya protein with polyvinyl fluoride (PVF) in the Shanghai Winshow Soybean-fire Industry Company's Winshow soy protein fibre (SPF). Substituting soya bean fibres for cashmere may help minimise desertification in China caused by the grazing goats. American research, much of it undertaken at the Georgia Institute of Technology, has focused on developing bicomponent monofilament fibres with soya bean protein and PVA or polyacrylic acid (Kotliar and Ghasemzadeh 1994; Zhang et al. 1999).

Peanut (groundnut) fibres

The American peanut fibre Sarelon does not seem to have been very successful. Imperial Chemical Industries (ICI) only produced the British version, Ardil, in 1951 although Astbury and Chiball patented their production process in 1936 (BP 467, 704). Production ceased in 1957 (Brooks 1993). Ardil links Britain's imperial past, using peanuts produced in African colonies, with the need for post-war redevelopment at home. It was intended as an alternative product for ICI's redundant explosives factories and was used for clothing (Fig. 5), curtains and carpets (Brooks 1991). An Ardil/cotton blend nightdress (York Castle Museum 431.78) bears a label reading 'This fabric contains Ardil. Styled by "Unique" from Potter's Ardingle. Wash as for wool' (Brooks 1991: 88). This suggests different manufacturers may have produced different blends.

ICI gave Ardil blend scarves as promotional gifts to British solders fighting in Korea. The donor of such a scarf in the Nottingham Museum of Costume and Textiles (1979.609) reported that it dissolved in saliva (Fig. 5).

Corn (maize) fibres

The US Department of Agriculture supported research into fibres made from zein, the protein in corn (maize) gluten. Only the Virgina-Carolina Chemical Corporation, purchasers of National Dairy's Aralac factory, appears to have developed it commercially. Vicara was used as blending fibre for wool, cotton and nylon while Zycon was made specifically for hat felts. Production ended about 1957. The Smithsonian Institute's Vicara samples include a woven automobile upholstery sample (Dirks 2000: 84). Nuno, the innovative Japanese textile company, is using corn fibre fabrics today.[5]

Why did azlons fail?

For a brief period in the mid-20th century, azlons seemed to be the fibres of the future. They had similar moisture-absorbing properties to wool, good handling properties and were thought to be mothproof as well as being more light-stable than the regenerated cellulosic fibres. The drawback was their lack of strength due to their disordered crystalline structure (Brooks and Garside 2005). Weaker than both natural fibres and the newly arriving synthetic fibres, an azlon fibre such as Aralac had less than half the dry strength of wool. Azlons were therefore generally used to bulk out more expensive fibres, such as wool, and in blends with linen, cotton and regenerated cellulosics to improve handle- and crease-resistance. Manufacturers filed numerous patents seeking to overcome problems resulting from their poor wet strength. One researcher memorably described them as 'flabby' when wet, soluble in saline solutions, acids and alkalis and brittle when dry (Traill 1951: 265). Azlons also required great care during domestic laundering. Aralac had a further distinctive and regrettable property: it smelt of sour milk when wet.

Azlons had an economic edge while wool was expensive but were not financially viable once wool prices dropped dramatically after the war. Nevertheless, both the milk and peanut fibres were made in some quantity. There were practical and economic reasons for the end of production – but why have they been so quickly forgotten? Where are surviving examples now?

Disappearing fibres?

Azlons may literally have disappeared because of their physical weaknesses. It is more likely, however, that they have 'disappeared' because they lacked a separate identity. As they were used as blends and marketed under unfamiliar trade names, they may be both physically and intellectually difficult to identify. Advertisements in *Vogue*, *Harper's Bazaar* and *California Stylist* provide details of the inventive names given to different fibre blends. Duvalara, Sutara and Lacara are all Aralac blends (the clue to identification being the ARA included in the fabric name). It therefore seems likely that some azlon textiles and garments have survived unrecognised as the fibres themselves have slipped from cultural memory. Tracing some of the reasons for this is interesting. Following the war, with rationing still in force, using food for fibres seemed wrong: 'Why use good food to make poor wool?' (Wormell 1954: xiii). Azlons were seen as inferior ersatz fibres and commercial failures – an experiment to forget rather than remember, linked with a period of deprivation and substitution. Certainly, association with these fibres had a depressing effect on some key researchers' careers; for example, after leaving Drackett, Boyer had a long period of rejection before going on to develop edible soya food products.

Michael Thompson's rubbish theory is a useful model to explore attitudes to collecting textiles from the recent past. Rubbish theory traces shifting attitudes to the value of objects: he explains it as the 'relationship between status, the possession of objects, and the ability to discard objects' (Thompson 1979: 1). It provides a framework for exploring why some objects lose their high value status and become unappealing before they regain interest and value. In consequence, they move from the (literal or metaphorical) 'rubbish dump' and become desirable and collectable. Azlons seem to fit this cycle closely: at first promoted as desirable and exciting and then forgotten before revision and re-evaluation. Schneider (1994) traces a similar cycle in her study of attitudes to synthetic fibres. These theoretical approaches provide helpful illumination as to why these fibres are rarely identified in textile and dress collections.

Implications for curatorial practice

American textile and dress curators surveyed in 2002 were fascinated by the azlon fibres but few reported examples in their collections. This is possibly because, unless manufacturers' names or labels provide a clue, there is little physically to suggest the presence of an unusual fibre. Coupled with the increasing absence of azlon fibres from many textile histories, the significance of unusual names may go unrecognised. Nevertheless, the speed with which knowledge about these fibres has been lost – some of which were being made within living memory – is a warning about the importance of contemporary collecting practice. Ideally, contemporary examples of new fibres, both in sample and garment form, should be acquired for museum collections. Given the speed with which technology changes, as much information as possible about manufacturers, marketing and, where relevant and possible, the owners' or wearers' attitudes and experiences should be recorded when items are accessioned.

Implications for conservation practice

Azlon fibres present conservators with some interesting problems. First, they need to be identified. Secondly, developing a conservation strategy for fibres for which there is little precedent in terms of intervention and preservation is difficult.

Identification

Regenerated protein fibres are not easy to identify as they are usually found as blended yarns or in mixed fibre fabrics. In order to establish their presence, it is necessary to sample yarns from both the warp and weft and to identify all other fibres which may be present. With contemporary fibres, it is necessary also to establish whether PVA is present.

At a microscopic level, fibres which have so far been available for examination, tend, like synthetic fibres, to be smooth and almost featureless. They are usually uniform and structureless but may have some longitudinal striations. Cross-sections tend to be round although this will alter depending on the nature of the spinnerets used in the extrusion process. Tests, such as those described by the Textile Institute (1975), may be used to establish the presence of protein although this, of course, does not distinguish a regenerated protein fibre from a natural protein fibre. Moncrieff (1975) reports that casein, peanut, soya bean and zein fibres will all give a yellow-orange coloration with cold Shirlastain A stain. Measuring fibre density used to be the established method of distinguishing azlon fibres. Moncrieff describes how measuring specific gravity enables Ardil (specific gravity 1.30) to be distinguished from casein fibres (specific gravity 1.29). More sophisticated methods such as conventional and polarised Fourier transform infrared (FTIR) spectroscopy are now available and will enable identification of specific types once comparative spectra of known fibres have been established (Brooks and Garside 2005; see also Garside and Brooks in this volume, pp. 67–71).

FTIR–ATR (attenuated total reflectance) spectra also enable the copolymers, such as polyacrylonitrile, found in modern azlon fibres to be identified.

Conservation strategies

Unfortunately, the lack of firmly identified mid-20th century azlon fibres means that it has not yet been possible to characterise a typical degradation pathway. What is known about their poor stability, however, allows some tentative recommendations to be made beyond the normal procedures for light and relative humidity (RH) levels for display (Museums and Galleries Commission 1998). Manufacturers' recommendations for cleaning the new mid-20th century regenerated protein fibres varied. Most suggested gentle washing as for wool while some, such as Ardil, were said to be stable in most organic solvents. Extreme caution would seem advisable, however, before attempting to wet clean any aged regenerated protein fibre as it seems likely that lack of stability to water, and possibly to solvents, will increase over time. The evidence of the Nottingham scarf, which was susceptible to saliva, suggests that enzyme treatments should be avoided. Until more is known about the behaviour of degraded mid-20th-century protein fibres, interventive treatments would seem to be undesirable.

What next?

The fate of these fascinating but short-lived mid-20th-century fibres hopefully will alert both curators and conservators to look closely at surviving garments from that period and undertake closer examination of their fibres. At present, the majority of data about these fibres are drawn from textual and visual sources. The lack of data from actual objects means that evaluating object-based data against documentary data – and vice-versa – is difficult. Only by identifying examples of these forgotten fibres can their history be recovered from the 'rubbish dump' and integrated into the broader framework of 20th-century social and technological history. The rapidity with which these fibres have been forgotten should also encourage full documentation of the next generation of regenerated protein fibres so future generations can identify and understand the textile innovations of the 21st century.

Acknowledgements

Thanks are extended to the Getty Conservation Institute for the author's 2002 Conservation Scholarship; Sandy Rosenbaum and Gail Stein, Doris Stein Research Centre, Los Angeles County Museum of Art; Denise Buhr, Research Librarian, Central Soya, Indiana; Patricia Starrett, Archivist, ICI; Katherine Dirks, Smithsonian Institution; Jennifer Harris and Ann Tullo, the Whitworth Art Gallery; Susan Mossman, the Science Museum. The author would also like thank colleagues at the Textile Conservation Centre, especially Dr Maria Hayward, Director and Dinah Eastop, Associate Director of the AHRC Research Centre for Textile Studies and Textile Research, Dr Paul Garside, Mike Halliwell, Cordelia Rogerson and Nell Hoare, Director of the Textile Conservation Centre for permission to publish.

Notes

1. The textile trade had great difficulty in finding an appropriate term for these new fibres. Terms such as 'natural protein fibres' and 'prolon' ('pro' from protein and 'on' from nylon and cotton) were rejected. It is not clear how the term 'azlon' for 'manufactured fibres in which the fibre-forming substance is composed of regenerated naturally occurring proteins' originated (Federal Trade Commission Textile Products Identification Act; see www.ftc.gov/os/statutes/textile/rr-textl.htm).
2. In the same spirit, the Arthur D. Little Laboratories later made a viable lead balloon.
3. See www.fabricha.it.
4. Chemurgy (meaning 'chemistry at work') is defined as 'a branch of chemistry that deals with industrial application of organic raw materials especially from farm products' (*Webster's Third New International Dictionary*, 1976: 384). The term 'biochemical engineering' is now used.
5. Nuno produces maize fabrics as part of its Eco Series, first developed in 2001; see www.nuno.com.

References

American Magazine (1945) 'Dresses from chicken', 10: 32–3 and 94–5.

Araki, S., Tarumi, K. and Iguchi, R. (1946) 'Artificial fibers from soybean protein', *Journal of the Society of Chemical Industry Japan* 49(33) reported in *Chemical Abstracts* 20 August 1948, 6120d.

Arthur D. Little Laboratories (1921) *On the Making of Silk Purses from Sows' Ears*. Cambridge, Mass.: Arthur D. Little Laboratories.

Brooks, M.M. (1991) 'Man-made fibres and synthetics in the Wallis Archive, York Castle Museum', in *Per una Storia della Moda Pronta, Problemi e ricerca, atti del V Convegno Internazionale del CISST, Milano, 26–28 February 1990*, 81–93. Florence: Edifir.

Brooks, M.M. (1993) 'Ardil: the disappearing fibre?' in *Saving the Twentieth Century: The Conservation of Modern Materials*, D.W. Grattan (ed.), 81–93. Ottawa: Canadian Conservation Institute.

Brooks, M.M. (2005) 'Soya bean protein fibres: past, present and future', in *Biodegradable and Sustainable Fibres*, D. Blackburn (ed.), 398–440. Cambridge: Woodhead Publishing.

Brooks, M.M. and Garside, P. (2005) 'Investigating the significance and characteristics of modern regenerated protein fibres', in *Art '05. Proceedings of the 8th International Conference on Non-Destructive Investigations and Microanalysis for the Diagnostics and Conservation of the Cultural and Environmental Heritage. Lecce (Italy), 15–19 May 2005*, C. Parisi, G. Buzzanca and A. Paradisi (eds), 1–14. Lecce: Italian Society for Non-Destructive Testing Monitoring Diagnostics, Ministry of Cultural Heritage and Activities and Central Institute of Restoration and Department of Materials Science, University of Lecce.

Dirks, K. (2000) 'Aralac: "The cow, the milkmaid and the chemist"', *Ars Textrina* 33: 75–85.

Jaumann, A. (1936) 'Carnofil and Marena fibres', *Kunsteide* 5(18): 109–13.

Koch, I.P-A. (1972) 'Regenerated protein fibres', *Ciba Geigy Review* 1: 25–33.

Kotliar, A.M. and Ghasemzadeh, S. (1994) *Novel Silk-like Bicomponent Textile Fibers Based on Soybean Protein and Poly(vinyl alcohol)*. USB/ASA Research Report Project Number E-27-663. Atlanta, GA: Georgia Institute of Technology.

Lundgren, H.P. and O'Connell, R.A. (1944) 'Artificial fibers from corpuscular and fibrous proteins', *Industrial and Engineering Chemistry* 36(4): 370–74.

Mana, D. 1998. Italy 1918–1938; www.fortunecity.com/tattooine/zenith/134/timelin1.htm

Moncrieff, R.W. (1975) *Man-made Fibres*, 6th edn. London: Newnes-Butterworths.

Museums and Galleries Commission (1998) *Standards in the Museum Care of Costume and Textile Collection*. London: Museums and Galleries Commission.

O'Brian, R. (1942) 'Wartime textile adjustments', *Journal of Home Economics* 34: 512–14.

Ramseyer, D.R. (1941) 'Ford develops soybean upholstery fiber', *Soybean Digest* 11(12): 12.

Schneider, J. (1994) 'In and out of polyester: desire, disdain and global fibre consumption', *Anthropology Today* 4: 2–10.

Textile Institute (1975) *Identification of Textile Materials*, 7th edn. Manchester: Textile Institute.

Thompson, M. (1979) *Rubbish Theory: The Creation and Destruction of Value*. Oxford: Oxford University Press.

Traill, D. (1951) 'Some trials by ingenious inquisitive persons: regenerated protein fibres', *Journal of the Society of Dyers and Colourists* 67: 257–70.

Wormell, R.L. (1954) *New Fibres from Proteins*. London: Butterworths Scientific Publications.

Zhang, Y., Ghasemzadeh, S., Kotliar, A.M., Kumar, S., Presnell, S. and Williams, D.L. (1999) 'Fibers from soybean protein and poly(vinyl alcohol)', *Journal of Applied Polymer Science* 71(1): 11–19.

The author

Mary Brooks trained at the Textile Conservation Centre after working in the book world and management consultancy. She has worked as a conservator and curator in Europe and America. At York Castle Museum, she jointly curated 'Stop the Rot', which won the 1994 IIC Keck Award for promoting public understanding of conservation. She is a member of the ICOM Conservation Committee's Task Force for raising awareness of heritage conservation. She has a special interest in the contribution that object-based research and conservation approaches can make to the wider interpretation of cultural artefacts.

Address

Mary M. Brooks, Reader, Textile Conservation Centre, University of Southampton, Park Avenue, Winchester, Hants SO23 8DL, UK (mmb1@soton.ac.uk).

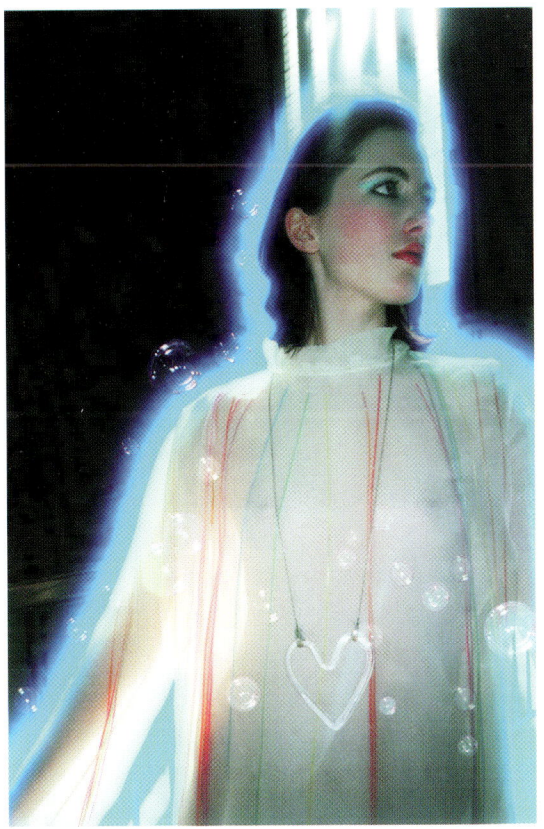

Plate 1 Smart second skin dress (Fig. 1, p. 4).

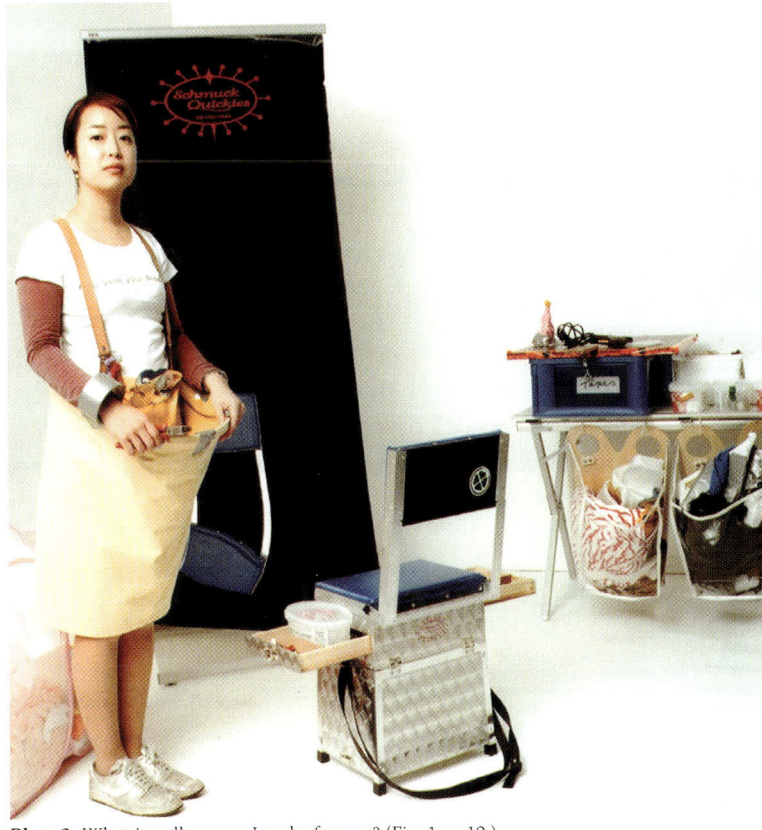

Plate 2 What jewellery can I make for you? (Fig. 1, p. 12.)

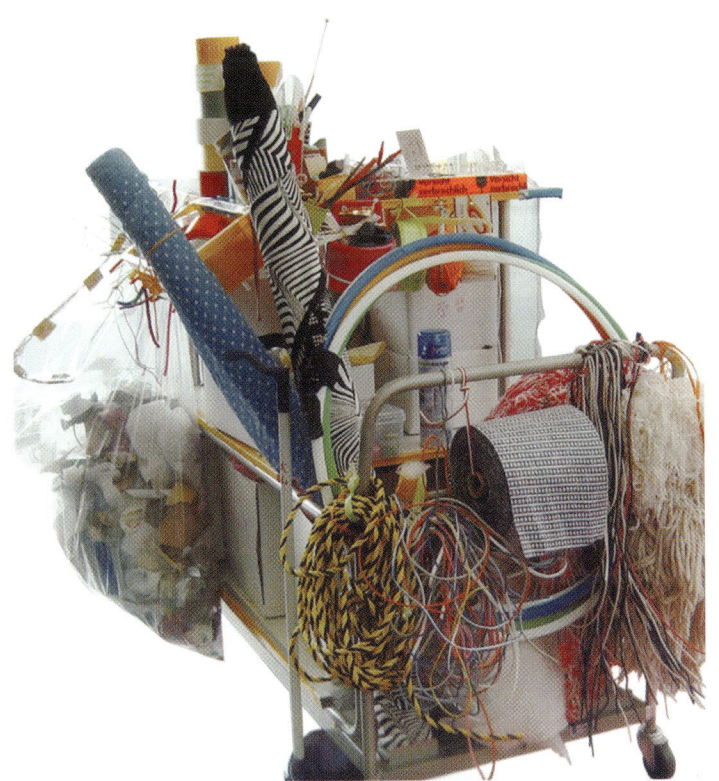

Plate 3 The cart of materials that forms the basis of Oyama's Schmuck Quickies (Fig. 2, p. 13).

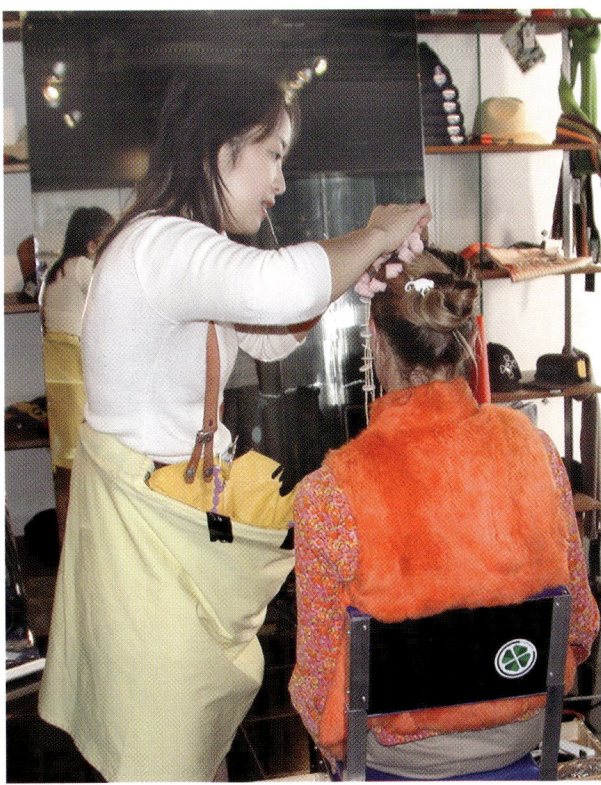

Plate 4 Schmuck Quickies performance in progress, Psyche, Middlesbrough, 26 October 2005 (Fig. 3, p. 14).

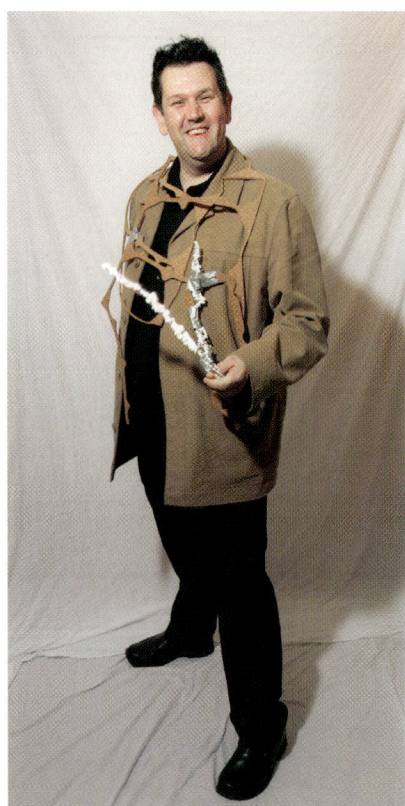

Plate 5 Schmuck Quickies: materials are selected to reflect the location and character of the wearer (Fig. 4, p. 14).

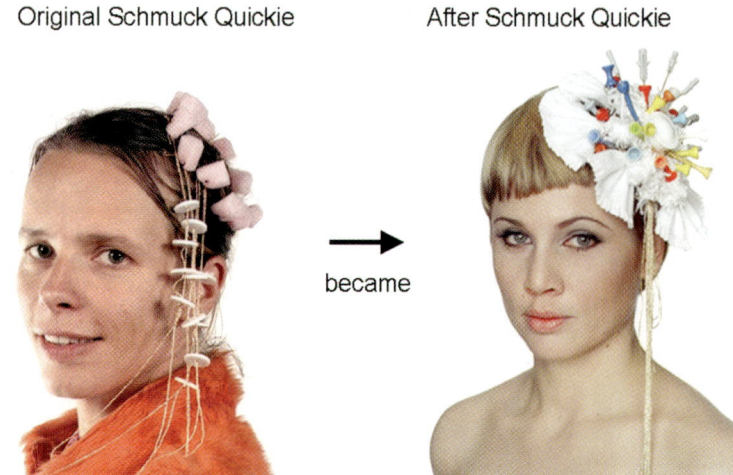

Plate 6 Schmuck Quickies and After Schmuck Quickies (derived from the former Schmuck Quickies) (Fig. 5, p. 15).

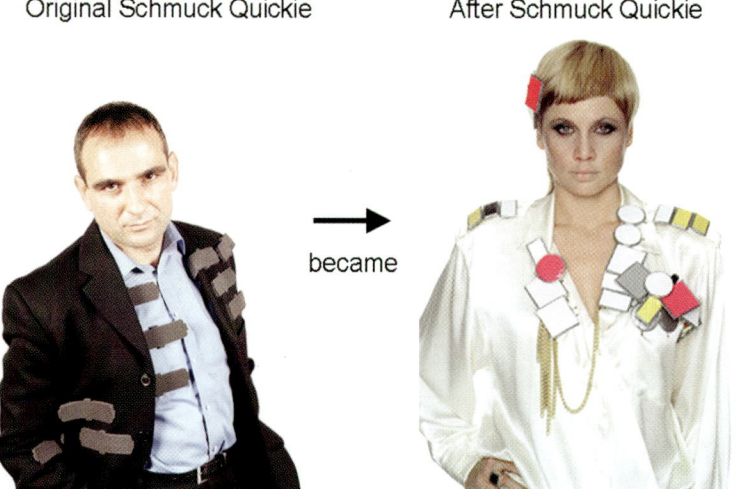

Plate 7 Schmuck Quickies and After Schmuck Quickies (derived from the former Schmuck Quickies) (Fig. 6, p. 15).

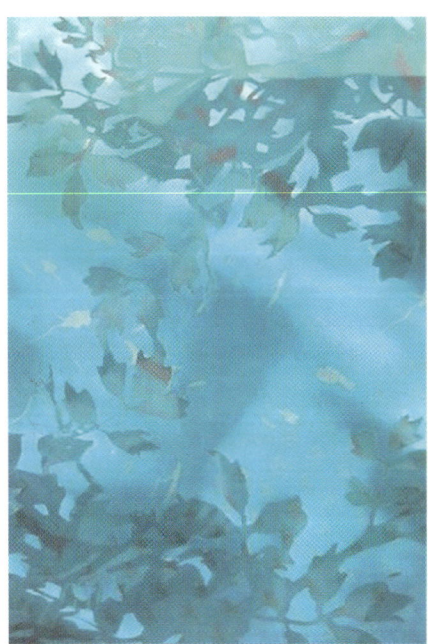

Plate 8 Silk satin viscose devoré print (Pippa Tinning) (Fig. 1, p. 19).

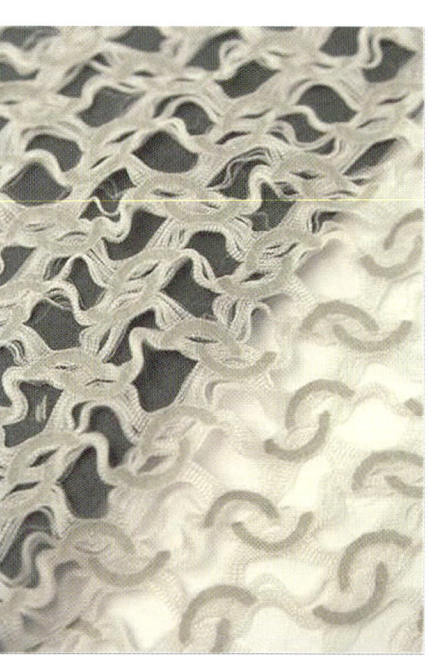

Plate 9 Latex printed silk cotton elasticated fabric (Claire Harbin) (Fig. 2, p. 21).

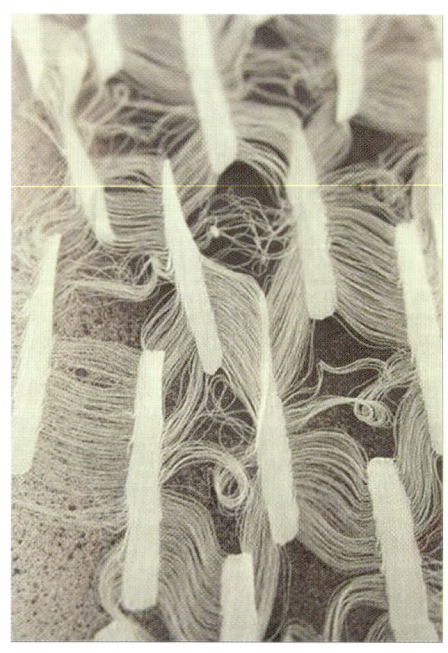

Plate 10 Silk viscose plain weave fabric devoré, printed. Sketchbook swatch (Claire Harbin) (Fig. 3, p. 22).

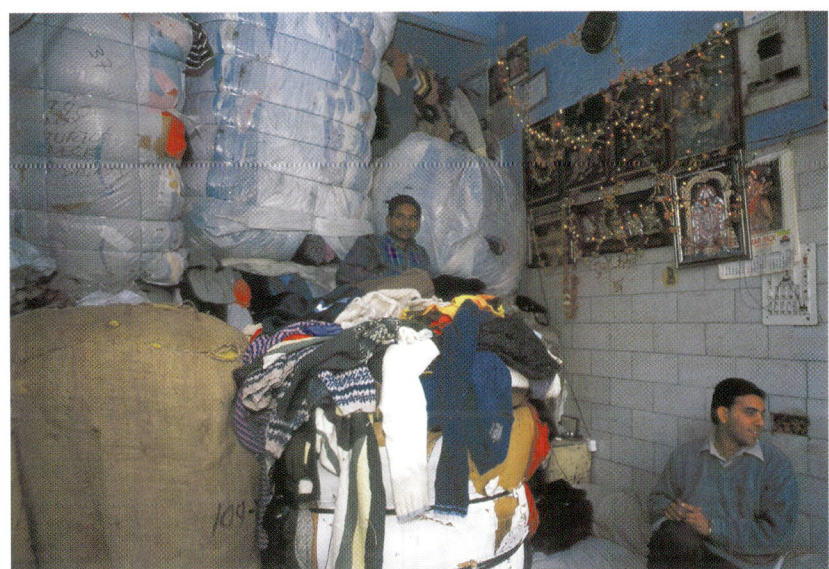

Plate 11 A bale of cast-off clothes sold for export (photo: Tim Mitchell) (Fig. 1, p. 25).

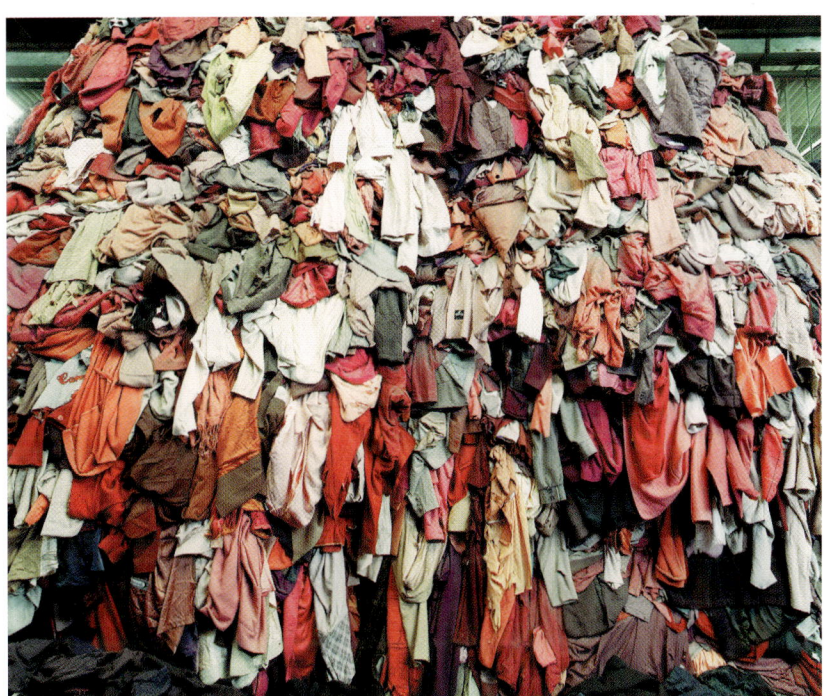

Plate 12 Piles of 'mutilated hosiery' sorted by colour and waiting to be processed (photo: Tim Mitchell) (Fig. 2, p. 26).

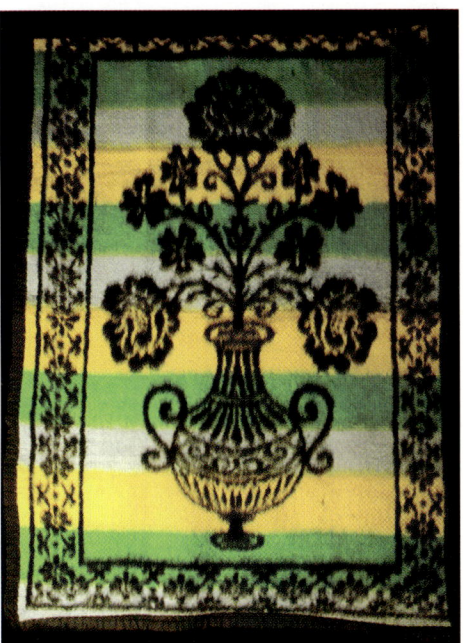

Plate 13 A baby's blanket with vase of flowers design made from recycled jumpers (photo: Lucy Norris) (Fig. 3, p. 28).

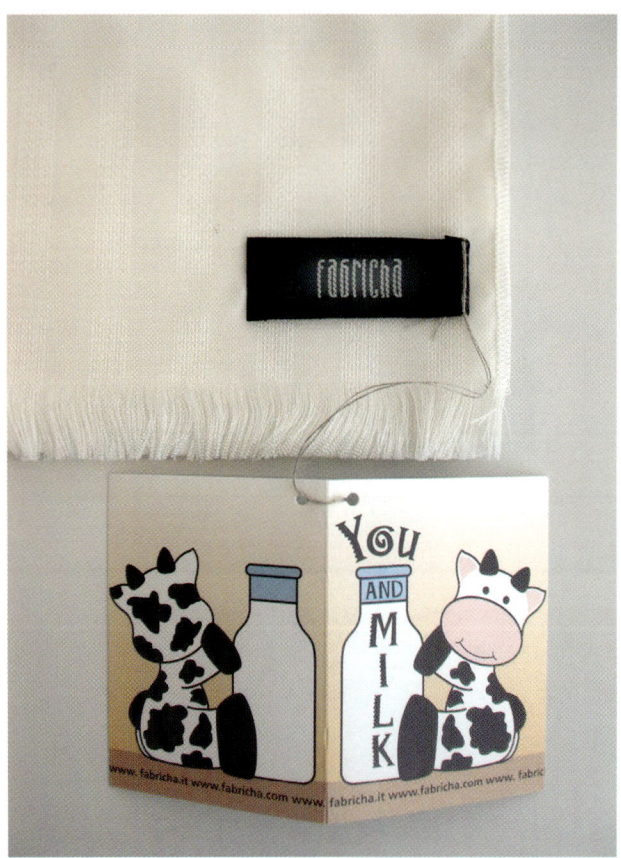

Plate 14 Fabricha milk and PVA scarf (author's collection) (Fig. 3, p. 36).

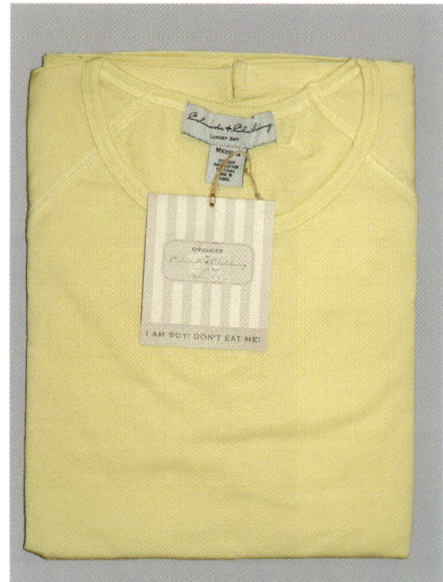

Plate 15 Colorado Clothing Company T-shirt; soya, cotton and Lycra (private collection) (Fig. 4, p. 37).

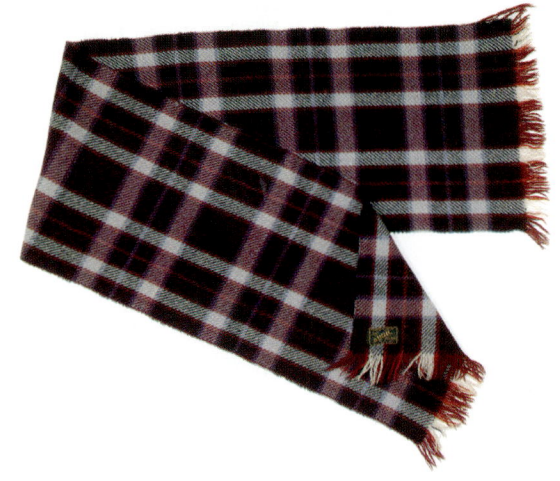

Plate 16 Ardil scarf (Karen Finch Reference Collection; reproduced with permission of the Textile Conservation Centre) (Fig. 5, p. 37).

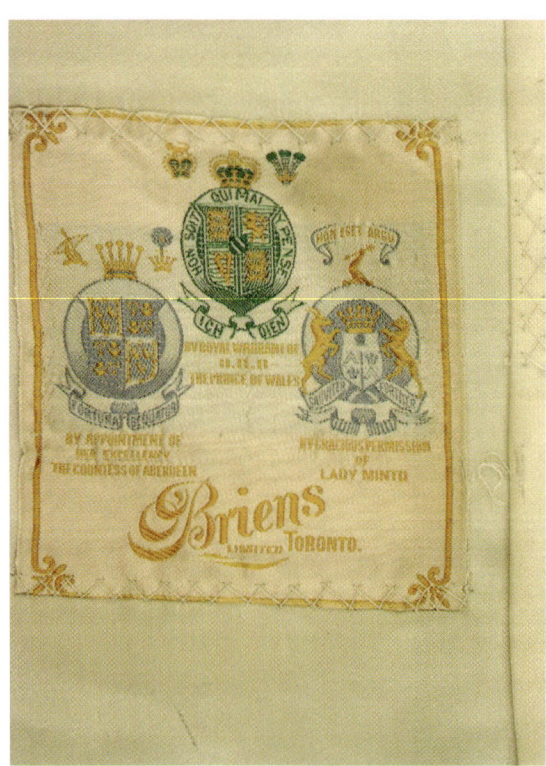

Plate 17 The Toronto dress house, O'Briens, was one of the leading Canadian dressmakers during the first quarter of the 20th century, as is reflected in the royal warrant and appointments displayed on the label (ROM 965.227.1ab, gift of Mr. John M. Morley) (Fig. 1, p. 42).

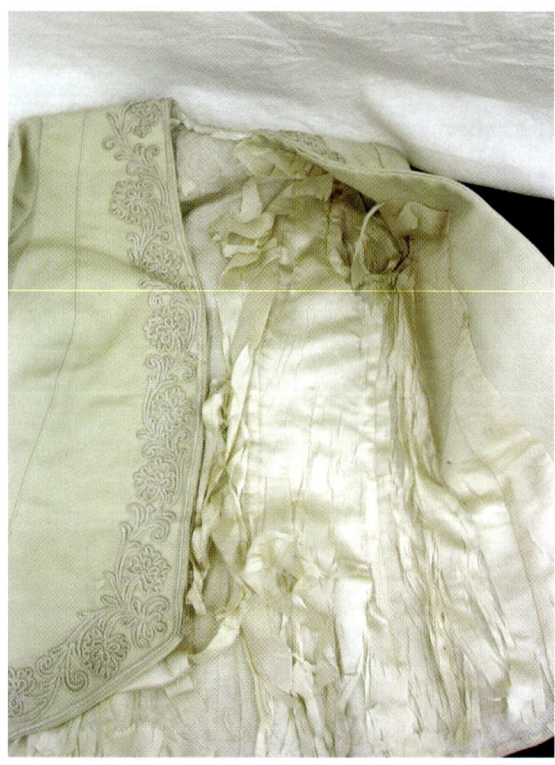

Plate 18 Woollen suit jacket (c.1909) with shattered silk lining that needs stabilisation or replacing for future display (ROM 959.43.24ab, gift of Mrs. F.W. Trusler) (Fig. 2, p. 43).

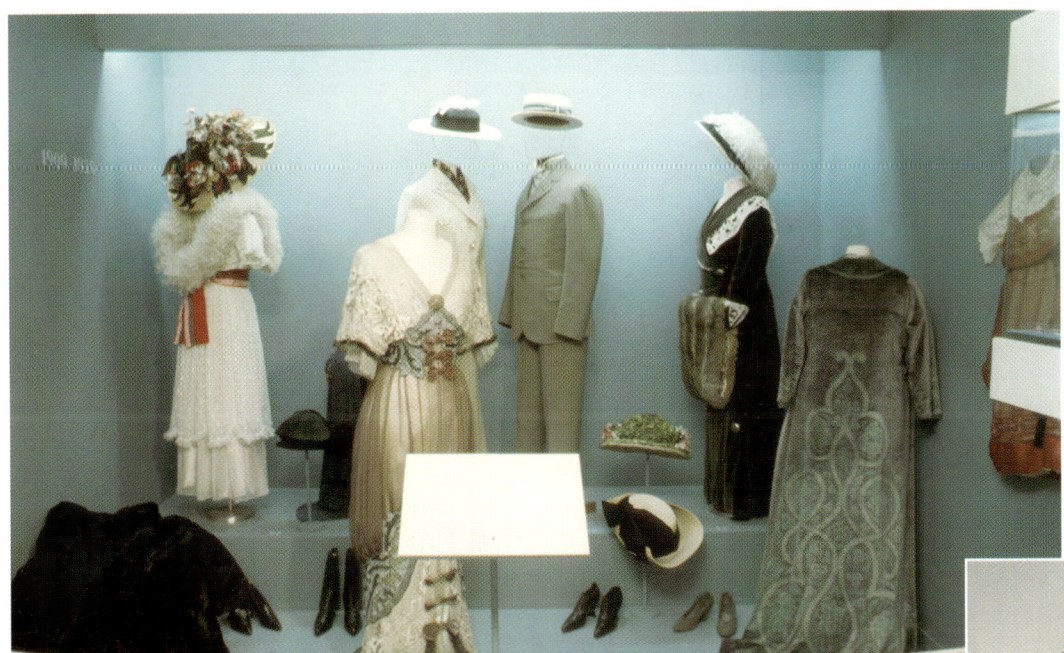

Plate 19 ROM's 1967 centennial year costume display *Modesty to Mod*. This is the last time that the *c.*1910 semi-formal Parisienne evening dress (front) was on public display. It is made of silk chiffon, beaded and trimmed with mole (ROM 952.203.1, gift of Miss Hattiemae Austin) (Fig. 3, p. 44).

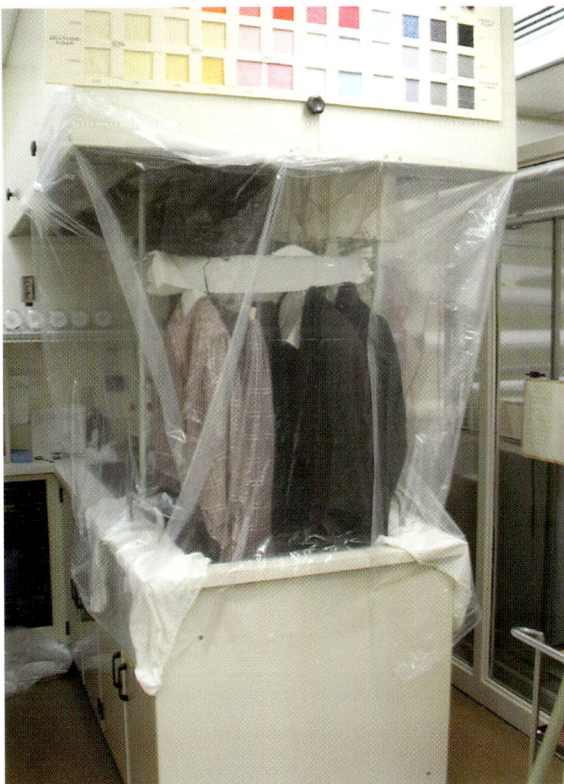

Plate 21 Canadian-designed red silk Directoire-style dress. The underdress of silk satin is made of weighted silk that is severely shattered (ROM 972.436.1, gift of Miss Kathleen Myers) (Fig. 5, p. 45).

Plate 20 Decontaminating chamber made for incoming mid-19th-century artefacts donated to ROM's collection in July 2005 (Fig. 4, p. 45).

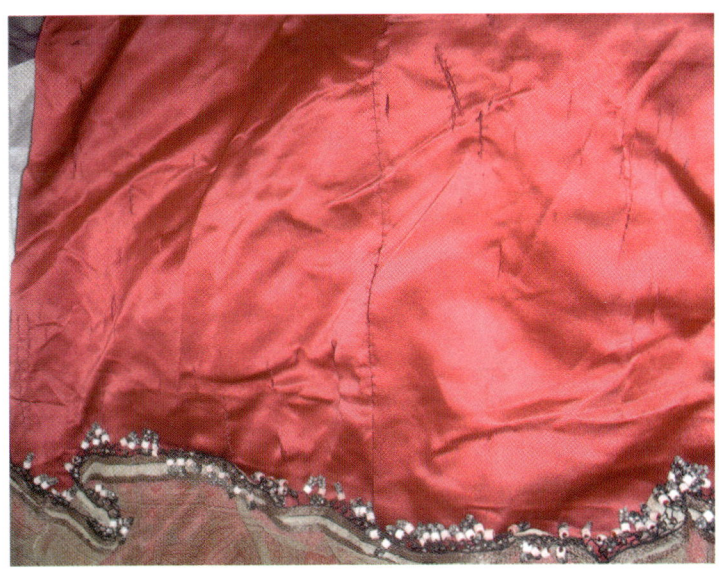

Plate 22 Detail of Plate 21 (Fig. 6, p. 46).

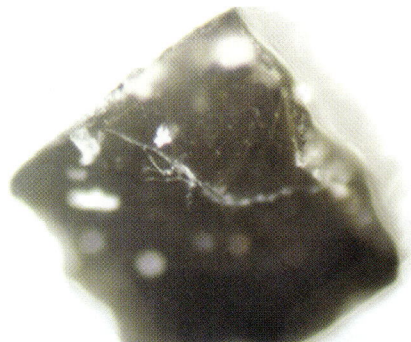 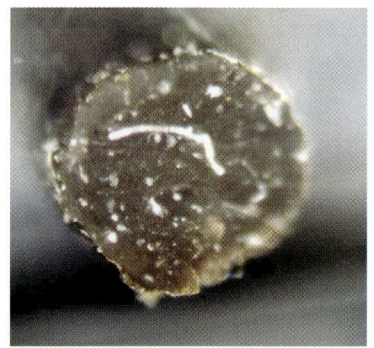 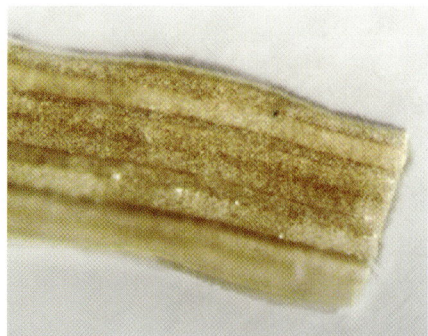

Plate 23 Cross-sectional view of a cut rubber thread (Fig. 1, p. 49).

Plate 24 Cross-sectional view of an extruded rubber thread (Fig. 2, p. 49).

Plate 25 Longitudinal view of a multifilament elastomeric yarn (Fig. 3, p. 50).

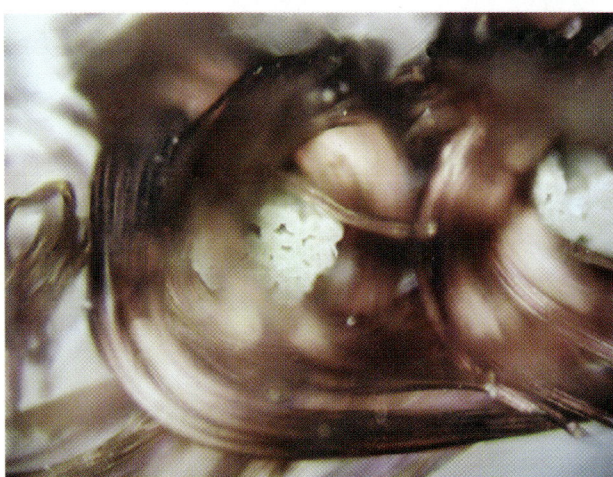

Plate 26 Cross-sectional view of a fabric with multifilament elastomeric yarn within its structure (Fig. 4, p. 50).

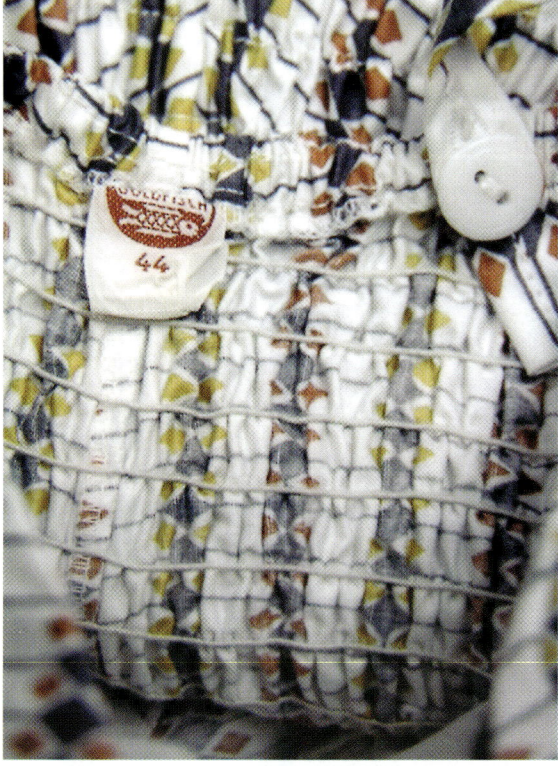

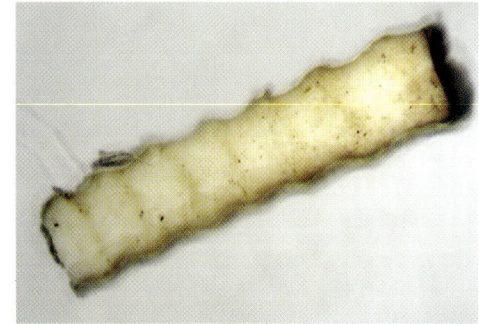

Plate 28 Section of a broken and stiff degraded natural rubber thread (Fig. 6, p. 51).

Plate 27 Swimming costume (German, 1950s): interior detail showing horizontal rows of elastic yarns (Fig. 5, p. 50).

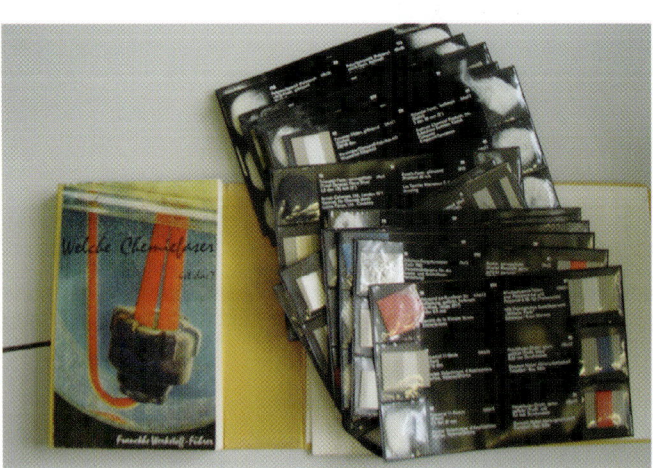 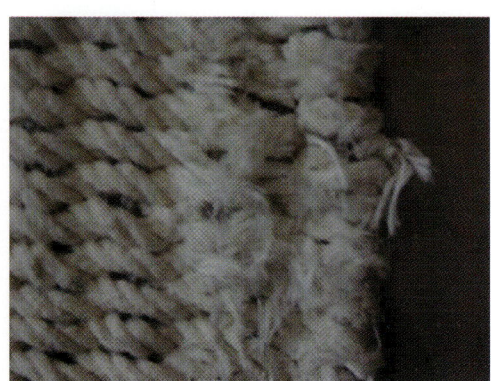

Plate 29 *Welche Chemiefaser Ist Das?* (Driesch 1962) (Fig. 1, p. 61).

Plate 30 *Fête II* (Will Fruytier, 1969): white fibres (Fig. 2, p. 62).

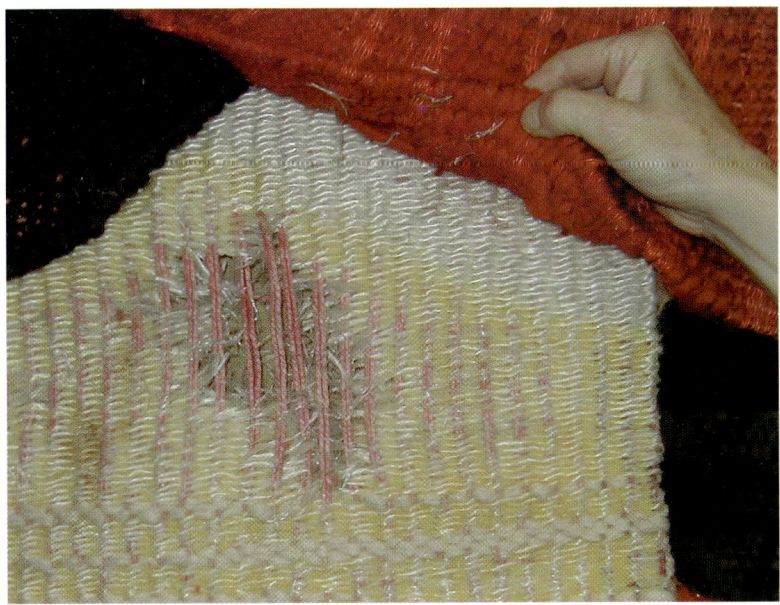

Plate 31 *The Knot* (Herman Scholten, 1968): degraded area of tapestry (Fig. 3, p. 62).

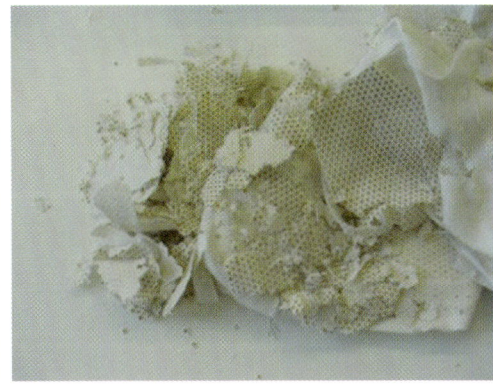

Plate 32 Relining fabric (1975) (Fig. 4, p. 62).

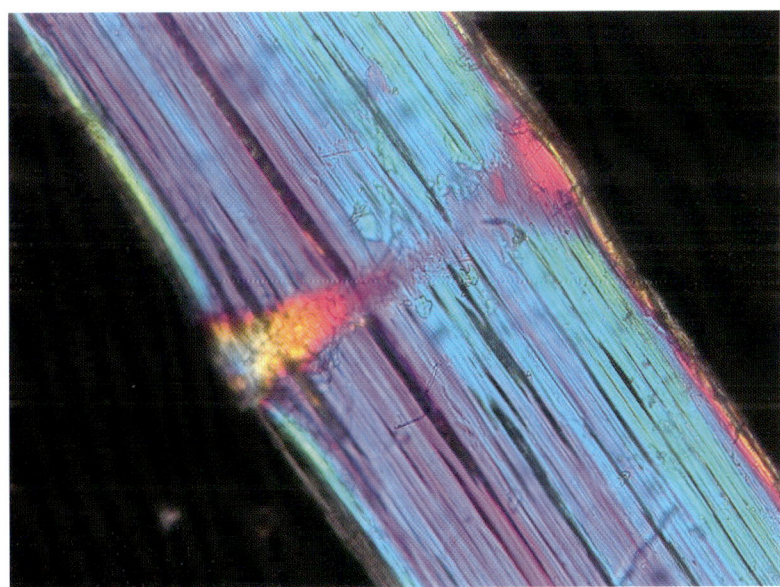

Plate 33 Polarisation microscope image of white fibre of *Fête II* (100×) (Fig. 7, p. 65).

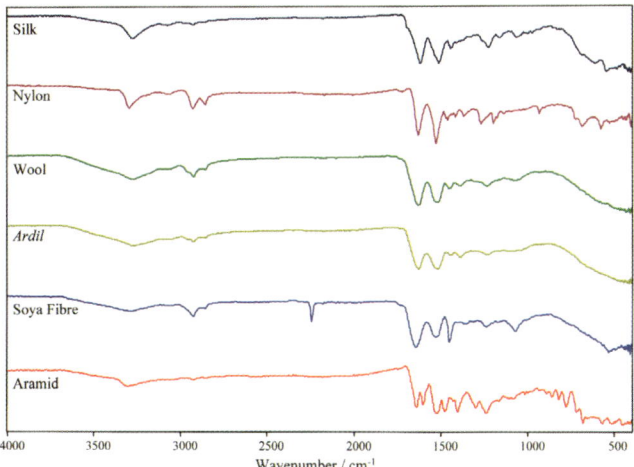

Plate 34 Typical ATR spectra for the experimental samples over the range 4,000–400 cm^{-1} (Fig. 3, p. 69).

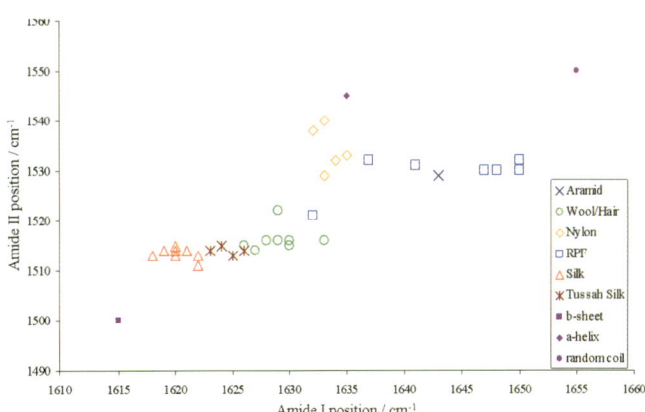

Plate 35 Plot of amide I and amide II positions for various proteinaceous and polyamide specimens (Fig. 4, p. 70).

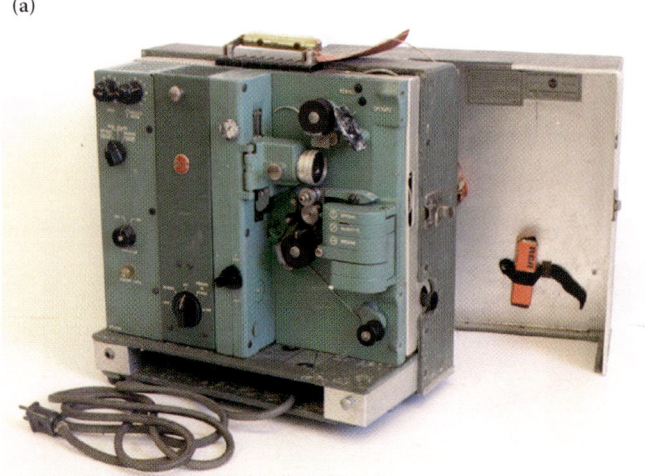
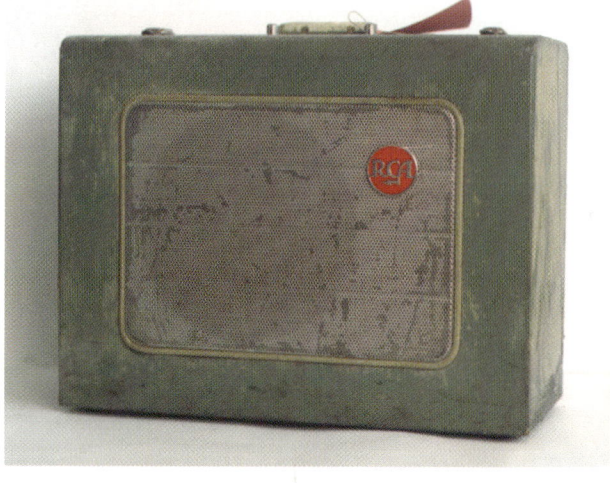

Plate 36 Film projector's case 1 (a) and case 2 (b) (Fig. 1, p. 73).

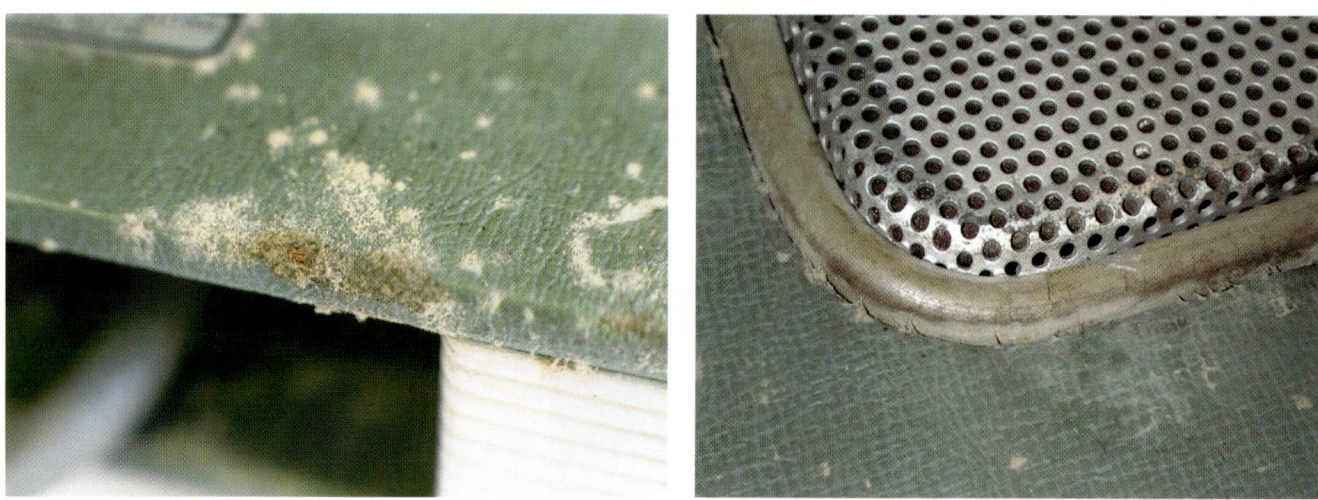

Plate 37 Morphology of the white substances appearing on the synthetic leather surfaces: (a) powdery appearance on case 1 and (b) filamentous appearance on case 2 (Fig. 2, p. 73).

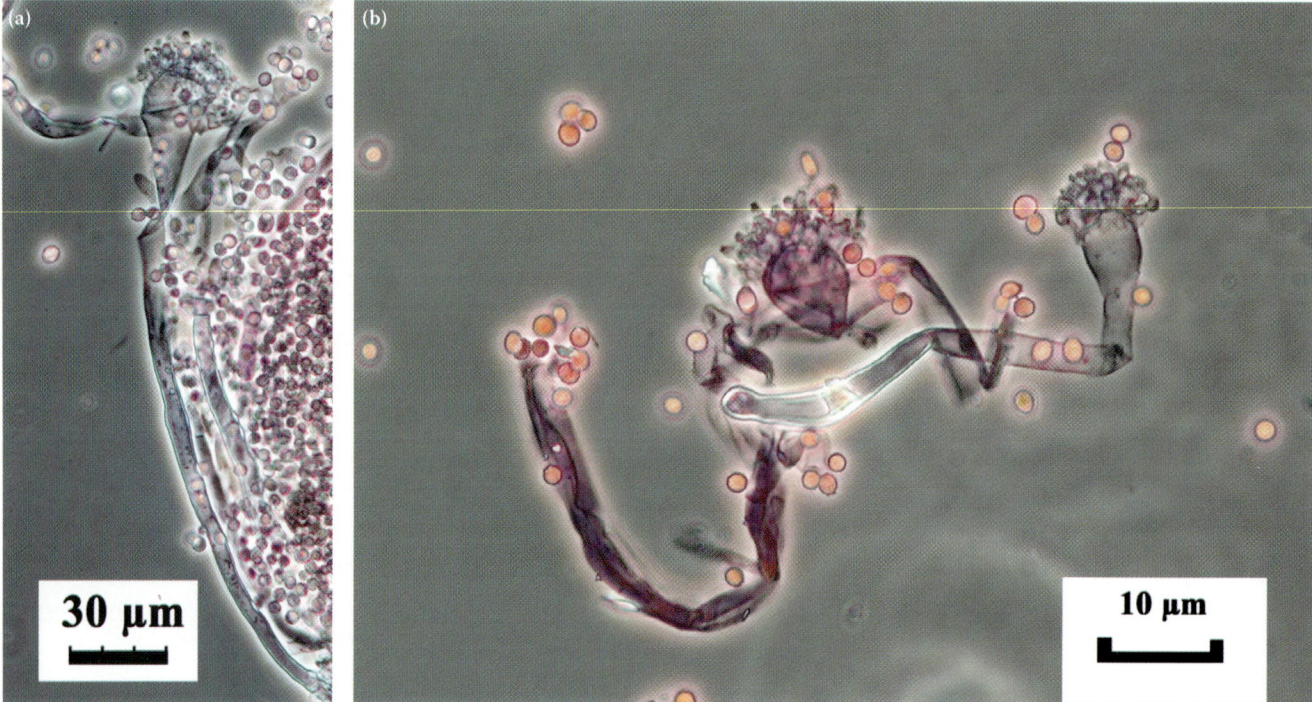

Plate 38 (a) and (b) Conidiophore and spores of *Aspergillus*, observed on microscope with a magnification of 40× (Fig. 4, p. 75).

'A bomb in the collection': researching and exhibiting early 20th-century fashion

Alexandra Palmer

ABSTRACT In *Twilight Memories*, cultural critic Andreas Huyssen noted that: 'the museum serves as both a burial chamber of the past – with all that it entails in terms of decay, erosion, forgetting – and as a site of possible resurrections' (Huyssen 1995:15). This project, an exhibition and publication, aims to resurrect the decaying artefacts of early 20th-century fashion. It investigates the significance and function of European and Canadian couture for the makers, merchandisers and consumers in Toronto. This research is dogged, however, by museological and conservation issues that are reframing the result.

This period has barely been researched. One reason may be the complex designs themselves that are highly experimental, coupled with their rapid deterioration that is leaving even less of a record for future research. Usually when an object becomes a museum artefact, the public expects it to be kept permanently. But, as conservators know only too well, many of our collections are mutating and rapidly decomposing. These costumes are a lively ticking time bomb of self-destructing material due to the late 19th and early 20th century popularity and use of weighted silks. Many pieces are already too far deteriorated to repair and languish in costly museum storage rooms as we wait for a more complete disintegration that would permit unequivocal de-accessioning. Each design raises ethical issues for conservation that strikes at the heart of the purpose of museums and collections.

This paper proposes an aggressive strategy for documenting and displaying selected designs from the collection of the Royal Ontario Museum (ROM) in an exhibition that would have two parallel strands: the sociocultural design history of these fashions and the fashion artefact as scientific specimen with its inherent difficulties of decomposition. Collaboration between the historian and the scientist could create a major exhibition of 'resurrected' early 20th-century fashion that exposes and questions the concept of permanence within museums and articulates the importance of museum research and exhibits for the public.

Keywords: weighted silk, 20th-century costume, fashion, exhibition, lining

Introduction

Dress and textile collections hold some of the most sensitive, decaying and labour-intensive materials in museums. The author's current research on early 20th-century couture in the collection of the Royal Ontario Museum (ROM) amplifies this reality. Even though the inherent conservation problems were known already, the author was unprepared for the full extent of the problems of the materials, in particular the explosive state of the weighted silks. The textiles, trims and linings all pose an enormous physical and intellectual challenge to the conservation of these costumes that were destined for photography and display for a publication and exhibition that retrieves the social, cultural and economic history of couture from this period. Cultural critic Andreas Huyssen has noted in his book *Twilight Memories* that 'the museum serves as both a burial chamber of the past – with all that it entails in terms of decay, erosion, forgetting – and as a site of possible resurrections' (Huyssen 1995: 15). The delicate state of the material has shifted the original historic project to a more complex one with a secondary contemporary and parallel theme of resurrection for these decaying artefacts. This paper, therefore, articulates the labour-intensive and costly reality of museum work and the complexity of dealing with collections of dress and textiles from the 20th century.

The project investigates European- and Canadian-produced couture from the point of view of the makers, merchants and consumers in Toronto in the early 20th century. Fashion historians have only superficially documented this period, one reason probably being the complexity of the designs themselves that are highly experimental. They are often asymmetrical, composed of multiple, diverse and conflicting materials; a factor that makes them seem messy to an orderly modernist eye and difficult for conservation. The period is often glossed over in favour of a simplistic comparison between the more clearly delineated Edwardian silhouette that is set in opposition to that of the post First World War 'new woman', despite the fact that c.1913–19 was the transition period. Another reason for the neglect of this era is that this was a difficult time for the first wave of fashion scholarship: the Cunningtons, Laver and Langley Moore. These scholars, who would have personally felt the effects of the war and the rebuilding of lives in its aftermath, have placed these years in the shadows as they preferred to discuss earlier or later times. Thirdly, there is a repeated tendency to negate the significance of fashion and style during wartime. The fashion press tends not to tout the latest designs by leading designers with the same bravado and designers cannot market their goods in the same manner. The complexity of our relationship to fashion and vanity and nationalism and war also accounts for the general lack of research on this period that has only recently begun to be addressed (Veillon 2002; Guenther 2004).

Scholars have paid attention to what we now consider the major European couturiers of this period such as Worth, Doucet or Poiret. But even cursory research shows that there were many more important houses at the time that are unrecognised today. While many Paris and London designers remain to be reinstated, the author is also salvaging the history and role of Toronto's leading dressmakers and merchandisers of style and culture who have left tantalising traces. The history is one that links Canadian fashion directly with Old and New World traditions. The significance of Canadian merchants and makers who were bastions of style and taste, such as Murray's, is reflected in the extant costumes held in museum collections across Canada. Murray's store interior, complete with an art gallery and tearoom decorated with faux garden decor, clearly demonstrates that this shop was an important rendezvous for women at the turn of the century. A postcard of this decorative setting is inscribed on the front 'This may give you some idea of the shops here', indicating the strength of the impression it made on one visitor. Such large-scale retailers displaced older 19th-century dry goods merchants such as John Catto, a Scot from Aberdeenshire. He arrived in Toronto in 1854 and within ten years was running his own dry goods business specialising in linens and tartans. He added a dressmaking department in 1894.

Another important Canadian dress house, O'Brien's, is also unknown despite the fact that its designs are found in collections across the country. O'Brien's held a royal warrant from His Royal Highness the Prince of Wales, an Appointment from Her Excellency the Countess of Aberdeen and displayed the coat of arms of Lady Minto 'by gracious permission' – all printed on the label (Fig. 1).

O'Brien's employed the top calibre of design, materials and techniques of international dressmaking of the time. The business began as a men's tailoring firm in operation from the 1880s; by the 1890s it was dressmaking for the most prominent Canadian families. Born in 1864, Donel O'Brien started out as a tailor's apprentice in Cork, Ireland and left at the age of 11 for another apprenticeship in London. In 1886 he took a job as a cutter in Toronto with the tailoring firm of Stovel and Armstrong, for which he received a secured paid passage and £4 sterling a week with four months guaranteed work. He was eventually given a partnership that he later bought out, creating O'Briens. The O'Brien family ran a smart dressmaking and retail clothing firm for several generations and finally closed in the 1990s.

Problems posed by the ROM costume collections

While patching together this rich history, large museological and conservation issues continue to hamper the process. The photographic records, newsprint and correspondence on fugitive carbon copies typed onto highly acidic onion skin papers in archives, require special permission to view them and leave small castles of crumbs on the desk afterwards. Even the microfiche onto which one journal is archived is becoming problematic as the Thomas Fisher Rare Book Room, University of Toronto, has only a few microfiche readers left and only one that prints. Thus it seems that the archival records are also disappearing.

As a museum curator, however, it is the fashions themselves that are the primary concerns and challenges. Usually when an object becomes a museum artefact the public expects it to be kept permanently. But, as conservators know all too well, many of our collections are mutating and rapidly decomposing. The fashions from this period in particular, are lively ticking time bombs of self-destructing material. Thus the theme of decay and deconstruction seems to be haunting all aspects of the research.

The key concern is that the degradation appears to be exponential once it begins. In many objects, observed for over 15 years, the deterioration has seemingly accelerated. This is far more rapid than in other areas of the collection where it is imperceptible. Pieces examined years ago, and known to be fragile, at the time were believed capable of being displayed, albeit carefully. Many of them are now far more degraded, however, requiring an enormous amount of conservation, while others are now too deteriorated to consider. The costumes that have already decomposed beyond repair languish in the storage drawers. The process will continue and a more complete disintegration will eventually result in unequivocal de-accessioning. While not all pieces are in dire condition, none of them are static and unchanging and none will

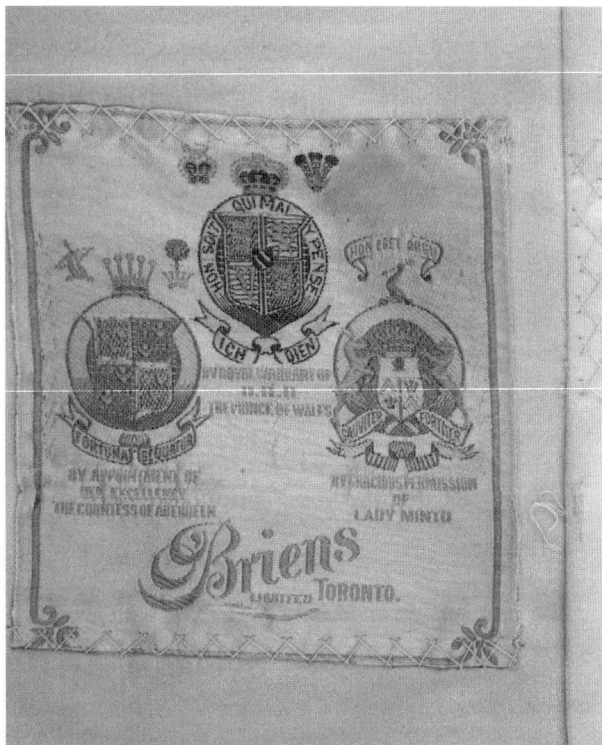

Figure 1 The Toronto dress house, O'Briens, was one of the leading Canadian dressmakers during the first quarter of the 20th century, as is reflected in the royal warrant and appointments displayed on the label (ROM 965.227.1ab, gift of Mr. John M. Morley) (Plate 17 in the colour plate section).

improve. Most dresses cannot be exhibited without the investment of hours of conservation work. Each raises interesting ethical issues that strike at the heart and purpose of museum collections.

The main problem is the late 19th and early 20th century popularity of weighted silk (tinned) that is now shattering or splitting (Fig. 2). The weighting process employed metallic particles, usually containing lead and arsenic, in order to add more body, texture and weight to the fabric. Over time, the damage of the metal salts on the fibres alters the mechanical properties of the silk, causing the weighted silks to shred. The process is commonplace for weighted silks and is frequently encountered by curators and conservators alike. The fragility of the textile is also compounded by the method of design and construction. Dresses from the period *c*.1913–19 tend to be constructed with lightweight materials at the shoulders that are then trimmed with heavy beading, velvets and even furs. This style of design makes the costume conservator's perennial struggle against gravity pulling on textiles when on display even more tricky than usual. Even in flat storage these artefacts are often extremely fragile and difficult to handle. Moving an artefact from storage drawer to table typically requires two people, four hands and once set out the piece usually reveals additional weak areas. Mounting a costume so that it resembles a corseted, three-dimensional draped form is often impossible without prior extensive conservation. The alternative is to leave it lying flat in the drawer. This may be interesting for the connoisseur but less so for the visiting public – fashion was never intended to be seen limp and lifeless.

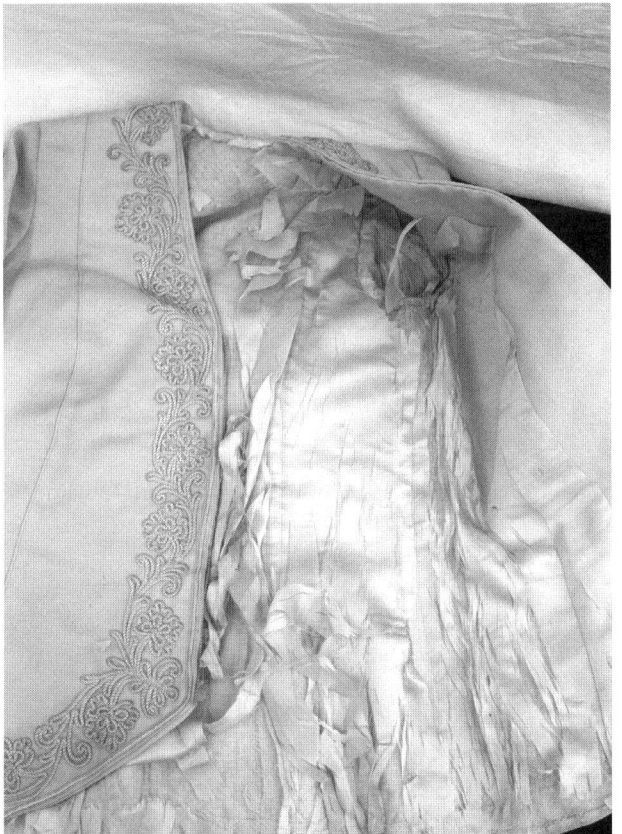

Figure 2 Woollen suit jacket (*c*.1909) with shattered silk lining that needs stabilisation or replacing for future display (ROM 959.43.24ab, gift of Mrs. F. W. Trusler) (Plate 18 in the colour plate section).

Given the number of pieces required for a major exhibition and the nature of the materials, however, the issue of how to make the costumes suitably attractive and safely mounted is foremost. It is impractical, if not impossible, to stabilise the materials using a stitched-on supporting fabric, as is common practice for textile conservation. Any stitching creates another split that further shatters the weighted textile. Adhesives are a possible solution that avoids stitching, but in many cases requires backing an entire dress, bodice, bodice lining or petticoat. It is a treatment that can substantially alter the drape and look of the design and is not a long-term solution. Adhesives were used in many of the garments in the exhibition, Shocking! The Art and Fashion of Elsa Schiaparelli held in Philadelphia and Paris (2003–04). This treatment was undertaken on the linings that were shredding in suits and coats, so it did not affect the outer look of the pieces. Textile conservator Sara Reiter noted that this definitely helped to stabilise the materials while on display but that the adhesive had weakened by the end of the tour, and in some instances the supporting material was no longer adhered to the lining (Reiter 2006). One was re-adhered for the Paris installation. This treatment was clearly not a final answer but a temporary one that served the immediate demands of the two displays. A solution is needed that takes into account the conservation and display requirements of the material and that also addresses the practical issues of staff time and costs that are required in order to realise the intended exhibition. The last time many of these garments were seen was in the large ROM exhibition, Modesty to Mod. Dress and Underdress in Canada 1780–1967 staged in the centennial year (Brett 1967) (Fig. 3). If these pieces are not presented to the public soon, this collection will be unsalvageable, and can be considered disposable.

In 1974, the exhibition The Destruction of the Country House at the Victoria and Albert Museum (London), very powerfully documented the enormous losses of over 300 English country houses and interiors, modest to lavish, dating from the 17th to the early 20th century. The exhibit was like a roll call with birth and death dates that was alarming, but it succeeded in many other homes still extant being placed on the endangered list – and many have since been saved. The house destruction can be likened to much of the late 19th and early 20th century extant fashions, many of which have reached this endangered stage. This is so acute that one dealer recently revealed that they will no longer purchase costumes with weighted silks as by the time they have been purchased, after the viewing and sale, the condition is often so poor they have difficulty realising a profit (Strong *et al.* 1974).

Along with other costume and textile curators, as well as private collectors, the author is faced with a beautifully decaying collection from this era that is not easily displayed. This issue raises conflicts between curators' and conservators' practice and even the fiscal responsibilities of museum management. There seems to be a long list of reasons for preventing a large exhibit of this early 20th-century material including:

- too damaging to the fragile artefacts;
- too time-consuming for conservation given other institutional pressures or even impossible because of the material;

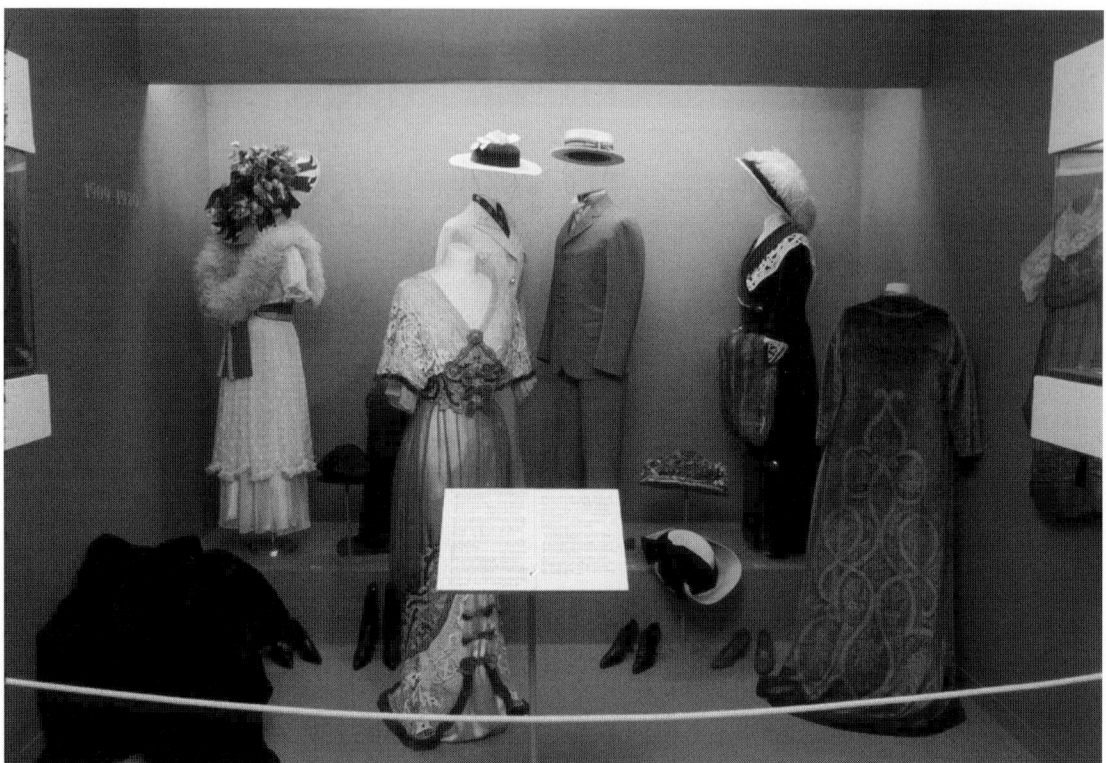

Figure 3 ROM's 1967 centennial year costume display *Modesty to Mod*. This is the last time that the *c*.1910 semi-formal Parisienne evening dress (front) was on public display. It is made of silk chiffon, beaded and trimmed with mole (ROM 952.203.1, gift of Miss Hattiemae Austin) (Plate 19 in the colour plate section).

- if the conservation was realised it would make the exhibit very expensive;
- the high cost of the exhibit would make it impractical given that major displays follow a profit-generating business model at the museum; this is compounded by the fact that the display would have to be short term because of the delicate material, and thus the time could not offset the costs.

Possible solutions

If exhibiting these dresses is impossible what is the point of keeping them in our overstuffed storage rooms? The need for a timely solution to this conundrum is especially pressing given our increasingly virtual world. Huyssen suggests that: 'What needs to be captured and theorized today is precisely the way which museum and exhibition culture ... can offer *multiple narratives of meaning* [author's emphasis] at a time when the metanarratives of modernity ... have lost their persuasiveness, when more people are eager ... to hear and see the stories of others, when identities are shaped in multiple layers' (Huyssen 1995: 34). The author proposes an exhibition with multiple layers.

Over the front door of the ROM was carved in stone in 1914: 'The record of Nature through countless ages. For the arts of man though all the years.' The ROM is still a museum of art, archaeology and science. To create an exhibition of these early 20th-century fashions, collaboration between its two intertwined strands, art (costume) and science (conservation) is suggested.

In addition to the sociocultural context, the exhibition and publication should expose the scientific aspect of collections and collections care; the preparation of artefacts for exhibition and the work that goes into daily museum life. Deconstructing the aura of museum work and the reality of the complexity of artefacts within the museum setting would be part of the exhibit itself. The physical, scientific effects of time on fashion history would be highlighted along with the new conservation record that has become a vital part of the garment's history. In this way the project would extend Kopytoff's discussion of the biography of an object (Kopytoff 1986). The biography would include the museum world in which the object continues to exist.

Interestingly, as Caroline Evans (2003) and others have noted (O'Neill 2001), contemporary fashion designers such as Martin Margiela make a fetish of deconstruction. Yet it is museums that deal with the detritus of historic and contemporary clothing on a daily basis (Fig. 4). This aspect of fashion is completely unheralded and silenced in fashion exhibitions. It is an intriguing story that lies at the heart of textile and costume displays – the museum visitor usually finds the 'behind the scenes' information fascinating. To show this within the context of the exhibition itself will make the museum visitor aware of the continuing movement and use of the costumes within the inner world of museum life.

By bringing to the foreground the nitty gritty of museum work in both exhibition and publication, this unglamorous aspect of the scientific exploding collection would be as privileged with public view as the history of the designs – the art. Such a programme that exposes the invisible to the public could be presented in an elegant and sophisticated manner. It would enable visitors, donors and sponsors to really

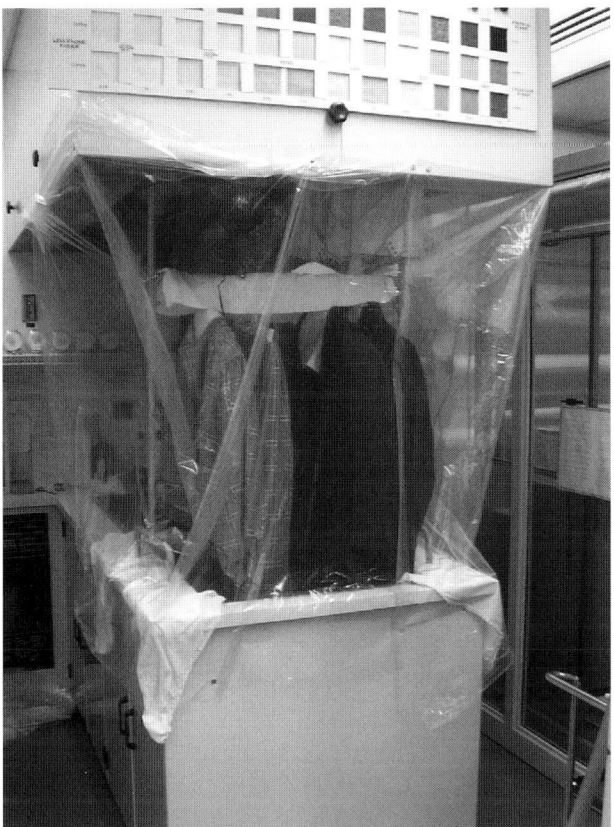

Figure 4 Decontaminating chamber made for incoming mid-19th-century artefacts donated to ROM's collection in July 2005 (Plate 20 in the colour plate section).

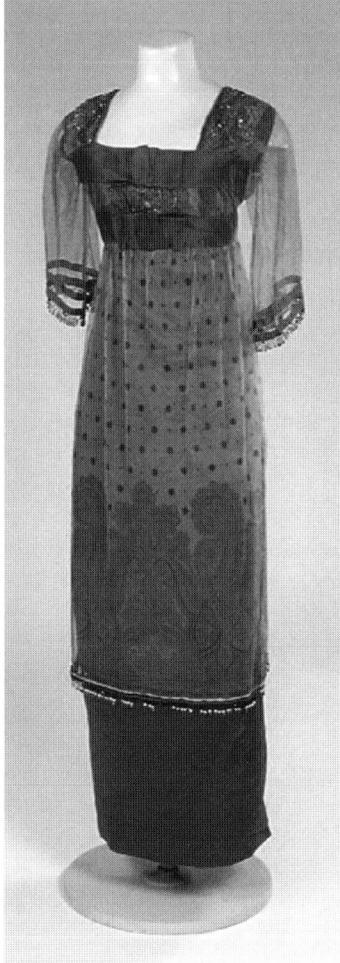

Figure 5 Canadian-designed red silk Directoire-style dress. The underdress of silk satin is made of weighted silk that is severely shattered (ROM 972.436.1, gift of Miss Kathleen Myers) (Plate 21 in the colour plate section).

understand the complexity of museums and the role of specialised staff, as well as highlighting issues related to the survival of clothing and textiles.

So what would it take? The selection of artefacts proposed for conservation and exhibition needs to be identified and is driven by the research. Each garment selected would be moderated by the conservation requirements and history. For instance, a deteriorating red Directoire-style dress (ROM 972.436.1) should be considered for display because it is Canadian and clearly is a wonderful example of the importance of Paris couture (Figs 5 and 6). Research links such a dress as this directly to the influence of Paul Poiret who stayed in Toronto during his North American tour in 1913. Poiret lectured and showed his collection at Eaton's department store where the audience laughed at the outrageousness of his designs, but he quipped back 'you'll wear it yet'. A design such as this proves that indeed Torontonians did (*The Globe* 1913).

The dress, however, is in poor condition. The visually disturbing stitching evident upon its fabric is an earlier conservation effort to stabilise it for display. It may have worked at the time, but the dress has continued to degrade and any more stitching would only further split the silk. Adhesives would alter the drape and, as discussed, cannot be considered a long-term solution. As it stands this dress is unacceptable for display because the damaged silk and the previous conservation treatment is too distracting and unattractive for the museum visitor. The costume needs to be presented as elegantly as it was intended to be seen. It is unreasonable to present it to the visitor otherwise and ask them to imagine and appreciate the original aesthetic that is now disguised by degradation. What are the options for such a piece? Replacing the damaged textile is the suggested solution here. Replacement of weighted silk material is a contentious issue as it is a radical intervention that is in opposition to conservation practice and is interpreted as a negative form of reproduction. In many instances, however, from a curatorial perspective the author considers it to be the only viable solution for the display of these pieces. There are existing precedents.

A privately owned Austrian 1906 tea gown designed by Ungar had a disintegrated weighted silk lining. This has been replicated, even to the extent of hand overcast finishing. It was sold for US $3800.[1] The restoration of this piece cost US $1544 including the $225 for materials. Most of the cost was in the handwork. Without relining, the dress was not saleable and held little value for the dealer who paid 'top dollar for it considering its condition'.[2] Linings are also an essential component of the period as they hold the structure, form and sometimes colour for the outer design of the garment. The solution of replication has also been undertaken within a museum. A 1908 wool and silk day dress by the couturier Lucile in the collection at the Museum of Costume in Bath (UK) was relined in order to make it possible to display. The conservation record notes that 'the overall condition of the dress was extremely poor ... and [it] could hardly be handled ... because possibly tin-weighted silk had been used and the lining had to bear most of the construction stress of the dress design'. The conservation record goes on to note that the dress was 'almost impossible to handle safely, nor could it be put on a mannequin. This meant that it was very difficult to imagine how it looked when it was worn.' The treatment was a relining

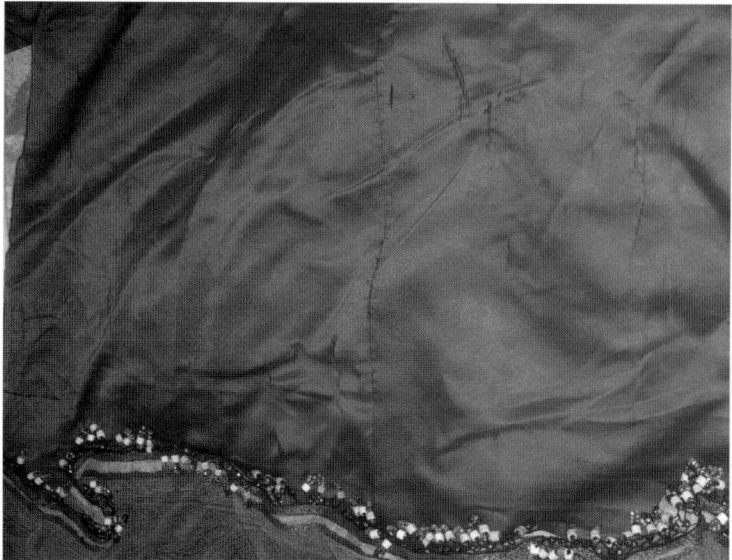

Figure 6 Detail of Figure 5 (Plate 22 in the colour plate section).

that was carefully documented and the garment was placed on exhibition. A sample of the original lining was kept.

In order to salvage the ROM collection, similar interventions are suggested. This may seem radical, by museum standards, as replacing materials goes against the grain of both curatorial and conservation practices. Costume exhibits are static formats, however, for objects that were animated by movement. They must look elegant, and costume exhibitions in particular need to move the museum visitor beyond merely shopping. So the difficulty is how to make a dress such as this look as elegant as it did originally, while retaining enough of the original design to be 'authentic'.

Replacing fabrics and especially linings is a practical solution that does not require slavish reproduction, stitch for stitch, but one that replicates the colour and effect of the original and provides a suitable support for the outer shell to be shown, pulling on gravity, for four to six months. Although this is costly in terms of time, it is less difficult for linings than the more delicate and time-consuming operation for pieces such as the red dress where the underskirt is part of the visible design. In this case, replacement may be appropriate because it is a plain colour and weave that has no complex pattern to imitate. The entire process and treatment of each object would be extensively documented (images, treatments and samples of the original saved), and all would be shown and discussed within the context of the exhibit itself, along with the sociocultural story of the artefact. This post-acquisition biography would be added to the public documentation. The lasting archival result would be images of these designs that are contextualised and recorded, as well as the detailed museum records of the conservation of each artefact. This will likely outlive the actual artefacts themselves as nothing can stop the deterioration.

Of course, the time and cost of this is a major stumbling block. ROM's textile conservator Shirley Ellis and the author have proposed a theoretical scenario for the time and cost required to exhibit a typical silk dress from this period. Ellis estimates that it would take approximately 44 hours to make an average deteriorating weighted silk costume safe for mounting (primarily the linings). Putting it on a form would take another few days. All the additional time required by technicians, database documentation, photography and sundry staff easily adds up to around 300 hours or two to three months per garment in order to get it ready for display. The financial cost from conservation to display is estimated to average *c*. $4500 per design. This quickly creates a sizeable budget for the exhibition, making it unaffordable for the existing ROM exhibition models.

In order to offset the cost, however, a strategy that addresses this directly could be implemented. The exhibition would have an interesting graphic format that would permit the object label to consist of two biographical parts: the cultural and historical information on the design itself and the before, during and after photographs of the conservation, complete with the time it took and the cost. The before-and-after of conservators' skills can indeed be staggering. Thus the exhibition would discuss and show the grand designs in their glory, but also acknowledge the challenges that were required to achieve this effect. It would create a two-for-one story that discusses the garment in pre- and post-museum context.

The obvious question is – is it worth it? The exhibition should contain around 75 designs, not all of which would require extensive work, as wool suits lined with weighted silk last better than the more glamorous evening wear that is often composed of many layers. The time frame required is long because of the conservation requirements and the resultant exhibit short because of the fragility of the costumes even after conservation. Once the entire scope of work is identified, however, it is possible to accelerate the time frame by creating internships and/or course credits with institutions that offer costume studies and conservation programmes, and with fashion schools that could assist in drafting patterns and making the new supporting structures.

Yet, the issue of funding still remains. Normally budgets are based on the fact that most artefacts are in good condition and require minimal conservation time. But, if the planning and exhibition foregrounds the conservation project, it could provide a unique opportunity for ROM to promote its collections and research in an innovative way and seek partnerships with the current leaders in textile science whose new textiles can assist our old ones.

Another initiative would be to orchestrate an 'adopt a dress' programme. Once the artefacts for exhibition and photography are identified, the project could be funded by virtually reselling the dress so the conservation of each design would be funded and recognised within the exhibition. In this manner, both moderate and more extensive conservation time and costs would be attached to each artefact slated for exhibition. The designs could be virtually 'sold' to patrons who would underwrite the cost. Their names would feature on the catalogue and object label as contributors to the longevity of the piece. The selection and cost of dresses would be wide: would you like to sponsor a Worth, O'Brien or Lanvin? Such a scheme of advertising and promoting the transformations in dress – social, historical and physical – relies upon public involvement and understanding of museum work and the exhibition process from its inception. If this were carefully positioned, ROM could lead the way in preserving and documenting this patrimony with a ground-breaking exhibition and publication.

Conclusion

In 1916, Sigmund Freud noted that: 'A time may come when the pictures and statues which we admire today will crumble to dust, or a race of men may follow us who no longer understand the works of our poets and thinkers … but since the value of all this beauty and perfection is determined by its significance for our own emotional lives, it has no need to survive us and is therefore independent of absolute duration' (Freud 1961: 304). Could Freud be correct? Certainly he captured his time. What about all the objects that will not survive and will crumble to dust? How do we validate the time they spend in museum collections?

It is clearly a museum's role to establish, save and communicate its material culture. As museum professionals we specialise in the real. The decomposition of these designs could not be more real. They will continue to degrade and without any intervention these pieces will remain in storage – flat, lifeless and useless, consigned to the cultural rubbish heap. Can we move beyond the decay and resurrect them before it is too late?

Acknowledgements

Thanks to the Catto and O'Brien families for kindly providing family histories. The author would also like to thank Sara Reiter for sharing her knowledge of the Schiaparelli costumes and to Ann French for information on the Lucile dress she worked on at Bath, as well as to Eleanor Summers at the Museum of Costume Bath for so generously sharing the information on the conservation of this piece.

Notes

1. www.antique-lace.com/date3/2557/2557.htm. See also www.antique-fashion.com.
2. E-mail correspondence between Karen Augusta and the author, 4 October 2005.

References

Brett, K.B. (1967) *Modesty to Mod. Dress and Underdress in Canada 1780–1967.* Toronto: Royal Ontario Museum, University of Toronto.

Evans, C. (2003) *Fashion at the Edge: Spectacle, Modernity and Deathliness.* New Haven and London: Yale University Press.

Freud, S. (1961) 'On transcience', in *The Standard Edition of the Complete Psychological Works of Sigmund Freud*, J. Strachey (ed.), Vol. 14, 303–7. London: Hogarth Press (original work published 1916).

The Globe (1913) 9 October: 5.

Guenther, I. (2004) *Nazi Chic? Fashioning Women in the Third Reich.* Oxford: Berg.

Huyssen, A. (1995) *Twilight Memories: Marking Time in a Culture of Amnesia.* New York: Routledge.

Kopytoff, I. (1986) 'The cultural biography of things: commoditization as process', in *The Social Life of Things: Commodities in Cultural Perspective*, A. Appadurai (ed.), 64–91. Cambridge: Cambridge University Press.

O'Neill, A. (2001) 'Imagining fashion', in *Radical Fashion*, C. Wilcox (ed.), 38–45. London: V&A Publications.

Reiter, S. (with H. Sutcliffe, B. Price and K. Sutherland) (2006) 'Second times the curse: the shattered silks of Schiaparelli', in *Textile Specialty Group Postprints*, AIC 33rd Annual Meeting, Minneapolis, MN. 9–12 June 2005. Vol. 15 (forthcoming).

Strong, R. *et al.* (1974) *The Destruction of the Country House, 1875–1975.* London: Thames & Hudson.

Veillon, D. (2002) *Fashion Under the Occupation.* Oxford: Berg.

The author

Alexandra Palmer is Senior Curator in the Textile and Costume section of the Royal Ontario Museum, Toronto, Canada. She has curated many costume exhibitions including Au Courant: Contemporary Canadian Fashion (1997), Unveiling the Textile and Costume Collection (2002) and Elite Elegance: Couture Fashion in the 1950s (2003). She is currently working on a new gallery at the Museum. Her research examines 20th-century fashion and textiles.

Address

Alexandra Palmer, Senior Curator, Textile and Costume Section, Department of World Cultures, Royal Ontario Museum, 100 Queen's Park, Toronto, Ontario, Canada M5S 2C6 (alexp@rom.on.ca).

Early elastic threads and fibres in clothing

Laura Petzold

ABSTRACT This paper provides an overview of the materials and processes used to create elastic fibres and the methods with which they are used to manufacture clothing, particularly swimwear and underwear. The development of the fibres encompasses natural rubber in the 19th century through to the synthetic fibres of the 20th century, including polyisoprene, nitrilo-butadiene rubbers and polyurethanes. The materials may be characterised by techniques such as Fourier transform infrared (FTIR) spectroscopy, visual examination and knowledge of the period of the garment. Elastic fibres are distinguished by their physical properties, most notably that they can be stretched to more than twice their initial length and return to their original dimensions. The materials degrade by a variety of mechanisms, principally oxidation and photo-oxidation, leading to the loss of their unique flexibility and elasticity. Following deterioration, options for interventive conservation are limited, so stable environments are the best strategy for the long-term preservation of these materials – ideally in the absence of light and with temperatures below 15 °C; anoxic conditions are also preferable. For display, the threads should not be not placed under tension and the items should be isolated from other objects to ensure that damage from off-gassing is minimised.

Keywords: elastic, elastomeric, natural rubber, synthetic rubber, segmented polyurethane, swimwear

Introduction

This paper (based on a diploma thesis) considers textiles that have elastic fibres incorporated within them. The principal aim of the paper is to provide a brief overview of the materials and the manufacturing processes used to create elastic fibres and the different ways of processing these yarns to manufacture clothing. In doing so, the incidence and construction of elastic threads may be better understood. Due to ageing, elastic yarns quickly lose their original properties, most importantly elasticity and flexibility. Recognising how they are incorporated into textiles will elucidate how these garments deteriorate. Information for this study was gathered by examining garments containing elastic threads in German collections and through a study of patents[1] and secondary written sources. Bathing costumes and underwear serve as notable examples for all elastic clothing since their function to fit closely around the body is an ideal application for elastic fibres, and these are therefore the predominant items with such fibres found in collections today.

Elastic threads or fibres are distinguished by the fact that they can be stretched repeatedly to more than twice their length and, after the tensile force is removed, they immediately revert to their original dimension. They should not be confused with highly crimped fibres, such as polyamide fibres in nylon stockings, which are sometimes also called 'elastic'. The two different terms 'thread' and 'fibre' indicate different diameters, threads being thicker than fibres.

This study is focused on the time period stretching from the development of the first elastic threads, using natural rubber, in the first half of the 19th century until the end of the 1960s, arguably the most active period of development and novel application for elastic fibres. The materials studied are the principal materials found in elastic clothing of this period: rubber threads of natural rubber (NR), synthetic polyisoprene rubber (IR), nitrilo-butadiene rubber (NBR) and polyurethane elastomeric fibres (PUR). The first representatives of polyurethane elastomeric fibres, also recognised by the terms Spandex (such as Lycra by DuPont) and Elastan (such as Dorlastan by Bayerfaser GmbH), are included within the study. The term 'elastomeric' is derived from the words elastic and monomer to reflect the all-important elastic property of these synthetically derived fibres.

Brief historical overview

A burgeoning desire for more freedom in clothing and an increased understanding of health issues brought about more research and use of elastic materials in clothing during the course of the 19th century. These were mainly threads and ribbons made of natural rubber. In 1844 several companies producing elastic woven goods already existed in the UK, and companies in France and the US were quickly established, promptly catching up with these initial developments. Unfortunately, few objects have survived from this time but one example is a pair of garters from 1875 in the collection of the Historisches Museum Frankfurt/Main, Germany (Clouth 1879; Heinzerling 1883).

In the early years the desired elasticity in garments was more or less achieved by the process technology, for example knitting, which quite naturally provides stretch to a textile structure. In Germany, however, knitted 'elastic' clothing was produced until the end of the Second World War, which effectively combined elastic threads within a knitted structure. Further developments permitted the production of even thinner elastic threads, such as Lastex threads, commercialised in the 1930s. These newer, more slender, threads allowed clothing to be 'tight without creases, so you would think it to be precision work, while the elasticity would cover the little irregularities (and the bigger ones) of the average body' (Anon 1934).

For swimwear, the introduction of elastic threads was very welcome. Until the development of elastomerics, knitted swimwear was usually made solely of natural fibres. An ordinary swimming costume made of wool or cotton, with an average weight of 300 g, would gain five times its weight when soaked with water, with the unfortunate propensity to become misshapen. With the addition of elastic fibres within its structure, swimwear could hold its shape far better and with the eventual introduction of synthetic fibres, was significantly lighter in weight and hydrophobic in nature (Bleise 1992).

The dependence of industrialised nations on crude rubber and the poor resistance of natural rubber to high washing temperatures, several chemicals and oxidation promoted research into synthetic rubber types. But it was not until the 1950s that rubber threads of synthetic rubber such as NBR and IR were manufactured on a significant scale. Since the extensibility of these new synthetics was not as high as that of NR, the natural product was still used in some instances. To combine the beneficial properties of the natural and synthetic rubber the two types were, at times, blended. Suspender belts are a typical application and one where long-term strength and elasticity are needed (Davis 1933; Hoffmann 1965).

Despite these technical advances, optimal properties had still not been achieved, and elastic fibres were the subject of much greater development, but now using synthetic materials. At the end of the 1950s, so-called Spandex fibres (better known as Lycra) were produced in the US after extensive research on segmented polyurethane elastomeric fibres. In the 1960s the former Farbenfabriken Bayer developed Dorlastan (Marshall 1961; Rinke 1962). These new fibres not only had a finer fibre diameter than rubber threads but also gave better resistance to washing, oxidation and ultraviolet (UV) radiation. Moreover, unlike rubber, they performed well at low temperatures. These properties were favourable for sportswear and underwear since the demands on the fibres for washing and use can verge on the extreme (Oertel 1965).

Composition and manufacture

The composition and hence properties of NR and IR threads are very similar. They are composed of almost 99% linear 1,4-cis-polyisoprene. The composition and properties of NBR are different because the material is produced by polymerisation of acrylonitrile and butadiene. In NBR, a higher content of acrylonitrile gives a higher fibre strength and resistance to ageing and a lower elasticity than NR and IR. Moreover, the increased acrylonitrile content of NBR enhances its resistance against several solvents as well as the effects of UV radiation. The elasticity of rubber, and therefore rubber threads, is generally obtained by the use of sulphur and heat in the vulcanisation process, which causes a change in the structure of the rubber from the plastic to the elastic condition. Cross-linking elastomer molecules by sulphur links gives the bulk material greater elasticity, making it harder, less soluble and more durable (Frötscher 1930; Korfmacher 1953; Hofmann 1965).

Generally, rubber threads can be produced in two major ways. The first is to cut strips from thin rubber sheets, the resulting threads having a square diameter.[2] The sheets from which the threads are cut can either be ready vulcanised, which produces rubber threads directly, or made of a mixture of rubber and sulphur, but not yet vulcanised (Fig. 1). Hot knives are used for cutting to vulcanise the threads in this instance. The second method involves squeezing a mixture of rubber and additives from nozzles or spinnerets to produce threads with a round diameter, which have slight longitudinal grooves (Fig. 2). Once extruded, the threads pass through a bath of coagulant, such as acetic acid, to 'set' the thread which is then washed in a water bath. Finally the threads are dried and heat-cured[3] (Korfmacher 1953; Förster and Wildfeuer 1953).

Elastomeric fibres consist of segmented polyurethanes. Their main characteristic is the alternating amorphous (soft

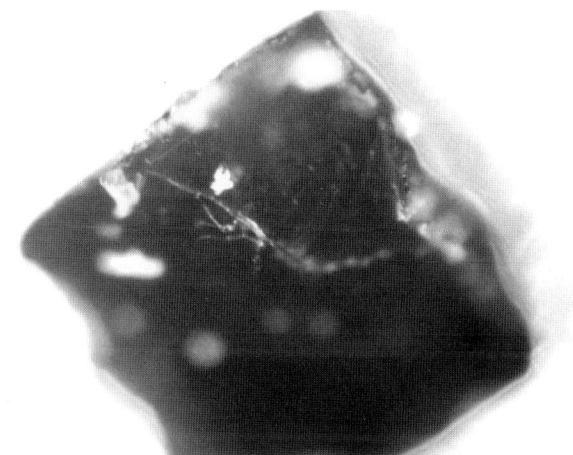

Figure 1 Cross-sectional view of a cut rubber thread (Plate 23 in the colour plate section).

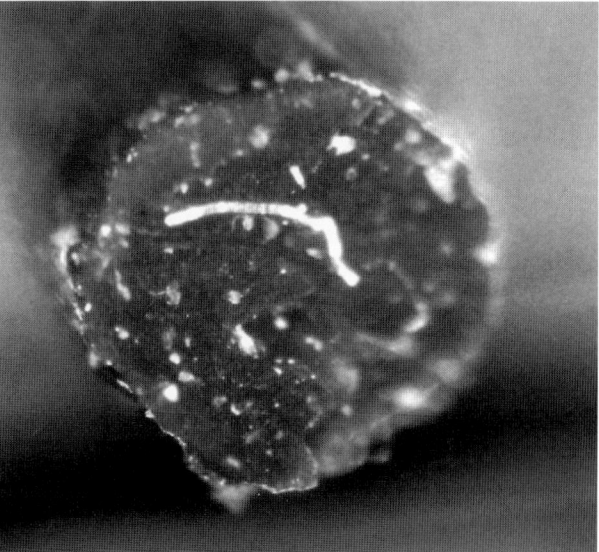

Figure 2 Cross-sectional view of an extruded rubber thread (Plate 24 in the colour plate section).

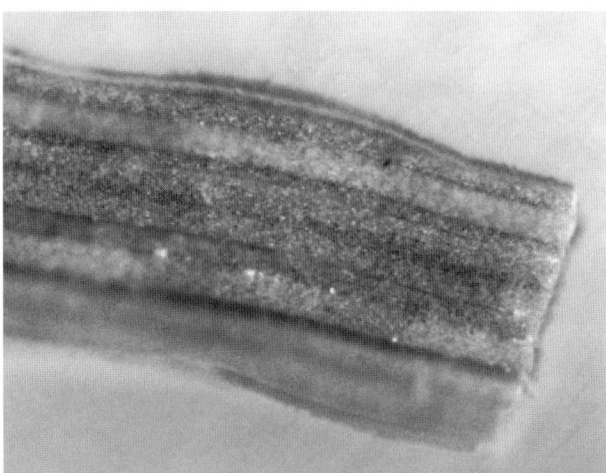

Figure 3 Longitudinal view of a multifilament elastomeric yarn (Plate 25 in the colour plate section).

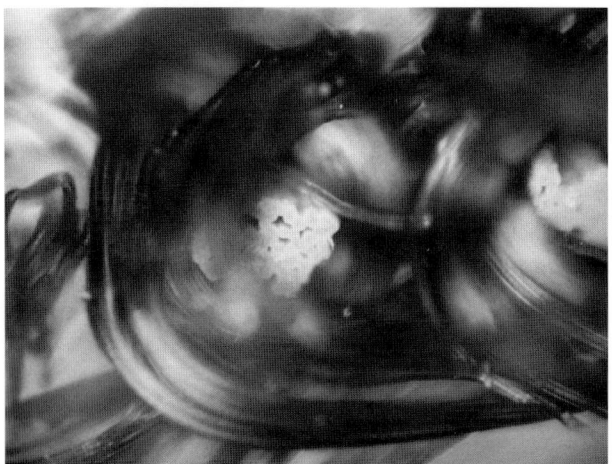

Figure 4 Cross-sectional view of a fabric with multifilament elastomeric yarn within its structure (Plate 26 in the colour plate section).

and flexible) and crystalline (hard and inflexible) segments, these fibres can be stretched by up to 400–800%. The high extensibility is due to the amorphous segments, while overall strength is provided by the crystalline regions. When the fibres are stretched, the molecules of the amorphous segments are aligned and hence become longer in length. For a successful elastomeric fibre to be produced, the number of amorphous segments has to be greater than the crystalline segments. Cohesion of the polymer chains is ensured by secondary valence bonds between the segments. When used in knitwear, elastomeric fibres are spun as multifilament yarns with several separate filaments placed together (Figs 3 and 4), and as monofilaments or a single thicker yarn, sometimes with a dumbbell-shaped cross-section. Monofilaments with a thicker diameter are also processed in woven textiles (Oertel 1965).

Elastic threads used in textiles

Rubber threads that were woven into textile structures were either encased by other textile yarns (covered) or left uncovered. A casing offers higher stability to the elastic thread because otherwise abrasion and loss of the surface occurs due to repeated mechanical strain. A casing also prevents the rubber from being overstretched. The coverings were frequently achieved by wrapping two layers of textile yarn around the rubber threads in different directions (Korfmacher 1953; Gayler *et al.* 1967). As for woven textiles, covered threads were sewn or stitched directly onto a fabric. They were usually applied to lie in the direction of the warp (Fig. 5).

To ensure fabrics were not overstretched when the elastic was elongated in use, the elastic threads were initially secured to the fabric while in an extended position. When constructing a garment, the elastic fabric was thus used with the warp lying appropriately to achieve elasticity in the direction needed, most often horizontally. In knitted fabrics, rubber threads were usually worked in conjunction with another textile thread (Korfmacher 1953; Gayler *et al.* 1967). PUR fibres could also be covered with staple fibres. The particular application of woven fabrics for clothing with elastic fibres within them was determined by the direction in which the PUR-elastomeric fibres had been processed. For example, warp-elastic woven fabrics were used for sports clothing to ensure a close fit, weft-elastic woven fabrics were used for ski garments and home clothing, such as so-called leisure suits, to give comfort and ease of movement. Weft-faced knitwear was used for outer and undergarments and the elastic threads were knitted either with or without other textile fibres as part of the same yarn (Marshall 1961).[4]

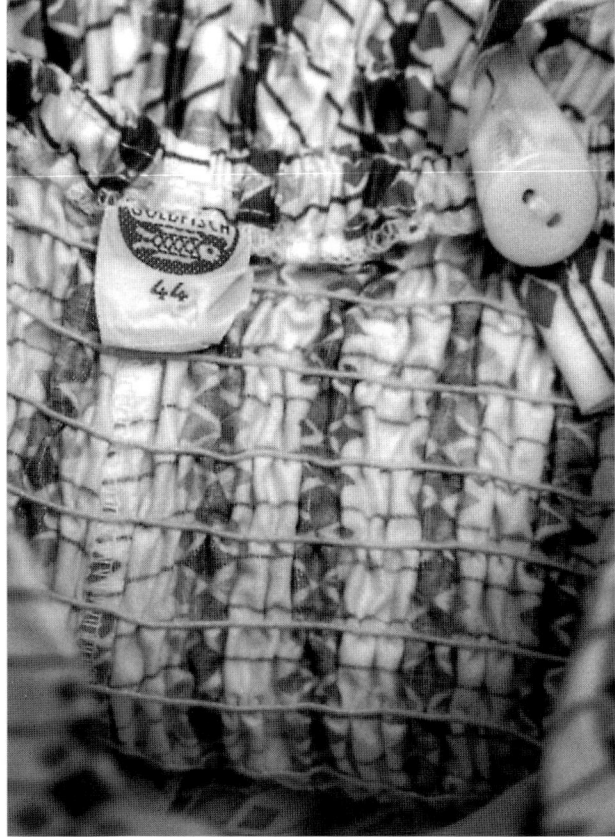

Figure 5 Swimming costume (German, 1950s): interior detail showing horizontal rows of elastic yarns (Plate 27 in the colour plate section).

Identification

Identifying the type of elastic thread present in a garment is beneficial for a conservator in order to understand any deterioration present and to propose treatment. Fourier transform infrared (FTIR) spectroscopy is an appropriate means to undertake such analysis using a sample taken from the object. Conducting FTIR analysis on elastic threads may not always be possible, however, since it is costly in terms of time and equipment; in these cases other means of identification have to be considered. Differentiation of rubber threads and elastomeric fibres is possible by dating the objects from a stylistic point of view, as PUR-elastomeric fibres were not manufactured in the clothing industry before 1958. If dating a garment proves difficult, a technical examination of the diameters of the threads or fibres present can broadly suggest time periods and, therefore, types of material present. Generally, earlier fibres tended to be more coarse while later technology produced finer threads that were used either as single strands or as multifilament yarns in configurations not encountered in the early history of elastic threads.

Characteristic damage found in conjunction with elastic threads

The most characteristic damage encountered with elastic threads is the loss of elasticity and flexibility. By association, the garments in which the threads are present are similarly affected. Swimwear, for example, will lose a degree of stretch when the elastomerics within it have deteriorated and the fabric may sag rather than conform to the original intended body shape.

The most harmful ageing processes that affect elastic threads are oxidation and photo-oxidation, which have been adequately discussed in other publications and are thus not described here (Loadman 1991; Blank 1988). The materials, discussed above as being typical of elastic and elastomeric fibres, are unfortunately susceptible to rapid degradation by these pathways. Garments that are used for their intended function will necessarily be exposed to both oxygen and light, as well as laundering and mechanical movement, which will further reduce their practical lifespan. Similarly, items permanently within collections will, in all probability, also be exposed to oxygen and to a lesser extent, light. Deterioration of garments in use is to be expected, therefore, but providentially many were only intended to be used for a moderately short-term period by their owners. Those garments in collections are also likely to degrade but, due to their place in museums, have greater expectations of longevity placed upon them and their deterioration is of greater concern. Indeed the poor resistance to ageing of early elastic fibres may go some way in accounting for the scant number of examples within collections. The situation is exacerbated by the fact that garments that were worn prior to being collected were probably suffering from early stages of degradation after being used, therefore their life in a collection is further truncated.

In order to investigate the degree of deterioration to an elastic thread, a textile can be stretched a little and then a basic

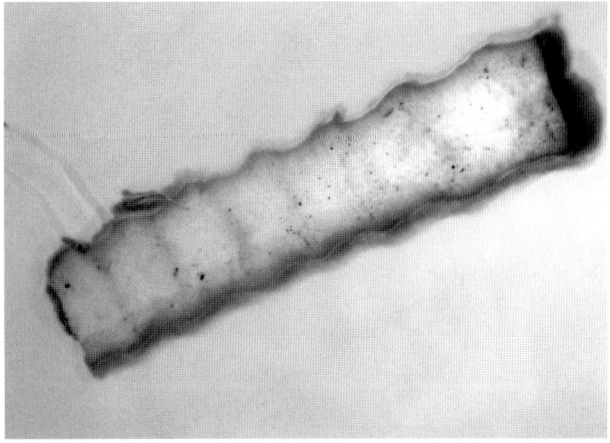

Figure 6 Section of a broken and stiff degraded natural rubber thread (Plate 28 in the colour plate section).

judgement can be made as to the extent of loss of elasticity. Obviously a much degraded thread will easily break when a force is applied. The degree of stiffness of a fibre can also give a clue to the extent of fibre damage. Hardening of elastomeric fibres was frequently encountered on garments in collections by the author. Another characteristic type of damage encountered during the course of the project, affecting rubber threads in particular, was deformation caused by other textile fibres. Tightly wrapped covering yarns may cause indents along the length of the elastic fibres (Fig. 6).

A noticeably sticky surface on elastic threads indicates softening of the materials and is another pathway for oxidative and photo-degradation. Furthermore, the threads will gradually turn yellow in colour as they degrade and become either rigid and hard or sticky and soft. PUR-elastomeric fibres are much more resistant to oxidative ageing. Antioxidants added to the material when the thread is manufactured can improve this further but as they are volatile and migrate out of the threads they can cause yellowing of surrounding textile fibres, particularly cellulose fibres.

Handling/storage

Options for interventive conservation treatment of elastic threads are limited. The degradation they suffer is irreversible and cleaning with either water or organic solvents is problematic due to the risk of swelling. Perhaps surface cleaning is the most prudent course of action to limit damage by removing potentially harmful soiling that could accelerate further degradation.

Even in the 1930s, the textile processing industry provided recommendations for storage conditions of elastic textiles because of their vulnerability to degradation. A dark environment with temperatures below 20 °C or ideally 15 °C was considered to be the optimum to reduce the rate of inevitable degradation (Anon 1958). These conditions can still be recommended today. A storage environment without oxygen would also help threads (particularly rubber) to remain in as good a condition as possible for as long as possible. Textile manufacturers also discovered that storage and display of garments with rubber threads, where they are put under tension, causes cracks along the thread. When exhibiting or storing elasticated clothing, mannequins and mounts used for storage or display should be 25–35% smaller than the original size of the intended body shape for a garment in order to ensure that the elastic threads are not permanently under tension and therefore liable to break. The numerical values correspond to the difference in size between a stretched piece of clothing and its un-stretched state. Moreover, textiles containing elastic threads should be stored and displayed separately from other garments in costume collections to prevent off-gassing of sulphur from rubber and volatile antioxidants from PUR-elastomeric fibres in particular, which are harmful to other textile fibres and can induce yellowing in them.

Conclusion

This paper has provided a brief overview of the material content, manufacture, use and preservation of elastic threads from their development until the 1960s. While few garments survive from earlier times, due to their susceptibility to degradation, examples do exist in collections and must be cared for appropriately. Furthermore, synthetic elastomeric threads continue to be used in fashionable and functional clothing today and in all probability examples will find a place in textile collections in the future. Curators and conservators need to be aware not only of the legacy of these types of threads but also the impending preservation problems newer ones may present. It is hoped that this paper will serve as a starting point for understanding the presence of elastic threads in garments and will raise awareness of their existence.

Notes

1. Patents examined covering elastic threads, fibres, yarns etc at the Deutsches Marken- und Patentamt, München (Patent Office, Munich) in 2002/2003: DBP 872268, Leverkusen 1953; DBP 888766, Leverkusen 1951; DRP 653329; DRP 662056; DRP 653329, Wien 1937, DRP 662056, Burkhardtsdorf/Erzgebirge 1935; DRP 662226, Münchberg/Bayern 1936; DRP 671473, New York 1935; EP 470722, 1938; EP 498851, St. Peter Port/Guernsey; USAP 1889102, London 1932; USP 2009361, Pawtucket/Rhode Island 1935; USP 2929804, Wilmington 1955; USP 2009361; USP 2929804.
2. United Kingdom Patent No. 470722, International Processes Ltd. Inventors: E.A. Murphy, R.G. James, 1938.
3. Deutsches Bundes Patent No. 872268, Bayer AG. Inventors: O.Bayer, W. Bunge, E. Müller, S. Petersen, H.-F. Piepenbrink, E. Windemuth, Leverkusen 1953.
4. Deutsches Reich, Reichspatentamt Patentschrift No. 662226. Inventor: H. Kalbskopf, Münchberg/Bayern 1936, published in *Melliand Textilberichte* (1939) 20: s. 603.

References

Anon (1934) Lastex für die Abendtoilette, Monatshefte für Seide und Kunstseide 39, Krefeld, 129.
Anon (1958) 'Gummifäden in der Wirkerei', *Melliand Textilberichte* 33: 65–6.
Blank, S. (1988) 'Rubber in museums: a conservation problem', AICCM *Bulletin* 14: 53–93.
Bleise, G. (1992) *Funktionelle Sportkleidung – eine Herausforderung der Textilfasern, Sportswear – Zur Geschichte und Entwicklung der Sportkleidung*, Exhibition Catalogue, Krefeld.
Clouth, F. (1879) *Die Kautschuk-Industrie oder Gummi und Gutta-Percha, ihr Ursprung, Vorkommen, ihre Gewinnung, Verarbeitung und Verwendung, Neuer Schauplatz der Künste und Handwerke – Mit Berücksichtigung der neuesten Erfindungen*. Weimar.
Davis, W. (1933) 'Properties of rubber threads used in textiles', *Journal of the Textile Institute* 24: T44–T53.
Ebblewhite, J.K. (1939) 'Elastische Waren', *Melliand Textilberichte* 20: 229.
Förster, H. and Wildfeuer, F. (1953) *Fabrikation Gummielastischer Bänder, Kordeln, Litzen, Galon- und Trikotageartikeln*. Saulgau/Württemberg.
Frötscher, E. (1930) 'Die Gummifäden', *Melliand Textilberichte* 11: 827–9 and 907–9.
Gayler, J., Kaminski, W. and Kantel, K. (1967) *Konstruktion elastischer Web- und Maschenwaren I–III*, Chemiefasern, 356–63, 590–96, 792–9. Frankfurt/Main.
Heinzerling, C. (1883) *Die Fabrikation der Kautschuk- und Guttaperchawaren sowie des Celluloids und der wasserdichten Gewebe*, Handbuch der Chemischen Technologie, Braunschweig.
Hofmann, W. (1965) *Nitrilkautschuk*. Stuttgart.
Korfmacher, P. (1953) 'Die Herstellung der Gummifäden', *Melliand Textilberichte* 34: 715–18.
Loadman, M.J.R. (1991) 'Rubber: its history, composition and prospects for conservation', in *Saving the Twentieth Century: The Conservation of Modern Materials*, D.W. Grattan (ed.), 59–80. Ottawa: Canadian Conservation Institute.
Marshall, T.B. (1961) 'Man-made elastic yarns', *Textile Industries* 125(8): 75–80.
Oertel, H. (1965) 'Elastomere Fäden auf Polyurethanbasis, ihr Aufbau, ihre Eigenschaften, ihre Verwendung', *Melliand Textilberichte* 46: 51–9.
Rinke, H. (1962) 'Elastomere Fasern auf Polyurethanbasis', *Chimia* 16: 93–105.

The author

Laura Petzold currently works as a Senior Textile Conservator at the Abegg Foundation. She studied textile conservation at the University of Applied Sciences in Cologne and completed these studies with her diploma thesis on 'Early elastic threads and fibres in clothing'. Earlier she trained as a tapestry weaver and then gained experience in textile conservation in the Textile Conservation Department of the Münchner Stadtmuseum, Munich.

Address

Laura Petzold, Senior Textile Conservator, Abegg Foundation, Werner Abegg-Strasse 67, 3132 Riggisberg, Switzerland (petzold@abegg-stiftung.ch)

Material challenges

Identifying modern materials: taking it to the collection

Paul Garside and Paul Wyeth

ABSTRACT Knowledge of the components of an artefact is key to its understanding and also to ensuring the best approach to its conservation. The authors are particularly keen to develop methodology that will permit on-site analysis for routine collection surveys and condition monitoring. To this end, they have investigated the value of Fourier transform infrared–attenuated total reflectance spectroscopy (FTIR–ATR) and reflectance near infrared spectroscopy (NIR) for characterising modern materials non-invasively. Both FTIR–ATR and NIR allowed the discrimination of five exemplar modern fibres: cellulose acetate, polyamide, polyester, polyacrylic and regenerated cellulose. An unknown could be matched straightforwardly to the appropriate reference within this limited spectral set. FTIR–ATR and NIR spectrometers are now readily portable so that analysis at collections is quite feasible.

FTIR–ATR, however, presents certain drawbacks as a technique for the direct, non-invasive analysis of artefacts, including the need to press the sample to the ATR crystal, which may damage fragile or deformable specimens. Access to the ATR top plate is also somewhat restricted. As ATR is a surface analytical technique, finishes may lead to confusion. Although NIR spectra are more complex, they are nonetheless characteristic. Contact is not required with the specimen to record a NIR reflectance spectrum, and excellent quality spectra can be acquired through a remote fibre optic probe allowing the interrogation of all sorts of objects. NIR may therefore be the technique of choice for the on-site analysis of modern fibres.

Keywords: non-invasive analysis, textile fibre identification, FTIR–ATR, NIR

Introduction

An appreciation of the make-up of an artefact is fundamental to its understanding and knowledge of the composition will, of course, influence decisions concerning its conservation. To ensure the best approach, identification of the components is therefore a necessary preliminary. Even when textiles are labelled by manufacturers or assigned by custodians, confirmation is still essential since the purported constituents are not always those revealed by analysis. Recently, for example, a wedding dress from a local collection believed to be polyamide was shown to be made from polyester fabric.

Traditional fibre microscopy is not particularly helpful in the characterisation of synthetic and man-made fibres.[1] The fibre cross-section and surface morphology, which are quite distinctive for natural fibres, are generally dependent on the extrusion process and the die used in manufacture rather than the constituent modern material. Staining, solubility and burn tests may offer some insight, but these can be somewhat laborious, have health and safety implications and the results may be equivocal. Another concern is the need to remove samples, albeit small amounts, to carry out the tests.

Mid-infrared spectroscopy offers a convenient means of differentiating many of the variety of modern textile fibres such as polyester and polyamide (Derrick *et al.* 1999), but in its routine application sampling is again usually required. The authors are particularly interested in developing practical, non-invasive analytical methodology which can be carried out on-site and have been investigating the value of Fourier transform infrared–attenuated total reflectance spectroscopy (FTIR–ATR) and reflectance near infrared spectroscopy (NIR) for distinguishing modern materials. Recent technological developments have led to portable instrumentation and, as detailed below, the results of the preliminary studies suggest that in both cases it should be possible to perform effective analysis at the collection.

Each of these two techniques is considered in turn, suggesting the appropriate protocol for their application to the identification of modern fabrics and concluding with a comparative assessment of their value as portable techniques for *in-situ* analysis.

Fourier transform infrared–attenuated total reflectance spectroscopy (FTIR–ATR)

To routinely record infrared (IR) spectra of fabrics it is necessary to use reflectance optics. The nature of the textile surface, however, generally results in a variable mix of specular and diffuse components (Fig. 1). The response is further complicated by the particular size of fibres, affording a surface roughness close to mid-IR wavelengths (2.5–20 μm), which can give anomalous spectra. In contrast, FTIR–ATR provides reproducible, albeit distorted absorbance spectra.

Five representative modern fabrics were selected from samples in the authors' collection: cellulose acetate, polyamide (nylon 66), polyester, polyacrylic and regenerated cellulose (viscose rayon). ATR spectra were acquired of the intact fabrics on a PerkinElmer Spectrum One FTIR instrument fitted with a single reflection ATR accessory, operating with PerkinElmer Spectrum software, over the range 4,000–400 cm^{-1} and at a resolution of 4 cm^{-1}, accumulating 32 scans. Spectra were subsequently processed with Thermo Galactic GRAMS 32/6 software.

The spectra of the five samples are quite distinct (Fig. 2). The differences arise, of course, from the diverse chemistry of the constituent polymers. For example, viscose is comprised of relatively short cellulose chains (Fig. 3a), and the mid-IR spectrum shows features, among others, arising from the fundamental stretching and bending vibrations of the hydroxyl (OH) groups. In contrast, polyester fibres are typically made from polyethyleneterephthalate (Fig. 3b), and the FTIR–ATR spectrum shows a dominant carbonyl stretching peak at 1,713 cm^{-1}. It is apparent that matching an unknown within a small spectral library should be straightforward. The composition of fibre blends, e.g. polycotton, might also be readily resolved.

The use of IR spectroscopy for the identification of small samples of modern materials has been appreciated for some time (Perry et al. 1985), but its direct application to artefacts is not routine. In this instance, some concern could be raised regarding the risk of damage to fragile or deformable specimens. The design of the PerkinElmer Universal ATR accessory, with the top plate proud of the instrument and the pressure arm able to swing over the top plate, enables relatively flat textile components to be interrogated without undue crumpling, although there is a limitation on the size of the artefact and on the site for interrogation. The sample must still be pressed to the diamond ATR crystal, but the applied force can be minimised by continuously monitoring the spectrum and it is often possible to obtain a characteristic pattern with just light pressure. The stainless steel anvil leaves no footprint on new nylon, for example, though there is a remnant impression (of 1 mm diameter) on a degraded sample.

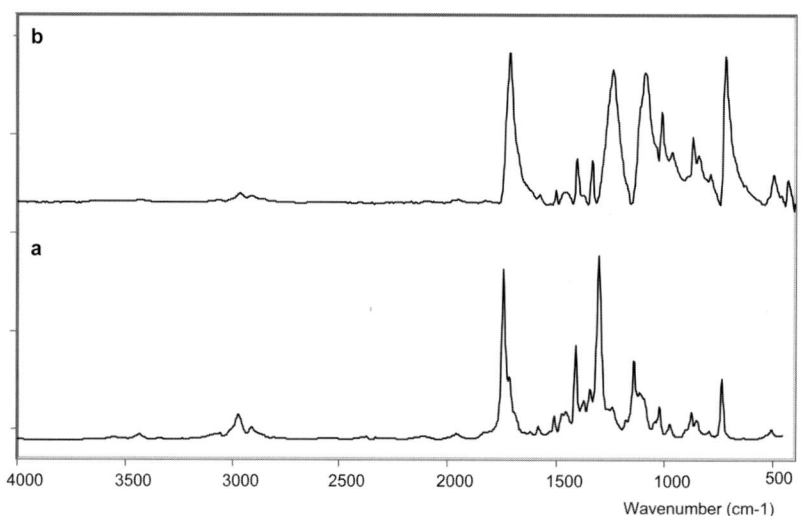

Figure 1 The mid-infrared reflectance spectrum (a) and attenuated total reflectance (ATR) spectrum (b) for a polyester fabric; both are presented as absorbance spectra.

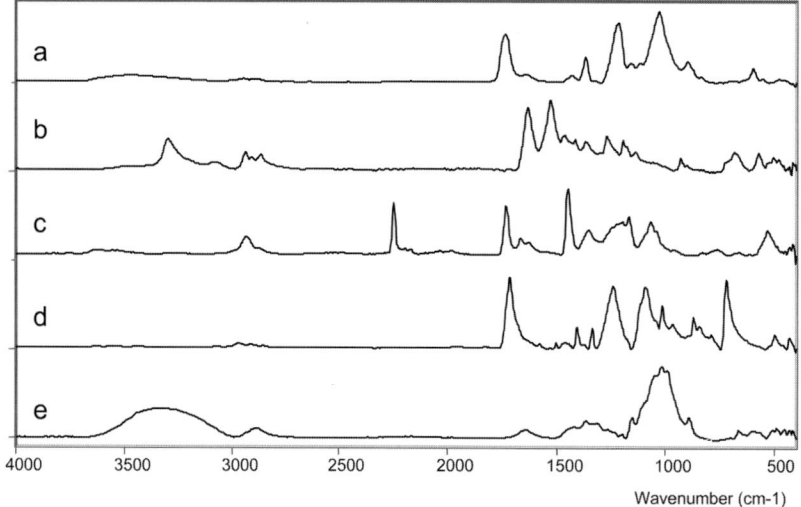

Figure 2 ATR mid-infrared absorbance spectra of the five modern fabrics: (a) cellulose acetate, (b) polyamide (nylon 66), (c) polyacrylic, (d) polyester, (e) regenerated cellulose (viscose).

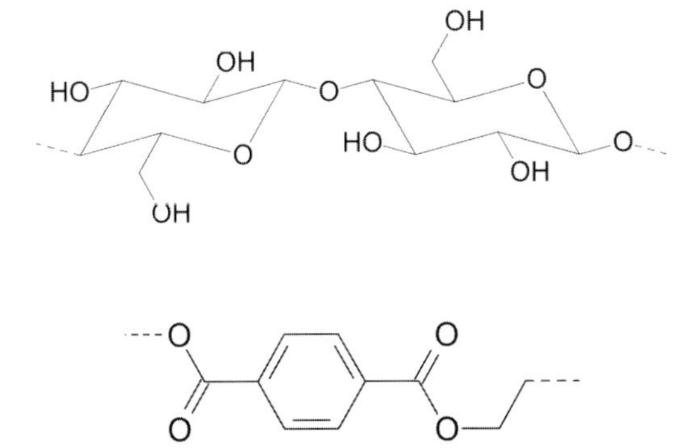

Figure 3 The repeat structural units of (a) cellulose and (b) polethyleneterephthalate.

For FTIR–ATR, the spectral data only come from the top few microns of the sample. A criticism can then be raised that representative data on the bulk are not obtained. For instance, for a viscose warp-faced fabric with polyester wefts, the spectrum of viscose only was obtained. Similarly, a surface finish might mask the substrate. In other circumstances, however, the limiting penetration may present an advantage, for example where a surface coating needs to be characterised (Chang and Wyeth 2002).

Another complication that can arise with FTIR–ATR is the inherent partial polarisation of the IR beam due to the reflectance optics. When there is polymer ordering within fibres, such as the axial alignment induced by extrusion and drawing, then orientation-dependent data can result. This is apparent for nylon 66, for example, in the altered spectra recorded for fibres lying perpendicular to and in the plane of the IR beam. Bands arising from amide groups in the ordered crystalline elements appear to change, notably the amide II peak at 1,530 cm^{-1}, which alters in intensity and shifts by four wavenumbers. Where there are significant changes in microstructure and chemistry consequent on degradation, these might be monitored by FTIR–ATR, though such orientation-dependent shifts could confuse matters.

So, although FTIR–ATR may have a place as an on-site characterisation technique for modern materials, a cautious approach is required and more fragile and larger artefacts are excluded. Use of an accessory that allows free access to the ATR top plate for flat textiles is to be preferred; and operating software that permits spectral monitoring, while pressing the fabric to the top plate, is essential to minimise the applied force and permanent local deformation. Where direct analysis of an artefact is appropriate, care must also be taken to fix the orientation of the material on the ATR top plate.

Near infrared spectroscopy (NIR)

As FTIR–ATR does not offer the complete answer to on-site textile fibre identification, the assessment has been extended to include near infrared spectroscopy (NIR). The findings suggest that, for modern materials analysis, NIR is certainly complementary and indeed may offer a more versatile approach.

The near IR region of the electromagnetic spectrum is that between the visible and mid-IR, 12,500–4,000 cm^{-1} (wavelength 800–2,500 nm), though the authors have focused on the range 8,000–4,000 cm^{-1} (1,250–2,500 nm), where the stronger absorptions arise. NIR spectra consist of overtones and combinations of molecular vibrational modes (the fundamentals of which are seen in the mid-IR), and are typically dominated by those arising from CH, NH and OH bonds, which show significant anharmonicity. So, for example, such cellulose OH features can be identified in the NIR spectrum of viscose (Fig. 4) around 6,500 and 4,780 cm^{-1} respectively. Because of the abundance of overlapping bands, the NIR spectra are rather more complex than the FTIR–ATR spectra, but are nonetheless characteristic, allowing the five exemplar fibres to be distinguished.

NIR spectroscopy is of course a well-established technique and details of the instrumentation and procedure can be found in various texts (Siesler *et al.* 2004). NIR spectra are about 20 times weaker and the peaks rather broader than those recorded in the mid-IR, so a somewhat coarser spectral resolution is

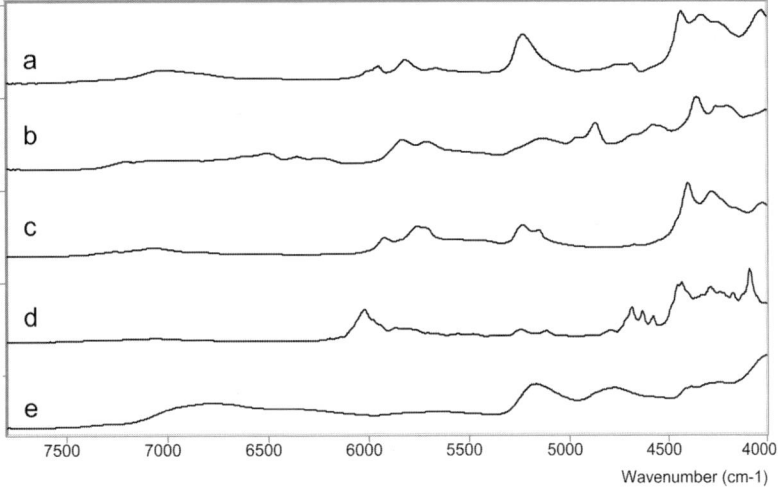

Figure 4 Near infrared (NIR) reflectance spectra (absorbance) of the five modern fibre types: (a) cellulose acetate, (b) polyamide (nylon 66), (c) polyacrylic, (d) polyester, (e) regenerated cellulose (viscose).

sufficient and good reflectance spectra can be acquired within a minute with a Fourier transform instrument or just a second using a photodiode array detector. The NIR spectra are diffuse as a consequence of the spectral selection rules, without the complication of specular components experienced for mid-IR reflectance.

To record the spectra of the various fabrics, samples were simply laid on the sapphire window of the integrating sphere near-IR accessory of the PerkinElmer Spectrum One FTIR instrument. The accessory has a built-in indium gallium arsenide (InGaAs) detector. Preliminary studies had shown that the orientation of the fabric on the window was not a concern. Spectralon, an efficient diffuse reflector, was used for the background reference; 256 scans were acquired at a spectral resolution of 8 cm^{-1}. Subsequent spectral processing was carried out with Thermo Galactic GRAMS 32/6 software. For the particular integrating sphere accessory used, the sampling area is a 13 mm diameter circle, but it is possible to select smaller areas with masks or using a remote sampling probe.

Typically, the second derivative of a NIR spectrum is generated to emphasise the minor contributions which are somewhat hidden within the spectrum, enhancing visual resolution (Fig. 5). So, for example, this highlights cellulose CH overtone bands in the spectrum of viscose at 5,816 cm^{-1}. Generating the second derivative also has the advantage of removing any spectral offset, which can be caused by a different weave or reflectance geometry and eliminating a sloping spectral background (Osborne *et al.* 1993) (the authors used the Gap method for generating derivatives, with a gap value of 19).

The second derivatives are routinely used for spectral matching and the identification of an unknown by comparison to a reference spectral library and can allow quite subtle distinction, for example of different nylons. This approach will also successfully distinguish the components of a fibre blend and, with appropriate standards, will quantify the proportions.

While the contributions of dyes to the NIR spectra of fabrics are likely to be minor, as they are present in very small

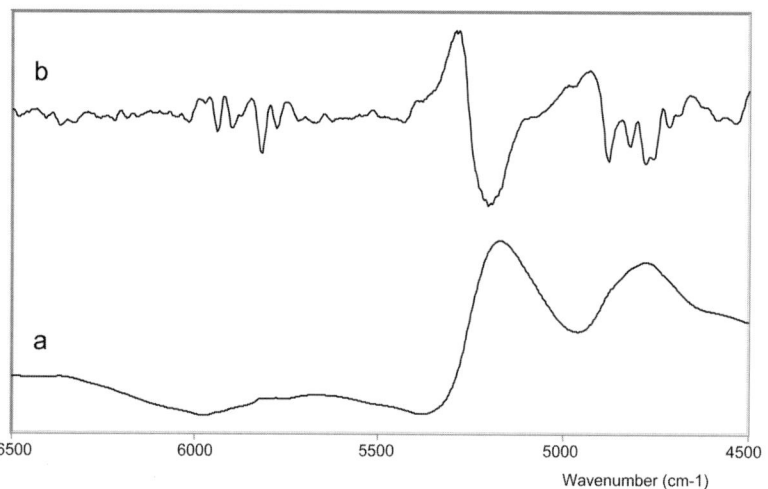

Figure 5 A comparison of a portion of the NIR spectrum of viscose (a) with the spectral second derivative (b), illustrating the enhanced visual resolution of minor components.

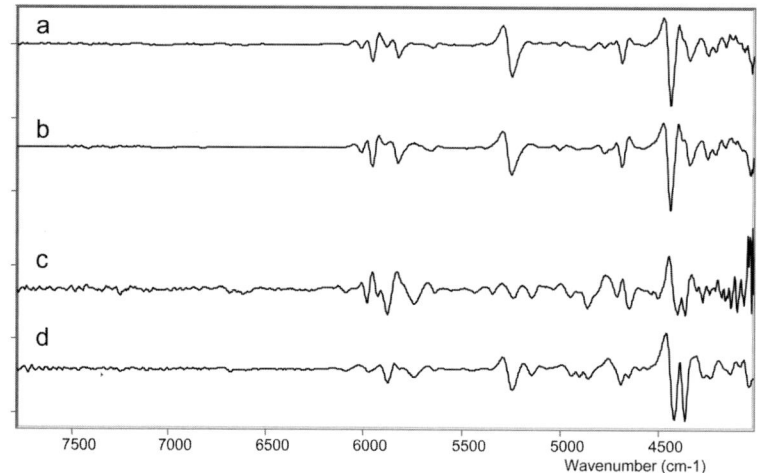

Figure 6 The NIR spectral second derivatives of (a) the lining of a grey jacket and (b) the lining with underlying foam shoulder pad. The difference (c) reveals the contribution from the foam pad, which gives a reasonable match to the spectral second derivative for an authentic polyurethane ester sample (d).

amounts, extended absorption from the visible region into the near-IR might be expected to cause problems with some coloured fabrics. So far though, significant interference has only been experienced with some grey-coloured materials, e.g. a grey acrylic scarf gave high absorbance for the darker grey fabric, and even then there was sufficient detail at lower wavenumbers to allow an assignment. Again, however, this was done more confidently from the second derivatives.

Trials on stacked, closely woven, thin samples have indicated a NIR sampling depth of 500 μm, so probing well into the bulk of a fabric. Although the information from the upper regions will tend to dominate the reflectance spectrum, surface treatments are unlikely to mask identification of the substrate. The effective depth of sampling suggests that it might be possible to interrogate multilayered objects non-invasively to reveal the identity of a hidden underlayer and this has been demonstrated successfully for the foam padding on a grey jacket.

The spectrum recorded through the lining at the shoulder of the jacket shows contributions from two components, as seen in the second derivative trace in Figure 6. Subtracting the spectrum recorded separately for the cellulose acetate lining leaves the weaker spectrum from the foam shoulder padding, but this difference is sufficiently intense to be able to identify the padding as a polyurethane ester. For this application, rather than the in-spectrometer NIR accessory, an Axiom fibre optic solids probe with a sapphire window was used and the probe held close to the specimen. Near-IR radiation is readily transmitted through silica glass fibre optics with little loss in energy so that good quality spectra can be acquired with remote examination. This is one of the significant advantages of NIR over mid-IR spectroscopy. In the latter case, expensive and stiffer chalcogenide glass fibres are required, but because of significant energy losses the resultant spectra may be of quite poor quality even for the best-behaved specimens. The NIR remote probe affords good quality spectra, for the authors' system the signal to noise ratio only being compromised at the two ends of the spectral range where the energy is comparatively low. The free access provided by the NIR fibre optic probe removes the limitation on the size and the three-dimensional nature of the object that can be interrogated.

As with FTIR–ATR, the value of NIR to the analysis of textiles is well known (Ghosh and Rodgers 2001) and it is the routine application to heritage artefacts that has yet to be developed. For the latter, NIR seems to offer much potential for on-site materials identification. It is a non-contact analytical technique and there is therefore no risk of stress damage. NIR radiation is quite penetrating so that information is derived from a sample depth of almost a millimetre, perhaps allowing the characterisation of multilayered artefacts. Furthermore, the use of a fibre optic probe realises excellent quality spectra, permitting the analysis of all accessible regions of an object, whatever its size and shape.

Just as FTIR–ATR may report on deterioration, so NIR appears to hold some promise for *in-situ* condition monitoring. Preliminary data suggest though that age-related changes to the spectra do not compromise material identification. For application to the identification of modern materials in collections, the remote solids reflectance probe can be used routinely. The authors have found with their Fourier transform instrument that adequate spectra for cross-matching can usually be obtained by accumulating scans over one minute, using a spectral resolution of just 16 cm^{-1}. With a thin cloth, the spectral quality is improved somewhat by placing a flat reflector, such as aluminium foil, behind the region of interest. The probe-to-fabric distance and geometry do not appear critical; generally the sapphire window of the probe is held close to the fabric surface, though reasonable spectra can still obtained from a distance of 1 cm. The nature of the weave and the orientation of the fabric seem inconsequential where a single fibre type is concerned.

Conclusion

Both FTIR–ATR and NIR spectroscopy allow the discrimination of the five modern fibres studied: cellulose acetate, polyamide, polyester, polyacrylic, and regenerated cellulose. In each case, an unknown can be matched to the appropriate reference within a spectral library. FTIR–ATR and NIR spectrometers are easily portable so that analysis at collections is eminently feasible. FTIR–ATR, however, presents certain drawbacks as a technique for the direct, non-invasive analysis of artefacts. Physical contact is required between a sample and the ATR crystal and the pressure required to achieve this can be damaging, especially for fragile objects. This can be alleviated somewhat by making use of operating software that allows the spectrum to be monitored as the force on the specimen is gradually increased. There are still limitations on the position and shape of the object that can be interrogated, however, imposed by the form of the ATR accessory and on the depth probed, which is determined by the sampling optics. Furthermore, the orientation of an ordered fibre sample will affect its spectrum.

NIR circumvents many of these constraints and, with the use of a remote probe, allows the acquisition of characteristic fabric spectra from all sorts of objects. It may be the technique of choice for the on-site analysis of modern fibres, perhaps with fibre sampling and on-site or laboratory-based FTIR–ATR as a back-up for the odd occasions when NIR presents equivocal results.

Acknowledgements

The authors gratefully acknowledge AHRC funding and the support of colleagues at the Textile Conservation Centre. Thanks are due to Robert Alexander, PerkinElmer Ltd. for recording mid-IR reflectance spectra.

Note

1. The distinction is made here among natural, man-made and synthetic fibres. Man-made fibres are taken to be those for which the constituent is a modified natural polymer, such as regenerated cellulose and cellulose acetate. Polyamide (e.g. nylon 66), polyester and polyacrylic, for example, are deemed to be synthetic fibres, i.e. ones for which the polymer is made entirely from petrochemicals.

References

Chang, L. and Wyeth, P. (2002) 'Chemical finishes on indigo-dyed cloth: characterization of Miao and Miao-related costume from Guizhou, China', in *Strengthening the Bond: Science and Textiles. Preprints of the North American Textile Conservation Conference April 2002*, V.J. Whelan (ed.), 25–34. Philadelphia, PA: NATCC.

Derrick, M.R., Stulik, D. and Landry, J.M. (1999) *Infrared Spectroscopy in Conservation Science*. Los Angeles, CA: Getty Conservation Institute.

Ghosh, S. and Rodgers, J. (2001) 'NIR analysis of textiles', in *Handbook of Near-Infrared Analysis*, D.A. Burns and E.W. Ciurczak (eds), 2nd edn, 573–607. New York: Marcel Dekker.

Osborne, B.G., Fearn, T. and Hindle, P.H. (1993) *Practical NIR Spectroscopy*. Harlow: Longman.

Perry, D.R., Appleyard, H.M., Cartridge, G. *et al.* (1985) *Identification of Textile Materials*, 7th edn. Manchester: Textile Institute.

Siesler, H.W., Ozaki, Y., Kawata, S. *et al.* (eds) (2004) *Near-infrared Spectroscopy*. Weinheim: Wiley VCH.

Suppliers

- Spectrum One and accessories: PerkinElmer LAS (UK) Ltd., Chalfont Road, Seer Green, Beaconsfield HP9 2FX, UK.
- GRAMS software: Thermo Electron Corp., Stafford House, 1 Boundary Park, Hemel Hempstead, HP2 7GE, UK.

The authors

- Paul Garside is a postdoctoral research fellow at the AHRC Research Centre for Textile Conservation and Textile Studies. He has developed significant expertise in applying analytical methodology to conservation science problems.
- Paul Wyeth is a Senior Lecturer in Conservation Science at the Textile Conservation Centre. His research interests encompass applications of microstructural and microspectroscopic analysis in the areas of conservation science, and the development of instrumental methods for the on-site condition monitoring of organic heritage.

Address

Corresponding author: Paul Wyeth, Textile Conservation Centre, University of Southampton, Park Avenue, Winchester, Hants SO23 8DL, UK (pw@soton.ac.uk).

Man-made fibres from polypropylene to works of art

Thea van Oosten, Ineke Joosten and Luc Megens

ABSTRACT While the potential of man-made fibres was realised when the first completely synthetic polymers were developed in the early part of the 20th century, many of these materials were not widely used until the 1960s and 1970s. From the early 1960s, polypropylene (PP) became increasingly common, due to the new polymerisation catalysts developed by Ziegler and Natta in 1954. PP possesses excellent properties that make it well suited to use as a fibre; however, it is also prone to oxidative degradation, accelerated by light, particularly in the ultraviolet (UV) region of the spectrum. A book published in 1962, entitled *Welche Chemiefaser Ist Das?* ('Which Synthetic Fibre is it?'), contains 126 samples of reference fibres, including polyester, polyamide, polyethylene, polypropylene and polyurethanes (Driesch 1962). At the time of writing this paper, all fibres in the book have been aged naturally for 43 years; some of them are discoloured and others have degraded entirely. The aim of this study is to investigate changes that may have occurred in the structure of the PP polymer in these reference samples and in PP fibres from two works of art and a relining fabric. Fourier transform infrared (FTIR) spectroscopy, polarised light microscopy (PLM) and scanning electron microscopy (SEM) were used.

The first of the works of art, *Fête II*, is a wall tapestry constructed from PP and PP/polyethylene (PE) copolymer fibres. It has hung for more than 34 years on open display and the component fibres exhibit degradation. The second piece, *The Knot*, has suffered damage due to both display conditions and conservation treatments; freezing, as a measure against possible insect infestation, resulted in the embrittlement of some of the white fibres. The relining fabric was heavily degraded. FTIR analysis of these fibres identified them as isotactic PP. While the normal version of the polymer can withstand temperatures of –30 °C, the isotactic form becomes brittle at –20 °C when under stress.

Keywords: polypropylene, synthetic fibres, freezing, Fourier transform infrared spectroscopy (FTIR), scanning electron microscopy (SEM)

Introduction

The work in this paper is based around a book published in 1962 entitled *Welche Chemiefaser Ist Das?* ('Which Synthetic Fibre is it?'), describing the history, chemistry, microscopic examination, manufacturing processes, applications and trade names of synthetic fibres (Driesch 1962). In addition to this information, the book contains 126 reference samples of these fibres, including polyester (PET), polyamide (PA), polyethylene (PE), polypropylene (PP) and polyurethanes (PUR) (Fig. 1). At the time of writing this paper, the fibres have been aged naturally for 43 years; some of them are discoloured and others have degraded entirely.

In addition to the investigation of the fibres in this book, the research also focused on the degradation phenomena observed in PP fibres found in two works of art, *Fête II* and *The Knot*, and on PP fabric used for the cold relining of paintings in the 1970s. The first of these works of art was the wall tapestry *Fête II*, made from PP rope by Will Fruytier, a Dutch designer, in 1969. This work was acquired by the University of Nijmegen, the Netherlands, in 1971. The tapestry was exhibited for more than 34 years in the hall of the university and white fibres within it were particularly degraded and brittle, making the work look messy (Fig. 2).

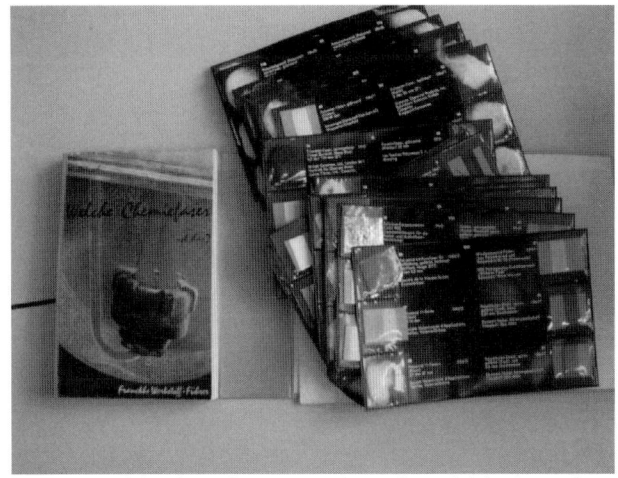

Figure 1 *Welche Chemiefaser Ist Das?* (Driesch 1962) (Plate 29 in the colour plate section).

The second item is a tapestry made from natural fibres entitled *The Knot*. It was made by the artist Herman Scholten in 1967 and acquired by the Rijksdienst Beeldende Kunst (RBK) (now the Netherlands Institute for Cultural Heritage (Instituut Collectie Nederland/ ICN)) in 1968. The history of the work is not known. In 2001, the tapestry was suspected to be infested with moths and was put in a freezer at −20 °C for three weeks as a remedial measure. Natural fibres withstand this treatment without damage, but in this item white, opaque fibres, which were only present in a rather small area of the work, became embrittled (Fig. 3). It was not previously known that a small proportion of the fibres in the carpet were synthetic.

The third item studied was a relining fabric, which was used for the nap-bonded cold lining of a painting in 1975 at the Central Research Laboratory (now part of ICN). According to the relining method, the linen of the painting was pre-treated with diacetone-alcohol (4-hydroxy-4-methyl-pentan-2-one) and water, then dried on a vacuum table followed by the application of Plextol B 500, an acrylic methacrylate copolymer emulsion adhesive (Mehra 1975). The relining has become extremely brittle and is partly pulverised (Fig. 4).

Microscopy, scanning electron microscopy (SEM) and Fourier transform infrared (FTIR) spectroscopy were used to compare the reference fibres from *Welche Chemiefaser Ist Das?* to the degraded fibres from the works of art and the relining fabric. FTIR analysis allows the composition and the degradation-related molecular changes of the samples to be studied. Birefringence colours, observed through polarised light microscopy (PLM), relate to variations in the crystallinity of the fibres, while SEM permits the examination of the fibre surfaces.

Polypropylene

The history of man-made fibres began at the end of the 19th century with the first semi-synthetic or regenerated materials (van Oosten 2002) and although the first completely synthetic polymers were developed in the early 20th century, many fibres that are now in common use were not fully exploited until the 1960s and 1970s. Currently, one of the most widely used fibres is polypropylene. The first commercially viable form of PP was developed in 1954 by the chemists Natta and Ziegler by using special catalysts (Gordon Cook 1993). Several different polymerisation methods can be used to produce PP, but all rely on exposing the propylene monomer to high temperatures and pressures in the presence of an active metal catalyst (Fig. 5). By selecting particular catalysts and polymerisation methods, the molecular configuration of the polymer can be directed to one of three forms: atactic, isotactic or syndiotactic (Fig. 6).

Atactic polymers are characterised by their tacky texture, amorphous behaviour and low molecular weights. Where present they act as a plasticiser by reducing the overall crystallinity of the PP. From a commercial viewpoint, isotactic PP is the most important form of the polymer. In comparison to the atactic and syndiotactic forms, isotactic PP has the greatest degree of stereo-regularity; as a result, a higher degree of crystallinity is achieved. Many of the mechanical properties and processing characteristics of PP are determined by the level of isotacticity and thus crystallinity.

The useful properties of the polymer, combined with its comparatively low price, make it suitable for a wide range of applications. PP is the lightest of all man-made fibres (0.91 g/cm^3) and has the lowest moisture absorption. Dirt will not penetrate the fibre and can easily be removed from the surface. Polypropylene fibres have the same tensile strength, wet or dry, and are also highly resistant to both acids and alkalis and to most organic solvents. The fibre does not rot and is not readily attacked by micro-organisms or moulds.

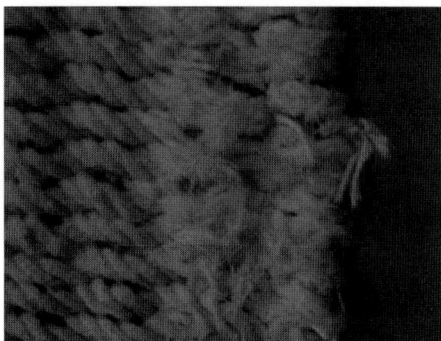

Figure 2 *Fête II* (Will Fruytier, 1969): white fibres (Plate 30 in the colour plate section).

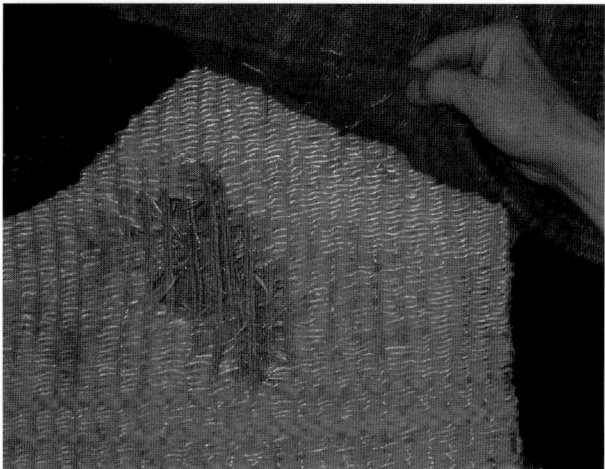

Figure 3 *The Knot* (Herman Scholten, 1968): degraded area of tapestry (Plate 31 in the colour plate section).

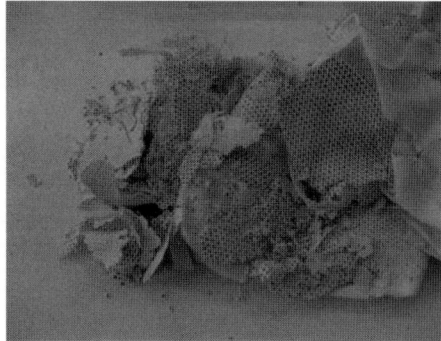

Figure 4 Relining fabric (1975) (Plate 32 in the colour plate section).

Production process

Most PP fibres are made by extruding the liquid material through the holes of a die plate, followed by spinning and stretching. The slit film method, however, in which the polymer is extruded through a small slit, was also used from the start of PP production. Slit film tape yarns were used principally in carpet backing and industrial sackings. Fibrillated yarns were developed in the 1970s. Nowadays, PP can be found as monofilaments, split or slit films, as either multifilament or staple forms. The polymer is usually pigmented before extrusion.

Modern PP comes in many varieties, ranging from tough rigid materials for outdoor furniture and car parts to soft, flexible fibres for clothing and babies' nappies. Some have good heat resistance, making them suitable for microwave food containers, while others melt easily and can be used for heat-sealable food packaging. Some are as clear as glass and others are completely opaque. Through research and development, the variety of materials available is increasing and polypropylenes are steadily replacing other polymers and traditional materials in many applications.

Degradation of polypropylene

The usefulness of polypropylene depends on the retention of its useful properties during a prolonged service life. Under mild conditions, unstabilised PP will retain these qualities for long periods of time. In most applications, however, exposure to heat and light will occur, accelerating oxidative degradation that results in a decrease in elastic properties, loss of flexibility, development of surface cracks and discoloration.

Photo-oxidation

Polypropylene readily suffers photo-oxidation, resulting in embrittlement, darkening and a decrease in molecular weight. It is considered that the induction period for these reactions represents the time necessary for the build-up of a sufficiently high concentration of peroxides, which subsequently decompose by a free radical mode and cause rapid oxidation (van Oosten and Aten 1996). The length of the induction period and the rate of subsequent oxidation depend upon the availability of oxygen. The induction time is inversely proportional to the square root of the oxygen pressure, the maximum rate of absorption increasing directly with the pressure.

Increasing the specimen thickness to the point where diffusion controls the availability of oxygen will prolong the apparent induction period. The presence of an antioxidant also prolongs this period by an amount directly proportional to its concentration within the polymer. An increase in temperature reduces induction periods and also increases the maximum oxidation rates. Both these functions give linear Arrhenius plots and have been used to determine apparent activation energies for the initiation and propagation stages of oxidation.

To retard oxidation and prolong service life, antioxidants and light stabilisers are normally added. These can be incorporated using any of the conventional thermoplastic compounding techniques. Pigments are the most effective UV absorbers, however, and a concentration of 5% of various organic and inorganic pigments has been found to restrict the degradation of PP to surface oxidation, thus increasing the time to failure. It has been demonstrated that channel black and furnace black also act as mild antioxidants for PP. Their effectiveness increases with concentration and there is a correlation with the concentration of surface groups. But they have a deleterious effect

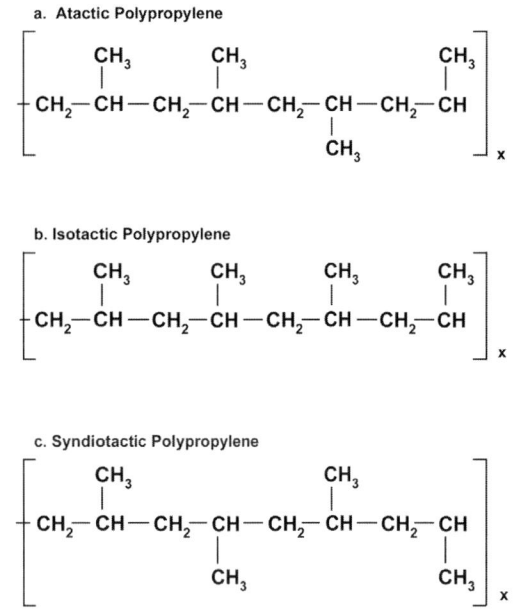

Figure 5 Monomer and polymer structure of polypropylene.

Figure 6 Three types of polypropylene.

on mechanical properties at concentrations greater than 5% and at this level there are no efficient stabilisers at processing temperatures so additional antioxidants may be required (Davis and Sims 1986).

Thermal degradation

Polypropylene tends to be stable to heat in the absence of oxygen; high temperatures are necessary to cause any significant decomposition. The activation energy, calculated from decomposition rates, is close to the value of 66.2 kcal/mol observed as the bond energy of a C-C single bond.

Photolytic oxidation

Although thermal oxidation at low temperatures is negligible, exposure to UV radiation in the presence of air can initiate chain scission, cross-linking and the formation of polar groups such as peroxides, acids, ketones and aldehydes.

Catalysed oxidation

The presence of metallic contaminants such as copper, copper compounds or Ziegler catalysts accelerates the thermal oxidation by catalysing the breakdown of hydroperoxides, producing a much higher concentration of deleterious radicals in the polymer.

Reactive stability

The initiation process of reaction is believed to begin with the 'weak bonds' and structural imperfections (Tennent 1998). The weakest bonds present in PP are those between carbon and hydrogen. Atactic PP is slightly inferior to isotactic PP in its resistance to oxidation at 140 °C. The isotactic configuration is sterically favourable for the interaction of a peroxide radical on a tertiary carbon atom with an adjacent tertiary carbon atom on the same chain to form a hydroperoxide and leave a hydrocarbon radical at the attacked site. This hydrocarbon radical then reacts with molecular oxygen to give a peroxy radical which can attack the next tertiary carbon atom along the chain, the process repeating zipper fashion to leave hydroperoxide groups on alternate carbon atoms.

Other chemicals and solvents

Polypropylene is highly resistant to chemical attack from most solvents and chemicals in very harsh environments. Contact with certain chemicals, however, such as liquid hydrocarbons, chlorinated chemicals and strong oxidising acids, can cause surface crazing and the swelling of the material. In general, PP is not susceptible to environmental stress cracking and it can be exposed under load in the toughest conditions.

Analytical research

Fourier transform infrared spectroscopy (FTIR)

To verify chemical composition and study molecular changes, FTIR spectra of all 126 fibres in the book and the fibres of the two works of art and the relining material were recorded. Spectra were recorded using a PerkinElmer Spectrum 1000 FTIR spectrometer combined with a Golden Gate single reflection diamond ATR unit (sampling area 0.6 mm^2), over the range 4,000–600 cm^{-1}, with 40 scans at a resolution of 4 cm^{-1}.

Polarised light microscopy (PLM)

The fibres were examined at 100 times magnification by PLM using a Zeiss Axioplan 2 imaging microscope equipped with an AxioCam MRc digital camera (1388 × 1040 pixels).

Scanning electron microscopy (SEM)

The samples were also examined by SEM, using a JEOL 5910 LV. The technique was used to study the surfaces of the fibres and to detect any deterioration phenomena. To avoid surface charging, the material was coated with gold using a JEOL JFC-1200 fine coater. The SEM was operated at an accelerating voltage of 2 kV.

Results and conclusion

FTIR

The infrared (IR) spectra of the three polypropylene samples from the book showed only isotactic carbon-hydrogen absorptions due to the backbone of the PP structure; no carbonyl absorption was present. It can be concluded that the PP fibres in the book are in perfect condition, probably due to the fact that they were never subjected to daylight. The book had been standing in the library, untouched, for more than 40 years. The IR spectra of the fibres from the tapestry *Fête II* showed a carbonyl (C=O) absorption in addition to the carbon-hydrogen absorptions, indicating that degradation of the polymer has occurred within the timespan of 34 years. IR spectra of the white and blue fibres from this work showed the presence of a polypropylene/polyethylene copolymer, while the black- and orange-coloured fibres contained only polypropylene.

The spectra of the yellowed white fibres from *The Knot* showed carbonyl (C=O) and C-OH absorptions as well as the expected carbon-hydrogen absorption of PP. Due to the oxidation of polypropylene, degradation products were formed and moisture absorption had occurred. This tapestry had then been put in a freezer at −20 °C for three weeks to prevent insect damage and therefore the C-OH absorptions were noticed. The IR spectrum of the relining fabric showed absorption bands of isotactic polypropylene. Carbonyl and water absorption bands are also present, indicating that the fibre is degraded and that oxidation has taken place.

PLM

The fibres from the artworks clearly show areas with varying birefringence colours. In addition, many cracks and tears are visible. The birefringence colours around a crack in a white fibre from *Fête II* differ from those in the main body of the fibre (Fig. 7).

SEM

The surfaces of the three polypropylene fibres from the book are either smooth or generally smooth with striations running along the length of the fibre. Four samples from the tapestry *Fête II* were examined. The white, blue and yellow fibres have a rectangular cross-sectional shape and appear to be very brittle. The surfaces show large cracks running parallel and perpendicular to the fibre axis. On a smaller scale, cracks running perpendicular to the fibre axis are also seen in the middle of the fibre (Fig. 8). The black fibres appear to be degrading differently from the others – instead of cracking, the surface seems to be flaking off (Fig. 9).

The fibres of the tapestry *The Knot* are rectangular in cross-section. There appears to be some degree of defibrillation at the edges of the fibres. Cracks can be seen running parallel to the length of the fibres and, in the surface striations, small cracks perpendicular to the fibre are visible. The relining material seems to be made up of short staple length fibres, some of which are cracked (Fig. 10). The results of the FTIR analyses, visual observations and SEM images of the fibres are summarised in Table 1.

Conclusion

The three PP fibres from the book are all in good condition – neither degradation nor yellowing was observed, which was to be expected as the PP fibres had been kept in the dark. In the tapestries *Fête II* and *The Knot*, however, exposure to UV radiation/daylight and the freezer treatment have caused the

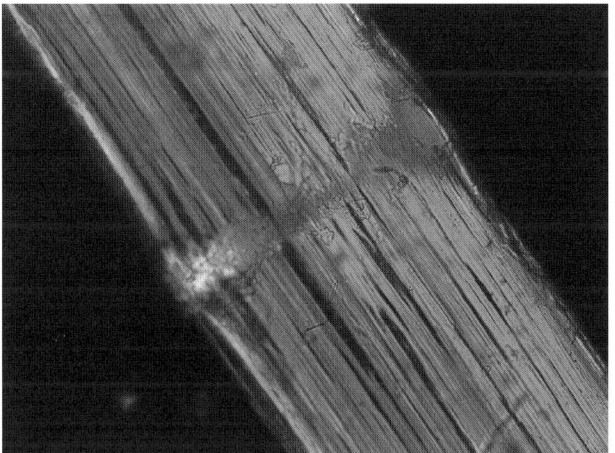

Figure 7 Polarisation microscope image of white fibre of *Fête II* (100×) (Plate 33 in the colour plate section).

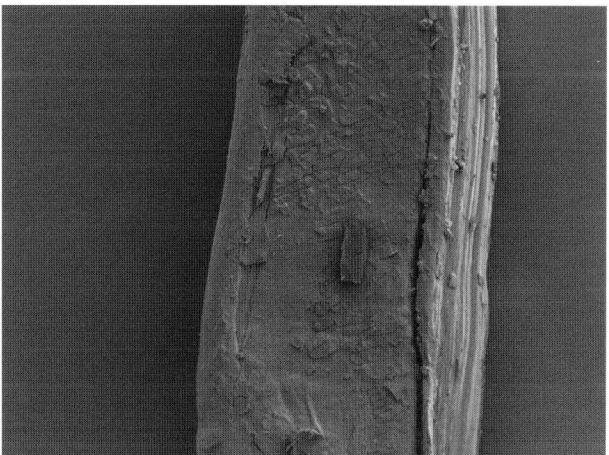

Figure 9 Secondary electron image of black fibre of *Fête II* (500×).

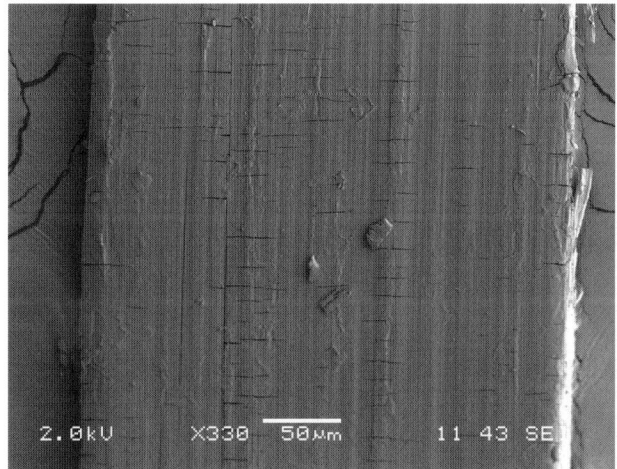

Figure 8 Secondary electron image of white fibre of *Fête II*.

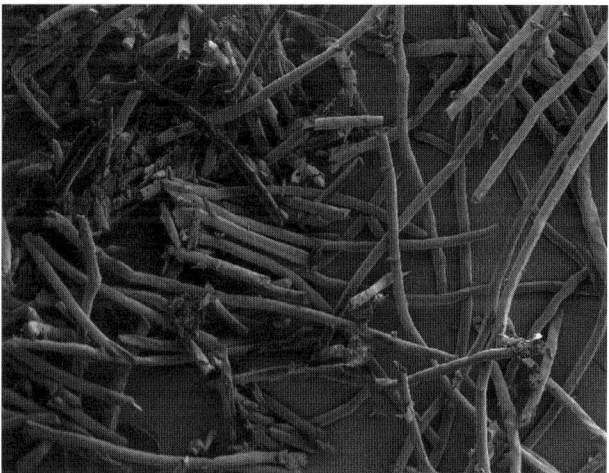

Figure 10 Secondary electron image from relining fabric (110×).

Table 1 Visual examination of fibres and FTIR and SEM results.

Sample	Polypropylene isotactic	Polyethylene	Carbonyl C=O	C-OH absorptions	Fibre or film	Yellowed/ brittle?	SEM
PP (book)	+	–	–	–	fibre	*	smooth surface
PP (book)	+	–	–	–	fibre	*	grooves
PP (book)	+	–	–	–	fibre	*	grooves
Fête II, white	+	+	+	–	film	brittle, not yellowed	cracks
Fête II, blue	+	+	+	–	film	brittle, not yellowed	cracks
Fête II, yellow	+	–	+/–	–	film	brittle	cracks
Fête II, black	+	–	+/–	–	film	brittle	surface flakes off
The Knot (yellowed area)	+	–	+	+	film	brittle, yellowed	cracks and fibres peel from the side
Relining fabric	+	–	+	+	fibre	brittle/not yellowed	cracks

+ present; +/– small amount present; – not present; * no changes serious degradation seen by yellowing and embrittlement of the PP. It is known that normal extruded PP can withstand temperatures of −30 °C, but isotactic PP under stress becomes brittle at −20 °C. The pulverised relining fabric was probably degraded by solvent (diacetone-alcohol) that was not allowed to fully evaporate from the old linen before the PP relining fabric with the adhesive was applied.

Overall, it can be stated that photo-oxidation has occurred in all cases where exposure to UV radiation or daylight occurred; although it is known that this can happen, it was not expected for these works of art over a timescale of only 40 years. The other degradation phenomenon, due to freezing, was not anticipated, as PP was not supposed to be present in a tapestry believed to be constructed from natural fibres. This highlights the importance of fully understanding the composition of an object before conservation procedures are carried out.

References

Davis, A. and Sims, D. (1986) *Weathering of Polymers*. London: Elsevier Applied Science Publishers.

Driesch, H. (1962) *Welche Chemiefaser Ist Das?* Stuttgart: Franckh'sche Verlagshandlung.

Gordon Cook, J. (1993) *Modern Materials. Handbook of Textile Fibres, II: Manmade Fibres*. Darlington: Merrow Publishing Co Ltd.

Mehra, V. (1975) 'Comparative study of conventional relining methods and materials and research towards their improvement', in *Interim Report to the ICOM Committee for Conservation, 4th Triennial Meeting, Venice, 1975*, Paper No. 75/11/5.

Tennent, N.H. (1988) 'An introduction to polymer chemistry relevant to plastic collections', *Preprints of Contributions to the Modern Organic Materials Meeting held at the University of Edinburgh, 14–15 April 1988*. Edinburgh: SSCR Publications.

van Oosten, T.B (2002) 'A survey of problems with early plastics', in *Contributions to Conservation, Research in Conservation at the Netherlands Institute for Cultural Heritage*, J. Mosk and N.Tennant (eds). London: James and James.

van Oosten, T.B. and Aten, A. (1996) 'Lifelong guaranteed: the effect of accelerated ageing on Tupperware objects made of polyethylene', *Preprints from ICOM-CC 11th Triennial Meeting, Edinburgh, 1–6 September 1996*. London: James and James.

The authors

- Dr Thea B van Oosten is a senior researcher at the Netherlands Institute for Cultural Heritage (ICN), Conservation Research Department. She is currently engaged in Fourier transform infrared spectroscopy (FTIR) and differential scanning calorimetry (DSC) analyses of organic materials, especially plastics in objects of cultural heritage and modern materials in modern and contemporary art objects. She also is member of the Directory Board of ICOM-CC.
- Dr Ineke Joosten is a researcher at the Netherlands Institute for Cultural Heritage (ICN), Conservation Research Department. She is currently involved in conservation research of metals and archaeological objects and is responsible for the scanning electron microscope equipped with an energy-dispersive spectroscopy system (SEM–EDS).
- Dr Luc Megens is a researcher at the Netherlands Institute for Cultural Heritage (ICN), Conservation Research Department. He is currently engaged in architectural paint research, especially pigment studies, and inorganic analysis in general, using x-ray fluorescence, diffraction and SEM–EDS.

Address

Corresponding author: Thea B van Oosten, Netherlands Institute for Cultural Heritage (ICN), Conservation Research Department, Gabriël Metsustraat 8, 1071 EA, Amsterdam, The Netherlands (thea.van.oosten@icn.nl).

Probing the microstructure of protein and polyamide fibres

Paul Garside and Mary M. Brooks

ABSTRACT The chemical, structural and physical properties of a range of proteinaceous and polyamide fibres, including silk, wool, nylon and various modern and historic regenerated proteins, have been assessed. The chemistry and microstructure of both pristine and artificially aged specimens of the fibres have been investigated using conventional and polarised infrared spectroscopy. Correlations have been drawn between these results and data derived from mechanical testing, demonstrating that the nature and state of such materials can be adequately characterised by analytical methods requiring minimal intervention. This will be of value in exploring the history and informing the identification, conservation, display and storage of artefacts containing these fibres.

Keywords: Aralac, azlon, protein fibre, polyamide, polarised infrared spectroscopy, regenerated protein, rubbish theory

Introduction

The aim of the research was to employ relatively simple spectroscopic techniques to investigate the microstructure of various proteinaceous and polyamide fibres, and to determine how structural information derived in this manner correlated with the bulk physical properties of these materials. In turn, it is hoped that this work will ultimately aid in the characterisation of regenerated protein fibres in collections, and thus help to inform decisions for conservation, storage and display strategies. Relatively few artefacts (garments, accessories or samples) from the mid-20th century containing these regenerated proteins have yet been located, so this is particularly pertinent.

Regenerated protein fibres

The work presented here developed from investigations into regenerated protein fibres. These are reconstituted from natural protein sources such as peanuts, soya, maize, egg, feathers and milk, in much the same way as viscose (regenerated cellulose) fibres are produced from waste cellulose such as cotton linters and wood pulp. These materials first found widespread use in the 1940s, when the fear of shortages of fibres such as wool, particularly during the Second World War, prompted the development of suitable alternatives – for example, Aralac (milk) and Vicara (maize) were developed in the USA, Ardil (peanut) in the UK (see Fig. 5 on p. 37) and Lanital (another milk fibre) in Italy. These materials were largely destined for domestic consumption, while the scarce supplies of wool were diverted for military use. The fibres had poor mechanical properties, however, particularly when wet, which ensured that they only found widespread use when blended with other, more durable, materials such as the natural fibres they were intended to replace, and so they quickly fell out of favour when shortages were no longer a problem. In the following decades various attempts were made to continue the development of these fibres, but they never achieved widespread usage. More recently, however, regenerated protein fibres have experienced a resurgence in popularity, in part because they can be marketed on their supposedly 'green' credentials. Advances in fibre technology have enabled the yarns to be produced with a range of copolymers and fillers (such as polyvinyl alcohol and polyacrylonitrile), intended to make good the technical deficiencies of their forerunners. Textiles containing milk, soya and maize fibres (generally blended with other fibres) can now all be found in commercially available clothing and other articles (Brooks 1993, 2005; Brooks and Garside 2005; see also Brooks in this volume, pp. 33–40).

In the experiments presented here, several of these modern regenerated protein fibres (milk and soya fibres) were examined, comparing them with both an historic regenerated protein fibre (Ardil) and other proteinaceous and polyamide fibres (silk, tussah silk, wool, nylon and aramids), chosen on the basis that they might be expected to share certain chemical or structural properties. The regenerated protein fibres are produced in a similar way to other regenerated or synthetic fibres, i.e. they are extruded from the bulk mixture and then drawn into fine filaments. As noted above, copolymers and other additives may be present.

Natural protein fibres

Silk is produced by the domesticated silkworm (*Bombyx mori*); the fibres are formed as two crystalline fibroin protein filaments, bound together by the amorphous protein sericin. In commercial use the fibres are often degummed, i.e. the sericin is removed. The fibroin crystallites are aggregates of anti-parallel ß-sheet protein structures, strongly aligned with the fibre axis during extrusion from the worm's spinnerets (Crighton 1993; Lewin and Pearce 1998; Merritt 1992; Otterburn 1977; Tímár-Balázsy and Eastop 1998; Wolfgang 1970). Tussah silk is produced by wild silkworms; its chemical composition differs slightly from that of domesticated silk, and it tends to have a slightly lower tenacity (Crighton 1993; Otterburn 1977; Tímár-Balázsy and Eastop 1998). Wool fibres are produced by many domesticated mammals, such as sheep, goats, and camels. Unlike silk, the fibres are cellular in nature, with durable external cuticle cells coated with waxes and proteins, and elongated cortical cells that form the bulk of the material. The proteins of the cortex have extensive α-helix structures (Lewin and Pearce 1998; Matthews 1947; Tímár-Balázsy and Eastop 1998).

Synthetic polyamide fibres

Nylon is a synthetic polyamide material principally derived from petrochemical sources. The structure of the fibres is imposed by the extrusion of the molten polymer, followed by cooling and then drawing, to lengthen the fibres and impart a greater degree of ordering (Lewin and Pearce 1998; Hatch 1993; Miller 1968; Morton and Hearle 1993; Nicholson 1991; Taylor 1990; Tímár-Balázsy and Eastop 1998). Aramids are also polyamides, but the simple hydrocarbon chains of nylon are replaced by aromatic groups, enabling the polymer to pack more tightly together, imparting a very high crystallinity at the molecular level and exceptional bulk mechanical properties (Taylor 1990).

Experimental rationale

The aim of this research was to determine the ways in which regenerated protein fibres differ from other, similar fibres (such as natural proteins and synthetic polyamides) in terms of their structural and chemical properties and their ageing behaviour. It was then hoped that this information would help to explain the generally poor physical properties of these materials, and ultimately lead to methods to characterise the fibres and to better inform conservation decisions for artefacts containing them.

Samples of natural protein, regenerated protein and synthetic fibres were subjected to accelerated thermal ageing, and then the mechanical properties of the unaged and aged samples were assessed, in both ambient and wet states. These data made it possible to gain an understanding of the ways in which these various materials behaved over time and under different conditions. The microstructures of the materials were then investigated, via spectroscopic techniques, to determine whether the differences in the measured physical characteristics could be correlated to differences in the conformation and ordering of the component polymers at the molecular level.

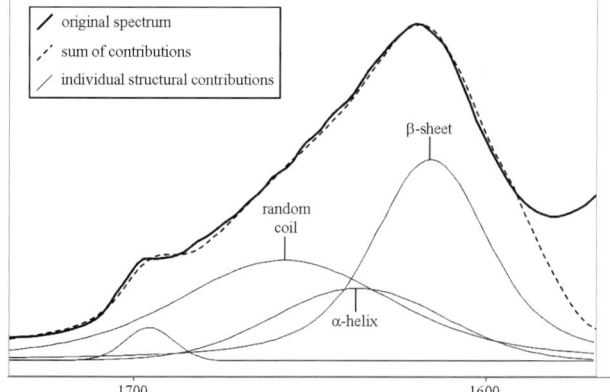

Figure 1 The amide I band from a protein spectrum, indicating the approximate contributions to the overall band from the ß-sheet, α-helix and random coil secondary structures.

Initially conventional attenuated total reflectance (ATR) spectroscopy was employed to investigate the presence of different protein secondary structural motifs, such as ß-sheets and ß-helices. Two of the more prominent features of a polyamide spectrum are the amide I and amide II bands (arising from the $\alpha(C=O)$ and $\alpha(N-H)$ vibrations, and occurring at approximately 1,635 and 1,520 cm^{-1}, respectively). Each of these bands is, in fact, composed of a set of superposed peaks, and each of these individual components can be assigned to a particular type of protein structure (Fig. 1). Therefore the precise positions of the composite bands depend on the structural composition of polymer itself and can gain an appreciation of the presence of different secondary structures found in the fibre.

Polarised ATR (Pol-ATR) spectroscopy was subsequently used to investigate the presence, or otherwise, of coherent long-range ordering in the fibres. This was done by exploiting dichroism in particular bands in the spectrum, i.e. the manner in which the intensity of the band changes with the orientation of the fibre when the spectrum is recorded using polarised radiation. Therefore, pairs of spectra were recorded, with the fibres aligned first parallel with ($F_\parallel$), and then perpendicular to ($F_\perp$), the electric vector of the incident polarised radiation. The use of techniques of this kind has already been demonstrated in monitoring the degradation of silk (Garside and Wyeth 2002; Garside et al. 2005). If the polymer chains are strongly aligned with the fibre axis, then it is expected that the two spectra ($F_\parallel$ and $F_\perp$) will have marked differences in the intensities of the dichroic bands, especially the amide I and amide II bands mentioned above. In the absence of a strong axial alignment of this type, however, the pairs of spectra will vary little.

Experimental method

Accelerated ageing

Samples of the eight fibre types (Ardil, modern soya fibre, modern milk fibre, silk, tussah silk, wool, nylon 66 and the aramid Nomex) were subjected to accelerated ageing by placing the materials in individually sealed hybridisation tubes, along with vials of saturated potassium chloride solution to ensure an atmosphere of 79±1% relative humidity (RH), before maintaining in an oven at 85 °C for three weeks. After this, the samples were removed and stored at 21±1 °C and 51±2% RH.

Mechanical testing

To determine the physical properties of the fibres, yarns extracted from the unaged and aged fabric samples were subjected to mechanical testing using an Instron 5544 machine fitted with a 100 N load cell, controlled by Instron Bluehill v1.4 software, and employing an extension rate of 10 mm/min^{-1}; the samples were tested to breaking point. The specimens were initially tested under ambient conditions (22 °C, 55% RH). Subsequently, similar sets of samples were tested after wetting by immersion in deionised water for a few minutes. Six replicate tests were carried out for each fibre type and condition. Average tenacities (breaking strength per unit linear density) were then calculated from these data.

Spectroscopy

The samples were then subjected to both conventional ATR and Pol-ATR spectroscopy. Absorbance spectra were recorded using a PerkinElmer Spectrum One FTIR spectrometer, fitted with a PerkinElmer Universal ATR accessory; spectra were captured over the range 4,000–400 cm^{-1}, using 32 scans and with a resolution of 4 cm^{-1}. Spectra were subsequently manipulated with Thermo Galactic GRAMS 32/6 software. ATR spectra were recorded from the bulk fabric specimens, laid square on the UATR accessory. Pol-ATR spectra of individual yarns were then acquired by placing a Specac wire grid IR polariser in the beam path before the specimen, aligned such that the electric vector of the radiation was perpendicular to the beam path. The yarns, and hence the component fibres, were carefully oriented such that they were either parallel or perpendicular to the electric vector, giving $F_\parallel$ and $F_\perp$ spectra, respectively.

Results and discussion

Mechanical testing

The unaged and aged tenacities of the various fibre samples, in both ambient and dry conditions, are presented in Table 1; these data are also shown graphically in Figure 2, indicating the percentage decrease in strength on ageing and wetting. It can be seen that of the specimens under examination, the regenerated protein fibres are both the most susceptible to ageing and possess the poorest wet properties.

Spectroscopy

Typical ATR spectra for the fibres under consideration are shown in Figure 3. As noted above, the positions of maximum

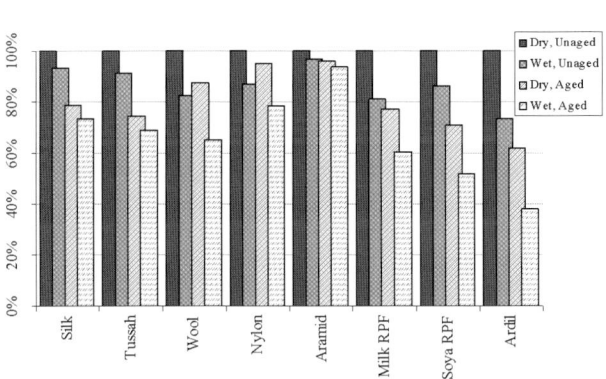

Figure 2 Percentage tenacities of the experimental samples, with reference to the unaged materials under ambient conditions.

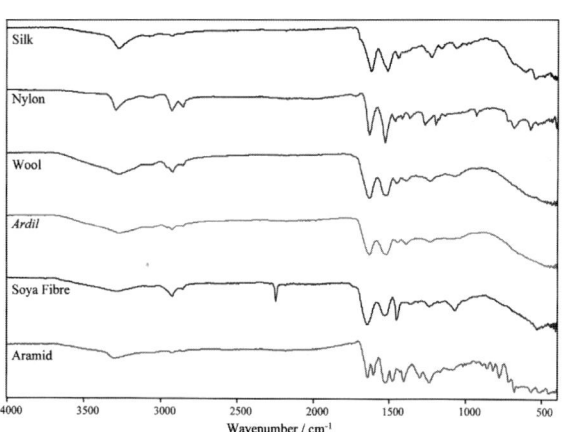

Figure 3 Typical ATR spectra for the experimental samples over the range 4,000–400 cm^{-1} (Plate 34 in the colour plate section).

Table 1 Average tenacities of unaged and aged fibre samples, under ambient and wet conditions, with standard deviations in parentheses.

	Average tenacities/cN.tex^{-1}							
	Unaged				Aged			
	Ambient		Wet		Ambient		Wet	
Silk	30.7	(3.1)	28.6	(3.9)	24.2	(1.3)	22.5	(1.6)
Tussah silk	24.3	(3.5)	22.2	(2.9)	18.1	(2.1)	16.8	(1.4)
Wool	10.3	(3.1)	8.5	(1.2)	9.0	(0.8)	6.7	(1.3)
Nylon	49.5	(5.5)	43.0	(5.9)	46.9	(4.2)	38.8	(2.9)
Aramid	164.8	(6.3)	159.5	(5.9)	158.3	(6.8)	154.3	(5.1)
Milk RPF	27.9	(1.9)	22.6	(1.6)	21.5	(1.1)	16.8	(0.5)
Soya RPF	24.1	(3.3)	20.8	(4.2)	17.1	(2.6)	12.5	(1.8)
Ardil	7.1	(0.2)	5.2	(0.2)	4.4	(0.2)	2.7	(0.2)

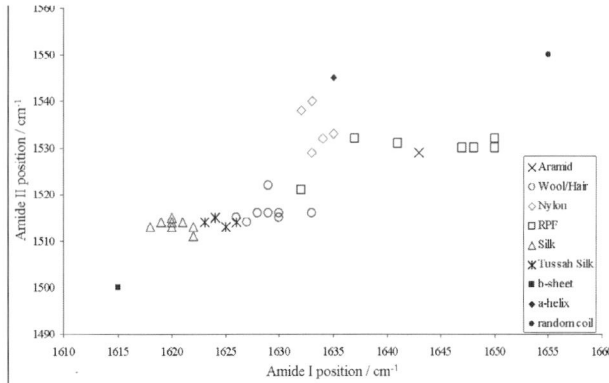

Figure 4 Plot of amide I and amide II positions for various proteinaceous and polyamide specimens (Plate 35 in the colour plate section).

intensity for the amide I and amide II bands are dependent on the relative presence of the various protein secondary structures. If these two band positions are plotted against each other (Fig. 4), along with the nominal positions of 'pure' ß-sheet, α-helix and random coil protein motifs (Jabs nd), it is possible to gain an indication of the occurrence of these microstructures within the fibre types under investigation. Materials that have a high degree of ß-sheet structures will be found towards the bottom left corner of this plot, whereas those which have a predominantly random secondary structure will fall towards the top right. Therefore, from this plot it can be seen that the silks, tussah silks and wools possess a high degree of polymer secondary structure, which the nylons, aramids and – most notably – regenerated protein fibres lack. This might be expected due to the different ways in which the fibres are formed: wools and silks are natural materials and so, in common with other natural proteins, a high degree of imposed protein conformational ordering is observed at the molecular level, whereas the formation of the synthetic fibres is a simpler process, with the bulk ordering of the polymer chain being afforded by extrusion and drawing. Although the regenerated protein fibres are derived from natural source materials, the manner in which they are processed and formed is analogous to that of the synthetic materials.

The pairs of Pol-ATR spectra ($F_\parallel$ and $F_\perp$) for typical fibre types are shown in Figure 5; the range of the spectra is limited to 1,750–1,400 cm^{-1}, highlighting the amide I and amide II bands. The degree to which these pairs of spectra exhibit dichroism (differing band intensities with fibre orientation) is related to the long-range alignment of the component polymer chains with the fibre axis: i.e. if the material possesses a strong degree of axial orientation, then the pairs of spectra will display marked differences in the intensities of these two bands; on the other hand, if the spectra are very similar then this extended, uniaxial orientation is not present. Therefore, from the spectra it can be seen that silk, nylon and aramid fibres possess this kind of ordering, as might be expected due the extrusion processes encountered in their formation, forcing the polymer chains into alignment. Wool, on the other hand, lacks this long-range alignment, again as might be predicted, as the ordering in wool occurs within the component cells rather than throughout the fibre as a whole. Neither the historic nor the modern regenerated protein fibres exhibit extended orientation of this type, however, even though they are also formed by extrusion and drawing like the other synthetics.

Conclusion

By comparing the results of the ATR and Pol-ATR experiments, the observed mechanical behaviour of the fibres can begin to

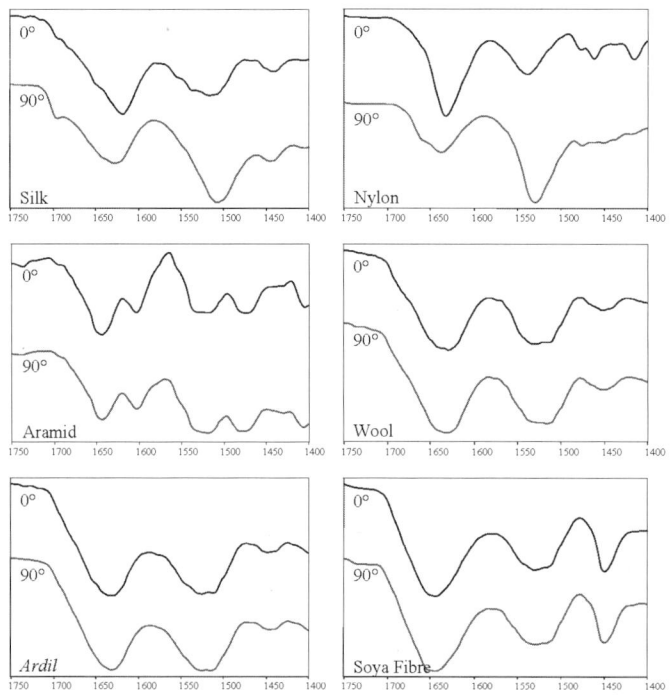

Figure 5 Pairs of $F_\parallel$ and $F_\perp$ Pol-ATR spectra over the range 1,750–1,400 cm^{-1}.

be explained by the differences in their microstructures: silk and wool fibres have localised polymer folding structures (ß-sheets and α-helices, respectively) that enhance their physical properties. Silk, nylon and aramid fibres have oriented crystallinity, with the polymer chains possessing long-range ordering along the fibre axis, which also serves to yield the desirable bulk characteristics of the materials. The regenerated protein fibres, on the other hand, possess neither localised structures nor long-range crystallinity, factors that help to explain both their poor mechanical performance and their particular susceptibility to wetting.

Thus it can be seen that it is possible to derive microstructural information from fibres using relatively simple, non-destructive techniques. These data can then be used to explain bulk physical properties. The authors hope to extend this work to allow the characterisation of these fibres as further garments and other textile artefacts are identified, and to apply additional analytical techniques to the problem; in particular, the use of liquid chromatography to assess both the degree of polymerisation of the polymers – a factor that will help to further understand their mechanical characteristics – and to monitor their degradation, and the amino-acid composition of the component proteins, thus allowing a greater appreciation of the various sources of these materials.

Acknowledgements

The authors would like to extend their thanks to the Getty Conservation Institute, University of California Los Angeles, Los Angeles County Museum of Art, Denise Buhr (Research Librarian, Central Soya, Indiana) and Patricia Starrett (Archivist, ICI) for information and access to collections, and DuPont for providing the Nomex sample. They would also like to thank colleagues at the Textile Conservation Centre and the AHRC Research Centre for Textile Studies and Textile Research, particularly Nell Hoare, Paul Wyeth, Maria Hayward, Dinah Eastop and Mike Halliwell.

References

Brooks, M.M. (1993) 'Ardil: the disappearing fibre?' in *Saving the Twentieth Century: The Conservation of Modern Materials*, D.W. Grattan (ed.), 81–93. Ottawa: Canadian Conservation Institute.

Brooks, M.M. (2005) 'Soya bean protein fibres: past, present and future', in *Biodegradable and Sustainable Fibres*, D. Blackburn (ed.), 398–440. Cambridge: Woodhead Publishing.

Brooks, M.M. and Garside, P. (2005) 'Investigating the significance and characteristics of modern regenerated protein fibres', in *Art '05. Proceedings of the 8th International Conference on Non-Destructive Investigations and Microanalysis for the Diagnostics and Conservation of the Cultural and Environmental Heritage. Lecce (Italy), 15–19 May 2005*, C. Parisi, G. Buzzanca and A. Paradisi (eds), 1–14. Lecce: Italian Society for Non-Destructive Testing Monitoring Diagnostics, Ministry of Cultural Heritage and Activities and Central Institute of Restoration and Department of Materials Science, University of Lecce.

Crighton, J.S. (1993) 'Silk: a study of its degradation and conservation', in *Conservation Science in the UK*, N. Tennant (ed.), 96–8. London: James and James.

Garside, P. and Wyeth, P. (2002) 'Characterization of silk deterioration', in *Strengthening the Bond: Science and Textiles. Preprints of the North American Textile Conservation Conference April 2002*, V.J. Whelan (ed.), 55–60. Philadelphia, PA: NATCC.

Garside, P., Lahlil, S. and Wyeth, P. (2005) 'Characterization of historic silk by polarized attenuated total reflectance Fourier transform infrared spectroscopy for informed conservation', *Applied Spectroscopy* 59(10): 1242–7.

Hatch, K.L. (1993) *Textile Science*. Minneapolis, MN: West Publishing Company.

Jabs, A. (nd) 'Determination of secondary structure in proteins by Fourier transform infrared spectroscopy (FTIR)', in *Image Library of Biological Macromolecules* (www.imb-jena.de/ImgLibDoc/ftir/IMAGE_FTIR.html).

Lewin, M. and Pearce, E.M. (1998) *Handbook of Fibre Chemistry*, 2nd edn. New York: Marcel Dekker.

Matthews, J.M. (1947) *Textile Fibres*. New York: John Wiley and Sons, Inc.

Merritt, J.L. (1992) 'Silk: history, cultivation and processing', in *Silk: Harpers Ferry Regional Textile Group, 11th Symposium*, 15–21. Washington DC: Harpers Ferry Regional Textile Group.

Miller, E. (1968) *Textiles (Properties and Behaviour)*. London: Batsford.

Morton, W.E. and Hearle, J.W.S. (1993) *Physical Properties of Textile Fibres*. Manchester: Textile Institute.

Nicholson, J.W. (1991) *The Chemistry of Polymers* (RSC Paperbacks). Cambridge: Royal Society of Chemistry.

Otterburn, M.S. (1977) 'The chemistry and reactivity of silk', in *The Chemistry of Natural Fibres*, R.S. Asquith (ed.), 53–80. New York: John Wiley and Sons, Inc.

Taylor, M.A. (1990) *Technology of Textile Properties*, 3rd edn. London: Forbes Publications.

Tímár-Balázsy, Á. and Eastop, D. (1998) *Chemical Principles of Textile Conservation*. Oxford: Butterworth Heinemann.

Wolfgang, W.G. (1970) 'Silk', in *Encyclopedia of Polymer Science and Technology 12*, 578–85. New York: John Wiley and Sons, Inc.

The authors

- Paul Garside is a postdoctoral Research Fellow at the AHRC Research Centre for Textile Conservation and Textile Studies. He has developed significant expertise in applying analytical methodology to conservation science problems.
- Mary Brooks trained at the Textile Conservation Centre after working in the book world and management consultancy. She has worked as a conservator and curator in Europe and America. At York Castle Museum, she jointly curated 'Stop the Rot', which won the 1994 IIC Keck Award for promoting public understanding of conservation and is a member of the ICOM Conservation Committee's Task Force for raising awareness of heritage conservation. She has a special interest in the contribution which object-based research and conservation approaches can make to the wider interpretation of cultural artefacts.

Addresses

- Paul Garside, AHRC Research Centre for Textile Conservation and Textile Studies, Textile Conservation Centre, University of Southampton, Park Avenue, Winchester, Hants SO23 8DL, UK (pg1@soton.ac.uk).
- Mary M. Brooks, Reader, Textile Conservation Centre, University of Southampton, Park Avenue, Winchester, Hants SO23 8DL, UK (mmb1@soton.ac.uk).

Investigating cellulose nitrate degradation caused by fungal attack

Margarida Silva

ABSTRACT This paper details the conservation problems associated with a portable film projector from the 1960s, contained in two cases. The machine is part of the collection of the Museu do Carro Eléctrico (Tramcar Museum), Porto, Portugal, which consists of a significant number of technical and engineering objects made of modern materials. The main goal of this paper is to discuss the fungal attack on the cellulose nitrate-based synthetic leather that covers the projector cases. In order to do this, it was necessary to identify the plastic and the particles found on its surface, using Fourier transform infrared (FTIR) microspectrometry and optical microscopy techniques. Although it is known that cellulose nitrate photographic films are attacked by fungi, it is not clear whether fungi are implicated in the degradation of this type of plastic. There is a common belief that the aggressive nature of nitric acid, a product of cellulose nitrate deterioration, prevents biodegradation from occurring. This paper confirms that fungal growth and attack on cellulose nitrate is possible.

Keywords: cellulose nitrate, synthetic leather, biodegradation, fungi, FTIR spectroscopy, optical microscopy

Introduction

The study of plastics is currently one of the most interesting topics in the field of conservation. This paper originated as the author's final project for the course on conservation and restoration at the Universidade Nova de Lisboa. Two case studies were chosen from the collection of the Museu do Carro Eléctrico (Tramcar Museum), Porto, Portugal: a natural rubber hose from a fire extinguisher and an example of synthetic leather from the cases from a 16 mm film projector (the latter being the subject of this paper).[1] This choice was based on the particular condition of the plastic that forms the exposed surface of the synthetic leather. These surfaces, on both the inside and outside of the cases, are characterised by the presence of powdery particles.

The first step was to identify the composition of both the powdery white particles and the constituent materials of the synthetic leather. Fourier transform infrared (FTIR) microspectroscopy and optical microscopy were chosen for this purpose. The results obtained from this analytical work revealed that the synthetic leather under study was composed of a cellulose nitrate-based plastic laid on a cotton textile. The particles were identified as fungal hyphae and spores.

In order to understand the mechanism of microbiological attack on such plastic, bibliographic research was undertaken. There is very little in the conservation literature on this topic, however, and even less that is helpful. There are some reports of cellulose nitrate photographic films being attacked by fungi, but in these cases it is thought that only the photographic emulsion is degraded. Studies found in the literature are not even in agreement that biological attack on cellulose nitrate is possible. There is a widespread assumption that the presence of nitric acid, one of the main products of cellulose nitrate degradation, prevents the growth of micro-organisms. There are also studies relating to the use of fungi to degrade cellulose nitrate explosives (Auer *et al.* 2005; Kim *et al.* 1998; Sharma *et al.* 1995; Rizvi and Zeto nd).

The main task of this paper is to discuss what made the biological attack on the synthetic leather possible.

Film projector

The portable 16 mm film projector, RCA 415R, is housed in two cases: the first contains the projector (case 1) (Fig. 1a) and the other, the loudspeaker (case 2) (Fig. 1b). This machine was produced by the Radio Corporation of America since 1949.[1] It was purchased c.1965 by the Porto Public Transport Company (STCP) for showing films about health and safety at work.[2] Both cases contain various polymeric materials (plastic, rubber, textile, wood, paper, paint and adhesive) and have green synthetic leather coatings and a transparent plastic handle with metallic fittings. Other materials include metal alloys, glass and ceramic. The main structure of the cases is made of wood; case 1 also has a removable metal lid. The synthetic leather is attached to the underlying surface with adhesive.

Synthetic leather

The synthetic leather plays both an aesthetic and protective role for the object: it provides an impermeable surface,

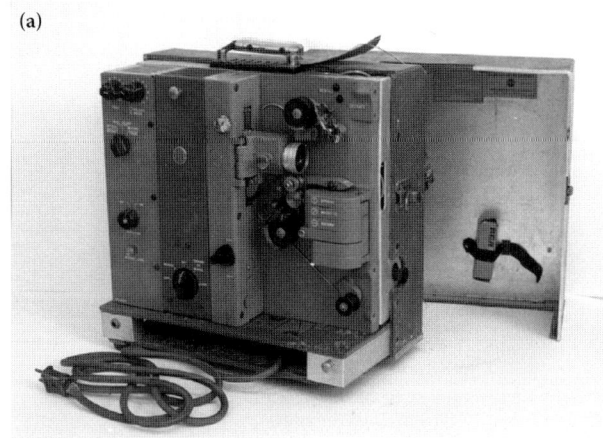

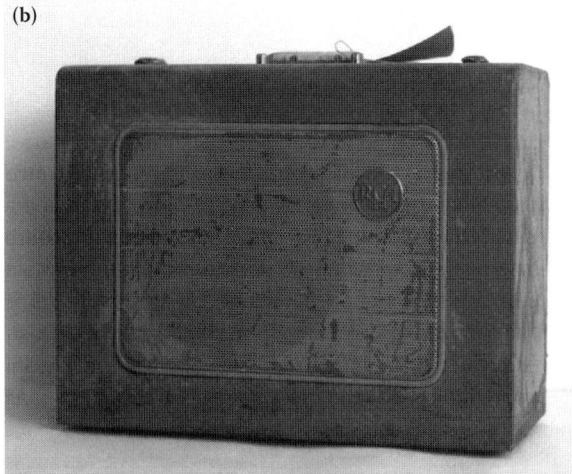

Figure 1 Film projector's case 1 (a) and case 2 (b) (Plate 36 in the colour plate section).

protection from both dust and humidity, and its colour and texture give a pleasing finish. The material is composed of a woven textile with a plastic coating; the example in this study was manufactured by a direct coating process. The materials that compose the synthetic leather were analyzed by FTIR (see 'Examination: methods and materials' below) and were found to be:

- plastic – cellulose nitrate;
- textile – cotton;
- adhesive – polyvinyl acetate.

Cellulose nitrate is a polynitrate ester of cellulose. It is obtained by adding a concentrated solution of nitric and sulphuric acids to some source of cellulose (usually cotton or paper pulp) at 20–40 °C (Selwitz 1988). The nitration rate on polymer prepared for plastics is around 10.5–11.5% (Williams 1994), which corresponds to an average substitution of 2.2–2.3 hydroxyl groups (OH) by nitrate groups ($-NO_3^-$) on the cellulose molecule.

This material is considered to be unstable due to the presence of the nitrate group and also the acidic residues that may remain from the production process. Its degradation produces acidic and nitrogen oxide gases. Cellulose nitrate's chemical and physical properties are improved, however, by adding plasticisers and stabilisers (Green and Bradley 1988). Plasticisers allow for more flexibility in the material as they permit a greater degree of movement in the polymer chains. Stabilisers act by absorbing degradation products released as cellulose nitrate ages, and by inhibiting light deterioration. Other additives may also be used to improve the qualities of the plastic. Colour is imparted by adding dyes and pigments.

The plastic was applied to the textile in a viscous liquid condition. In the case of the projector cases, the underlying fabric was a 2.1 z-twill, weft-faced material. The direct coating process involves applying the plastic directly on the fabric; a blade ensures an even coating thickness. After drying in a hot chamber, the surface can be given the desired texture by pressing the synthetic leather on a hot calender (mechanical press) with the appropriate pattern.

Condition

Preliminary observation revealed a white substance on the surface of the synthetic leather of both cases. This substance appears as small circular spots and is mainly distributed on the peripheral areas of the cases. There is some spreading of the substance on case 2, however, mostly on its sides, probably due to a past attempt at cleaning. The substance has two distinct morphologies: one is powdery (Fig. 2a), and the other is filamentous with a certain flexibility (Fig. 2b). The origin of this material could be inherent to the deterioration of the

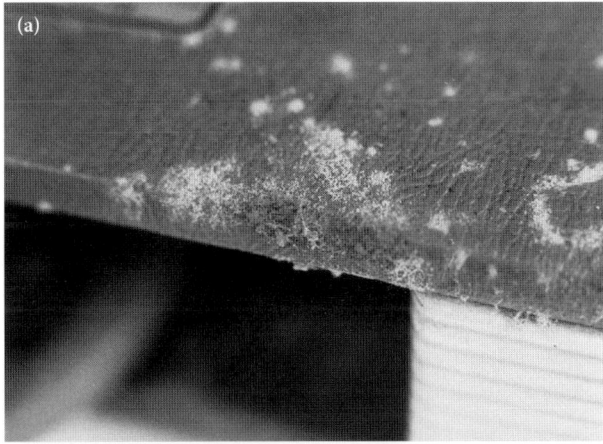

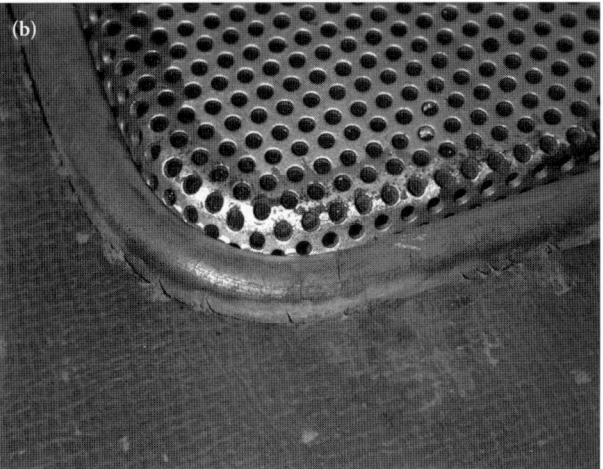

Figure 2 Morphology of the white substances appearing on the synthetic leather surfaces: (a) powdery appearance on case 1 and (b) filamentous appearance on case 2 (Plate 37 in the colour plate section).

plastic, such as a polymer decomposition product, deposition of acidic or salty products of cellulose nitrate, migration of some additive, or even caused by an extrinsic factor from some inorganic or biological agent.

In addition to this deposit, other signs of deterioration are apparent: some yellowing is visible on exterior surfaces and there is browning on top of the cases. The plastic displays gaps, little cracks, rips, scratches and abrasions. Another sign of the ageing of the material was the distinctive odour, concentrated in the cupboard where the machine was provisionally stored. This odour may result from the loss of nitrogen dioxide (NO_2) from the cellulose nitrate or the evaporation of some additive to the plastic. The FTIR spectra of the cellulose nitrate reveal an absorption band at 1,720–40 cm^{-1} (Fig. 3), which may be due to the increasing concentration of carbonyl groups (C=O), a result of the formation of nitrogen oxides (Shashoua et al. 1992). This band may also be attributed to some unidentified additive (Shearer and Doyal 1991) or to camphor (Derrick et al. 1993), used as a plasticiser of cellulose nitrate. The gases, if they are indeed nitrogen oxides, would be likely to endanger other objects, since they have a highly corrosive effect on a variety of materials.

Examination: methods and materials

Samples of the white powdery substance on the surface of the leather were taken for examination. These samples were assessed by three different techniques: microscopic observation, culturing and FTIR.

Microscopy

The samples were examined by an Olympus microscope using transmitted light. Specimens were placed on glass slides, stained with fuchsine dye (prepared as 0.1 g of fuchsine and 100 ml of lactic acid) and covered with cover slips; this dye improves the contrast of the sample and enhances the visibility of fungal structures.

Culturing

Two samples were collected with a scalpel and placed in Petri dishes prepared with MYP substrate (malt extract (0.7%), yeast extract (0.05%), peptone (0.25%), agar (1.5%) and water (97.5%)). These were then incubated at 23±2 °C, with indirect sunlight.

FTIR microspectroscopy

One sample of the particulate material was chosen for FTIR analysis in order to characterise its nature. This sample was compressed within a diamond cell and two transmission spectra were recorded, using a Nicolet Nexus FTIR spectrometer, employing a Nicolet Continuµm microscope with a Reflachromat 15× objective lens and a liquid nitrogen-cooled MTC detector. Each spectrum was recorded over the range 4,000–650 cm^{-1} at a resolution of 4 cm^{-1}, averaged over 128 scans.

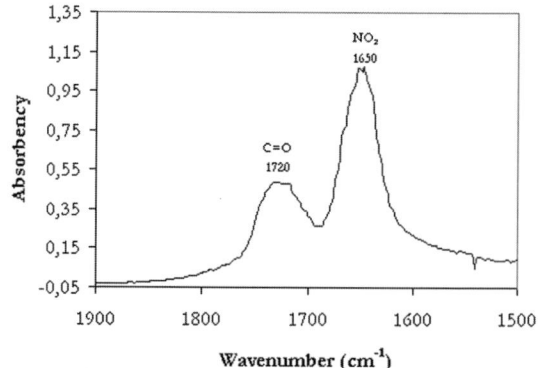

Figure 3 Cellulose nitrate FTIR spectrum showing absorbing band at 1,720 cm^{-1} from C=O vibration, possibly due to polymer degradation or some plastic additive related to the 1,650 cm^{-1} band, resulting from NO_2 stretching.

Results and discussion

As has already been discussed, the origin of the particles could be intrinsic or extrinsic to the nature of the plastic itself. The macroscopic appearance of the particles, however, indicates that the substance in question is fungal, as the circular nature of the spots is characteristic of fungal colonisation.

Microscope observations

The powdery and filamentous substances were identified as various fungal structures including hyphae, conidiophore, conidia and spores. This observation permitted the identification of the genus as *Aspergillus* (Fig. 4). It was impossible to identify the species because of the relatively poor and dry condition of the fungal structures.

Culturing

Two samples were taken for culturing in the hope that this would permit the identification of the fungal species present on the synthetic leather. At the end of the incubation period, however, no growth was observed. The MYP substrate is considered to be very versatile as it permits the growth of many types of micro-organisms. *Aspergillus* is one of the most common fungi; they appear in many different substrates and can survive under various conditions. The fact that no samples registered growth after a period of four weeks may be due to the poor condition of the fungi. This was confirmed by microscopic examination, which suggested that the fungi are probably inactive or even dead.

Table 1 Absorption bands registered on fungi FTIR spectra.

Functional group	Wavenumber (cm^{-1})
Protein amide II – N-H	3,300
Lipid = CH_2	3,100–3,000
Lipid – CH_2, -CH_3	3,000–2,850
Protein amide I – C=O	1,660
Protein amide II – N-H	1,540
Lipids, proteins CH_2	1,460
Phospholipids P=P	1,250
Carbohydrates C-O-C, C-O	1,200–1,000

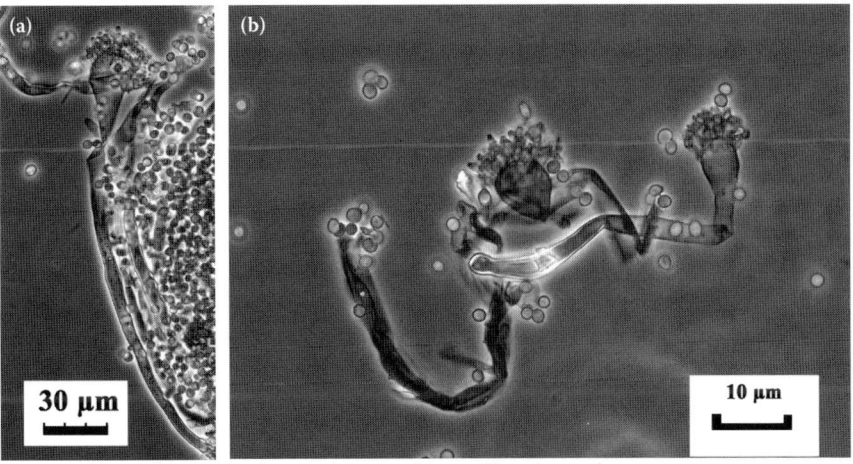

Figure 4 (a) and (b) Conidiophore and spores of *Aspergillus*, observed on a microscope with a magnification of 40× (Plate 38 in the colour plate section).

FTIR analyses

The spectra of the fungi reveal basic components of biological cells such as proteins, carbohydrates, lipids, water, etc. (Fig. 5). The main characteristic bands of fungi are listed in Table 1 (Yu and Irudayaraj 2005).

Why did fungi develop in this plastic?

Fungi are everywhere – indoors and outdoors. They are present in the air and can be found mixed with dust on the surfaces of objects, where they may then grow. The genera of fungi most commonly found are *Cladosporium*, *Penicillium* and *Aspergillus* (Florian 2002). Conidia and spores (~1–100 μm) are transported in the air and may settle on exposed surfaces. Fungi may then develop in humidity levels of more than 65% RH, and the surface can provide suitable nutrients. In favourable conditions, depending on the level of colonisation, fungal growth may be observed in 24 hours. *Aspergillus* is considered to be 'the common surface fungi found on artefacts and archival materials' (Florian 2002). Colonisation begins with conidia that germinate on the surface of a material and develop hyphae, initiating vegetative growth. A mycelium (a group of hyphae) produces asexual conidia, which are then transported by air to continue their spread. Cellulose nitrate is a potential source of all of the nutritional components required by fungi: they can derive nitrogen from the nitrate and carbon from cellulose, both of which are essential elements for their vital functions.

Until 2002, the projector cases were exposed to unknown environmental conditions. After that, the projector was incorporated into the museum collection and placed in storage. Currently the storage conditions vary slightly with outdoor climate: the RH is 51±5% and the temperature, 17.6±0.8 °C.[3] Dehumidification is the only active climate control within the storage area. At some time, the projector was located in a place contaminated by fungus spores or conidia. High humidity levels, in addition to other favourable environmental conditions, then promoted their growth.

Moisture is essential for fungal development on any kind of substrate, but also promotes acidity on cellulose nitrate surfaces. In the presence of moisture, nitrous oxides released by chemical degradation may form acidic products, such as nitric acid (HNO_3). Acidity improves fungal growth on a substrate. The surfaces of the suitcases are acidic (pH 4.5±0.5), presenting an optimum pH range for fungal development (Florian 2002). Considering that acidity is a result of the release of nitrogen oxides, this might indicate that the absorption band at 1,720–40 cm^{-1}, observed in the cellulose nitrate FTIR spectra, is due to an increase in the presence of carbonyl groups.

Fungi attack at a molecular level, using enzymes that are secreted into the material and break bonds between the macromolecules. In the case of cellulose nitrate, enzymes act by hydrolysis, breaking nitrate end-groups (exoenzymes) and cellulose inner bonds (endoenzymes). Although cellulose nitrate may be a nutrient source for fungi, their development could also have been promoted by the presence of some non-identified additive. Plasticisers, for instance, are generally 'low-molecular-weight organic liquids or solids' (Shashoua *et al.* 1992), which may also have nutrient value for fungi. Alternatively, it may be that some sort of protective coating over the cellulose nitrate was targeted by the fungi. There is no noticeable coating, however, and this was not a common procedure in the manufacture of this type of synthetic leather; nor was the presence of an additional material detected using any of the analytical techniques already discussed.

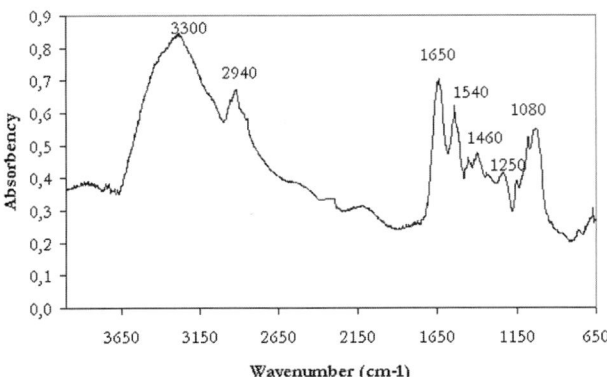

Figure 5 Fungal FTIR spectrum.

The fungal attack on this synthetic leather is clear and preferential, since it did not spread to the painting or wood with which it is in contact. After removing these micro-organisms, however, there is not always visible loss material from the surfaces in question. This may be due to a cessation of the fungal attack as a result of changes in environmental conditions, such as removal to a dry location, which could explain the inactivity of the micro-organisms and their poor condition.

Conclusion

Although it can be concluded that fungal attack does occur on the synthetic leather surfaces, for the time being it is not possible to identify whether this attack degrades the cellulose nitrate itself or, instead, targets some additive to the plastic. This deterioration, associated with the nature of the material, was almost certainly promoted by a combination of agents, including high humidity levels and the location of the objects in a contaminated environment.

As yet, it has not been possible to determine exactly how deeply the fungal attack has affected the plastic or how it has contributed to the chemical and physical deterioration observed. There are some places where the micro-organisms can be seen to have attacked exposed areas of the underlying textile. This may have been due to the loss of plastic caused by some physical damage in those areas, such as abrasion, rather than the complete decomposition of the cellulose nitrate by the fungus (Fig. 2a).

Therefore, this work cannot be conclusive. There are still many questions to answer and much work to be done. This work represents an initial study, which provides the basis for further work investigating the identity of fungi involved in attack on cellulose nitrate and the nature of the degradation that occurs as a result. Further research that would aid in this understanding may include the controlled culturing of fungi on polished cellulose nitrate surfaces, followed by microscopy and other analytical techniques to study the resulting damage.

Acknowledgements

The author would like to thank Professors Ana Ramos, Maria João Melo, José Paulo Sampaio and Miguel Lourenço for access to experimental work and scientific support.

Notes

1. List of vintage movie cameras, projectors, precinema, etc. www.xs4all.nl/~wichm/cinelisc.html.
2. Letter dated 8 June 1965 discussing the projector's characteristics for acquisition. This letter is kept by the Museum's Management Collections Department, with the projector's instruction guide.
3. These data were monitored (November 2004–May 2005) by a Tinytag Ultra data logger, produced by Gemini Data Loggers (UK) Ltd. This period registered low humidity levels in Porto.

References

Auer, N., Hedger, J. and Evans, C. (2005) 'Degradation of nitrocellulose by fungi', *Biodegradation* 16(3): 229–36.

Derrick, M., Stulik, D. and Ordonez, E. (1993) 'Deterioration of cellulose nitrate sculptures', in *Saving the Twentieth Century: The Conservation of Modern Materials*, D.W. Grattan (ed.), 169–82. Ottawa: Canadian Conservation Institute.

Florian, M.-L.E. (2002) *Fungal Facts: Solving Fungal Problems in Heritage Collections*. London: Archetype Publications.

Green, L. and Bradley, S. (1988) 'An investigation into the deterioration and stabilization of nitrocellulose in museum collections', in *Modern Organic Materials*, 81–95. Edinburgh: SSCR Publications.

Kim, B.J., Alleman, J.E. and Quivey, D.M. (1998) *Alkaline Hydrolysis/Biodegradation of Nitrocellulose Fines*. Technical Report, no. 98/65. Champaign, IL: Construction Engineering Research Laboratory.

Rizvi, A. and Zeto, C. (nd) *Proposal for In-Situ Sediment Bioremediation of a Complex Mixture of Nitroglycerin, Nitrocellulose, Ammonium Nitrate and Fuel Oil* (http://wvlc.uwaterloo.ca/biology447/modules/intro/assign2/447.htm).

Selwitz, C. (1988) *Cellulose Nitrate in Conservation*. Research in Conservation, Vol. 2. Los Angeles, CA: Getty Conservation Institute.

Sharma, A., Sundaram, S. T., Zhang, Y.-Z. *et al.* 1995. 'Nitrocellulose degradation by a coculture of *Sclerotium rolfsii* and *Fusarium solani*', *Journal of Industrial Microbiology and Biotechnology* 15(1): 1–4.

Shashoua, Y., Bradley, S.M. and Daniels, V.D. (1992) 'Degradation of cellulose nitrate adhesive', *Studies in Conservation* 37: 113–19.

Shearer, G.L. and Doyal, S. (1991) 'Use of FTIR in the conservation of twentieth century objects', *Materials Research Society Symposium Procedures* 185: 813–23.

Williams, R.S. (1994) *Display and Storage of Museum Objects Containing Cellulose Nitrate*. CCI Notes 15/3. Ottawa: Canadian Conservation Institute.

Yu, C. and Irudayaraj J. (2005) 'Spectroscopic characterization of microorganisms by Fourier transform infrared microspectroscopy', *Biopolymers* 77: 368–77.

The author

Margarida Silva is currently completing a graduate course in conservation and restoration at the Universidade Nova de Lisboa. Since 2001, Margarida has been developing the Conservation Department of the Museu do Carro Eléctrico in Porto.

Address

Margarida Silva, Museu do Carro Eléctrico, Alameda Basílio Teles, 51-4150-127 Porto, Portugal (silva.margarida@gmail.com).

Polyurethane foam: investigating the physical and chemical consequences of degradation

Paul Garside and Doon Lovett

ABSTRACT Polyurethane foams have been widely used in the fashion, upholstery and motor industries since the 1950s and therefore artefacts containing these materials are found in a broad range of museum collections. The foams are particularly susceptible to degradation, compromising their physical integrity and this may lead to the loss of form or meaning in these objects. A series of experiments was carried out to assess the differences in susceptibility to degradation between the two major types of polyurethane foam, namely polyester polyurethane (PU-PES) and polyether polyurethane (PU-PEt). The principal mechanisms of deterioration are believed to be oxidation and hydrolysis, so to investigate these modes, samples were placed in sealed environments in which the humidity and oxygen content were controlled (0, 50 or 80% relative humidity and ambient or nitrogen atmospheres, respectively); these samples were then subjected to accelerated ageing via either elevated temperature or irradiation with simulated sunlight.

Following ageing, the properties of the foams were assessed by mechanical (compression recovery) testing, microscopy and spectroscopy. The results of these analyses allowed the nature of the foams, and their particular susceptibilities, to be more fully understood and methods of non-destructive characterisation to be developed for testing objects *in situ*. This knowledge will assist conservators and curators in identifying these materials in collections and in providing optimal display and storage conditions.

Keywords: polyurethane foam, accelerated ageing, Fourier transform infrared (FTIR) spectroscopy, near-infrared (NIR) spectroscopy, mechanical testing, *in-situ* analysis

Introduction

Since their advent in the mid-1850s, synthetic polymers have become pervasive, replacing traditional materials and appearing as mouldable forms, fibres, adhesives, paints and foams. Whether found in everyday objects or works of art, they are being collected and appear in a vast range of museum collections including costume, social history and transport. Ironically, the 'everlasting' image of some of these plastics is not borne out by experience, and compared with naturally occurring materials and fibres, they have relatively short lifespans.

The deterioration of plastic artefacts was first reported in the literature in 1988 and prompted self-examination within the heritage sector (Pullen and Heumann 1988). A survey at the Victoria and Albert Museum (V&A) in London identified five general classes of plastics that were both in poor condition themselves and also a potential danger to other objects. The five classes were cellulose nitrate, cellulose acetate, polyvinyl chloride (PVC), rubber and polyurethane; of these, polyurethane was declared the 'most serious conservation problem' (Keneghan 1999: 357).

'Polyurethane', as a class, covers a wide range of materials. Polyurethanes were first developed in Germany in 1937, but were not commercially employed until the early 1950s. The class is characterised by the presence of the urea group (-NHCO-), formed between an isocyanate molecule and a hydrogen donor, most commonly either an ester or an ether species; the polymerisation reaction produces a polyester polyurethane (PU-PES) or a polyether polyurethane (PU-PEt), respectively.[1] By altering the choices of these initial components it was possible to produce polyurethanes with a wide variety of physical and mechanical properties, and therefore their suitability to different end functions. One author lists 294 different applications (Uhlig 1999).

It was not until 1957 that polyurethanes were produced as a foam, and it is in this form that the materials are considered in this paper. To produce any foamed polymer, a gas must be trapped during the polymerisation reaction: for polyurethane this is done by carrying out the reaction with an excess of isocyanate and water, which leads to the formation of carbon dioxide (Lovett 2003). This new form of the polymer rapidly revolutionised several industries, most notably the upholstery and motor industries, due to its cheapness and versatility. The possibilities of this new material were rapidly embraced by designers and artists alike, and the 1960s and 1970s saw an explosion in the use of foams to create both artwork and everyday items. For example, the Italian firm Gruppo produced a range of quirky furniture such as chairs designed to look like stones and grass. The American sculptor John Chamberlain, perhaps better known for his works utilising scrap metal, made a range of pieces using polyurethane foam between 1966 and 1972, and Gilardo created fantastic naturalist carpets. In addition to its more obvious use in objects, polyurethane has also been used as a significant 'hidden' element in many items, functioning as backings, paddings and interlinings.

The survey at the V&A (noted above) also highlighted two additional factors relating to the presence of these foams in museum collections: first, how difficult it was to trace objects containing synthetic elements, as the use of these materials was rarely recorded on documentation; secondly, that plastics appeared in all of the collections assessed – from transport to jewellery and science to sculpture. Furthermore, the cellular 'foamed' structure that gave these materials the physical properties that made them so popular also led to their vulnerability, due to the very large surface area compared to the actual mass of polymer.

The main pathways of degradation for all forms of polyurethane are a combination of oxidation and hydrolysis; it is now accepted that oxidation predominates for polyether polyurethanes, whereas hydrolysis is the principal cause of deterioration for the polyester variety (Horie 1987; Williams 1998). Exposure to conditions likely to promote degradation leads to reactions at susceptible bonds within the polymer, such as urethane, ester, ether and amide groups, causing chain scission and cross-linking. This leads to physical changes such as discoloration and the loss of mechanical properties: strength, toughness and resilience decrease, factors which become manifest initially as an impaired ability to recover from deformation, with consequent changes in shape and loss of structure, and ultimately by the disintegration of the material.

An initial study, preliminary to this work, was carried out on four foam-laminated dresses belonging to the Museum of London, dating from the 1960s (Lovett 2003; Lovett and Eastop 2004); the focus of this work was the unusual foam interlining, which was investigated to determine its identity, condition and the factors involved in its deterioration. Spectroscopic analysis revealed the foams to be polyester polyurethanes in all four cases; Oddy tests confirmed that the foam was releasing or off-gassing volatile organic acids as it degraded, gases that can potentially damage sensitive neighbouring materials and act in an autocatalytic role in the deterioration of the foam itself. The experimental phase of the project involved the accelerated ageing of samples of new polyester polyurethane under different relative humidities. These experiments suggested that a low RH is beneficial for storage as high humidity accelerates degradation causing discoloration and a loss of strength and elasticity. This research has been taken as a pilot study on which the work presented here has developed and expanded.

The authors were particularly interested in developing methods to identify initially the presence of these foams within objects, ideally by *in-situ*, non-sampling, non-invasive methods, and subsequently to characterise the chemical and physical state of these materials, using similarly non-interventive techniques. In order to achieve this, samples of both PU-PEs and PU-PEt foams were subjected to accelerated ageing, either by means of elevated temperatures or by intense illumination, and under a range of humidities and levels of oxygenation. This provided a range of materials exhibiting different degrees and types of deterioration. The physical properties of these foams were then assessed using mechanical testing and recorded both near- and mid-infrared (IR) spectra. The mid-IR technique is more sensitive to subtle chemical differences and the resultant spectra are more amenable to interpretation, but the near-infrared (NIR) approach has the advantage that it can be used with a fibre optic probe, and is thus better suited to *in-situ* analyses of objects that may be too fragile or bulky to be manoeuvred to the sampling window of the spectrometer itself. The spectra were analyzed both to identify characteristic differences between the two types of foam, and to find appropriate spectral signatures that reflected the physical state of the materials.

Finally, spectroscopic analyses were carried out *in situ* on clothing and other artefacts held in the Hampshire County Council Museum Services collections. This enabled the authors to judge the use of the technique in identifying the presence of these foams within items, particularly where the materials are found as paddings or interlinings, and thus are concealed beneath additional layers of fabric.

Method

Accelerated ageing

Samples of PU-PEt and PU-PEs were aged using either elevated temperature or intense illumination; under each of these two ageing regimes, specimens were placed under atmospheric or anoxic atmospheres, or in environments of high, medium of low humidity, as described below. For the thermal ageing experiments, samples were sealed in hybridisation vials (heavy, screwtop glass vials with Teflon seals, of suitable size to contain the foam specimens); for the light ageing, sealable transparent polyethylene bags were used.

The samples in anoxic environments were prepared in a glove bag, an inflatable polyethylene chamber with integral gloves and ports that can be connected to vacuum or gas lines. The samples were placed in the bag, which was then purged and repeatedly flushed with nitrogen (generated from liquid nitrogen), before sealing the specimens in the appropriate containers. The humidity was controlled at roughly 0, 50 or 80% RH using vials of molecular sieve type 3A (to maintain an atmosphere of approximately 0% RH) and saturated salt solutions (cobalt chloride, 52±4% RH at 80 °C, and potassium chloride, 79±1% RH at 80 °C). These samples were then placed in an oven maintained at 80 °C, and removed after 35, 49 and 63 days, or in a light-ageing chamber (using ten General Electric F20W/AD simulated daylight fluorescent tubes) and removed after 14, 21 and 35 days. After removal from the ageing environments, the samples were stored at 21±1 °C and 51±2% RH.

Mechanical testing

To determine the physical properties of the foams, the aged and unaged samples were subjected to mechanical testing using an Instron 5544 machine fitted with a 100 N load cell, controlled by Instron Bluehill v1.4 software. The system was programmed to compress the sample at a rate of 1 mm/min^{-1}, to half its initial thickness; the sample was then maintained at this degree of compression for 5 min; finally the compression was relaxed, again at 1 mm/min^{-1}. The testing was carried out under ambient conditions (21 °C, 51% RH). Three replicate tests were carried out for each sample.

Spectroscopy

The samples were then subjected to both mid- and NIR spectroscopy. Mid-infrared Fourier transform infrared (FTIR) spectra were recorded using the attenuated total reflectance (ATR) technique, with a PerkinElmer Spectrum One spectrometer, fitted with a PerkinElmer Universal ATR accessory; spectra were captured over the range 4,000–400 cm^{-1}, using 32 scans and with a resolution of 4 cm^{-1}. NIR spectra were recorded using the Spectrum One spectrometer fitted with a PerkinElmer NIRA accessory; spectra were captured over the range 7,800–3,900 cm^{-1}, using 16 scans and a resolution of 8 cm^{-1}. Spectra were subsequently manipulated with Thermo Galactic GRAMS 32/6 software.

In-situ *analyses*

The *in-situ* NIR analyses were carried out on foam padding from two garments from the Hampshire County Council Museum Services, namely the interlining of a tweed dress and a foam brassiere pad. The spectra were recorded using the Spectrum One spectrometer fitted with a PerkinElmer fibre optic interface and NTS fibre optic probe; spectra were captured over the range 7,800–4,000 cm^{-1}, using 16 scans and a resolution of 8 cm^{-1}. Spectra were recorded of both the underlying foam through the outer layer of fabric and of the outer fabric on its own.

Results and discussion

Accelerated ageing

As the samples are subjected to ageing, they undergo a physical deterioration and a discoloration; the rate of these reactions is dependent on both the nature of the foam (PU-PES or PU-PET) and the ageing condition employed (Fig. 1 a and b); it can be seen that the more extreme conditions lead to the catastrophic deterioration of the material, causing the structure of the foam to break down and disintegrate. The yellowing that accompanies the degradation is not an inherent property of the foam, but rather is due to the reaction of excess isocyanate species within the material.[2]

Mechanical testing

A typical load/extension curve from the mechanical testing is shown in Figure 2. From these data, it is possible to derive a range of potentially useful information, including the maximum force required to compress the foam, the modulus during compression, the drop in applied force while the foam is

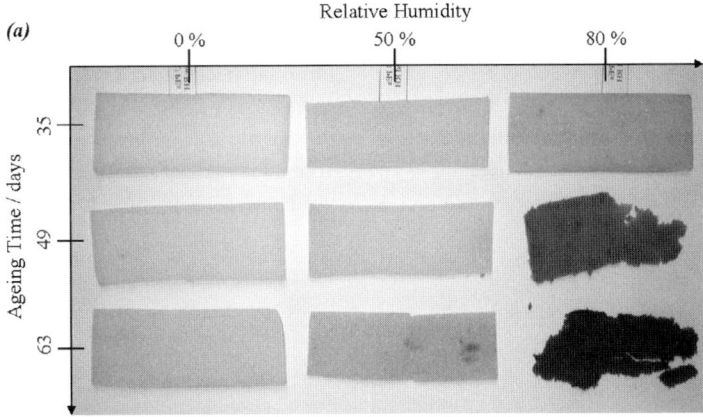

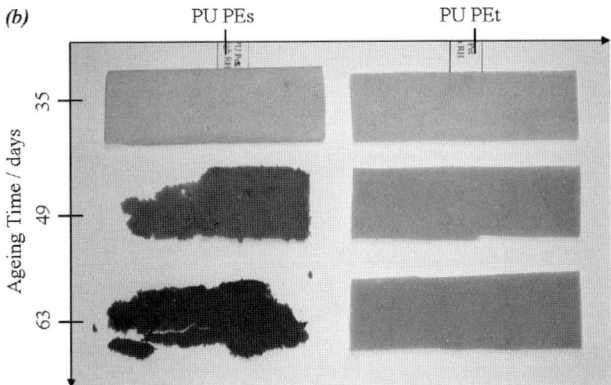

Figure 1 (a) Thermally aged PU-PES foam, under different humidities and (b) PU-PES and PU-PET foam thermally aged under 80% RH.

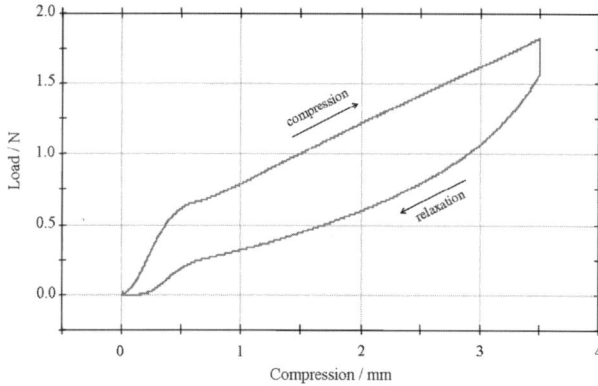

Figure 2 A typical load/extension curve for the compression and relaxation of a polyurethane foam sample.

held under constant compression, and the degree to which the foam returns to its original dimensions after relaxation, among others. It was found that all of these values changed with ageing; for the purposes of this research, compression recovery was chosen as a representative method of monitoring physical deterioration (Table 1).

From these data it can be seen that PU-PEs foams appear to degrade most rapidly under conditions that promote hydrolysis (that is, high temperature, high humidity atmospheres); although the literature suggests that PU-PEt foams are more susceptible to oxidative degradation, this was not observed in these experiments, probably due to the limited period of ageing. It is also apparent that for the ageing conditions and times employed, the PU-PEs foam samples have deteriorated much more significantly than the PU-PEt samples.

Spectroscopy

The initial examination of the FTIR and NIR spectra of the unaged foams shows that both of these techniques allow an adequate characterisation and differentiation of the two types to be made (Fig. 3). A further analysis of the spectra, including those of the artificially aged materials, can determine whether spectral signatures can be derived, corresponding to the physical changes characterised above. If the FTIR spectra of the PU-PEs foams are considered, it can be seen that a variety of changes can be observed, including the appearance of a broad, weak peak in the region 3,700–2,300 cm^{-1}, other new peaks at 1,410, 1,275, 920, 896, 680 and 510 cm^{-1}, along with more subtle changes in peak shape and position at approximately 1,710 cm^{-1} (ν(C=O)) and 1,065 cm^{-1} (ν(C-O-C)) (Kischel et al. 1998). From these observations, it is possible to derive a degradation parameter, X, calculated as the ratio of the intensities of the bands at 1,065 and 1,125 cm^{-1}, measured above a baseline drawn from 1,340 to 890 cm^{-1} (Fig. 4a):

$$X = I_{1065} / I_{1125}$$

If this value is plotted against one of the measures of physical deterioration, such as compression recovery, it can be seen that there is a good correlation between the two (Fig. 5), thus providing a potential method of following the ageing of PU-PEs materials. The NIR spectra of these foams reveal less obvious information, but nonetheless changes are observed, particularly in the growth of a new peak at approximately 5,190 cm^{-1}, which progressively obscures the original features observed at 5,235 and 5,150 cm^{-1} as the material ages, and which also appears to correlate with the loss in mechanical integrity (Fig. 4b).

Unfortunately the FTIR and NIR spectra of the PU-PEt materials are less amenable to interpretation in this manner; a simple analysis of the data revealed no obvious spectral changes that might readily be used as a degradation marker. It may be that this is simply due to the lesser degree of physical deterioration observed in the PU-PEt foam when compared to the PU-PEs, and that with more fully aged samples an appropriate spectral ageing signature would become apparent; this is an avenue for further research.

Table 1 Compression recovery for aged PU-PEs and PU-PEt foams.

		Compression recovery/%				
		Thermal ageing			Light ageing	
Conditions	Time/days	PU-PEs	PU-PEt	Time/days	PU-PEs	PU-PEt
Unaged	0	95.3	94.0	0	95.3	94.0
Atmospheric air	35	96.2	95.7	14	93.3	96.9
	49	96.0	95.3	21	92.8	94.7
	63	93.8	95.3	35	89.3	92.1
N$_2$ atmosphere	35	94.7	96.7	14	93.6	96.7
	49	95.6	96.3	21	94.3	96.6
	63	96.0	95.3	35	91.3	96.1
0% RH	35	93.3	96.4	14	95.8	96.3
	49	93.6	95.6	21	94.2	94.1
	63	91.4	96.2	35	94.8	95.8
50% RH	35	94.0	96.4	14	93.6	96.1
	49	87.3	95.6	21	89.9	95.3
	63	83.3	96.2	35	85.3	95.7
80% RH	35	87.1	96.4	14	88.3	95.6
	49	64.3	96.1	21	84.4	93.4
	63	53.3	95.0	35	79.2	92.0

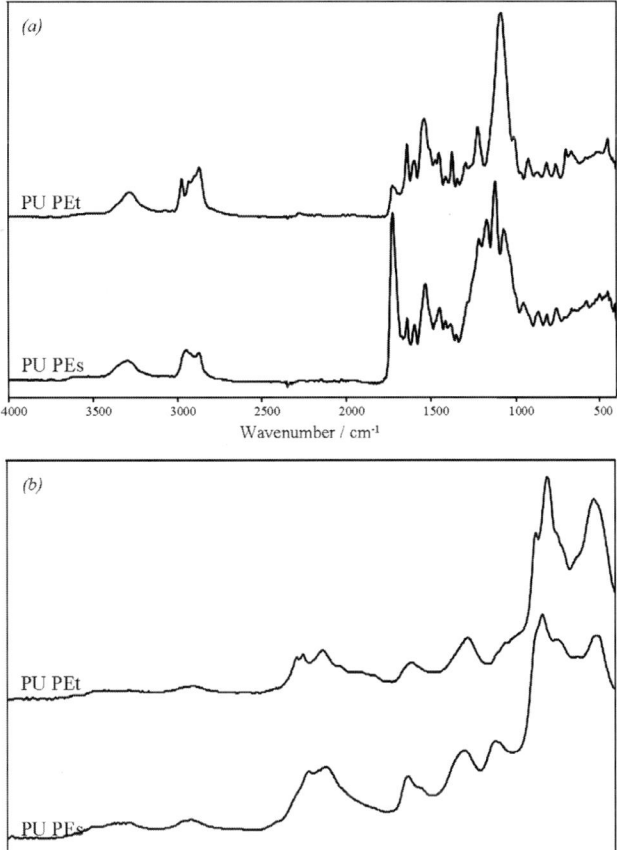

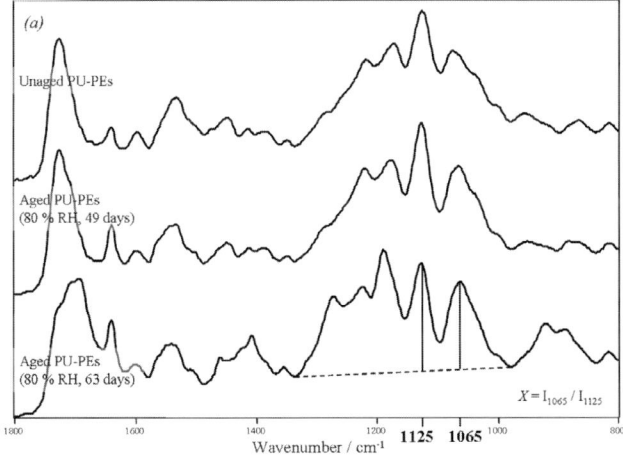

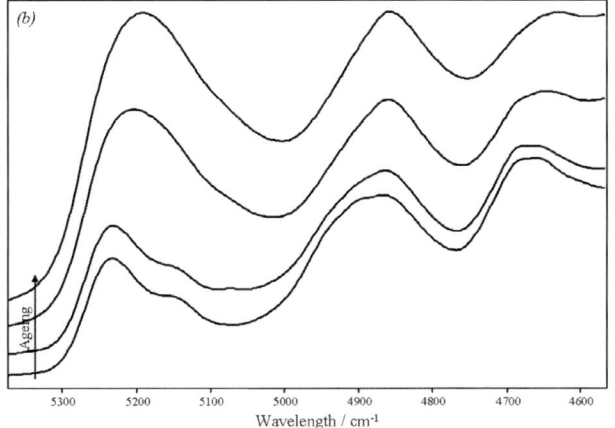

Figure 3 (a) FTIR and (b) NIR spectra of PU-PEt and PU-PEs foams.

Figure 4 Changes in the spectra of PU-PEs foam on ageing: (a) FTIR–ATR spectra, detailing the calculation of the degradation parameter, $X = I_{1065} / I_{1125}$ and (b) NIR spectra.

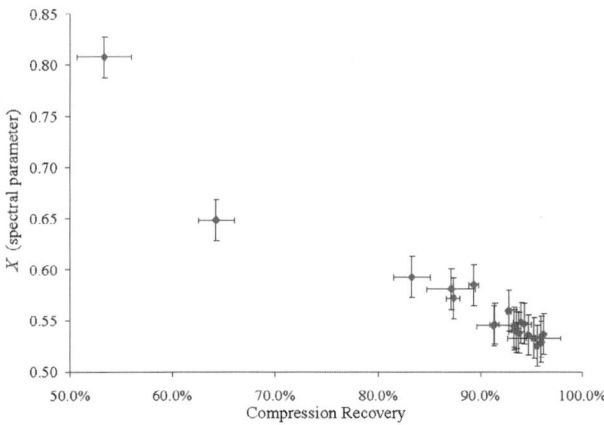

Figure 5 Correlation of compression recovery and the spectrally derived degradation parameter, X.

In-situ *analyses*

By employing spectral subtraction techniques on the spectra of the foam through fabric and of the fabric itself (see Garside and Wyeth in this volume, pp. 55–60), it was possible to accurately identify the nature of the underlying foam. This can be seen in the examples presented in Figure 6 (a and b), employing suitable reference materials, in which it can be seen that the foams are identified as PU-PEt and latex rubber, respectively.

Conclusion

The results of these experiments show that it is not only possible to accurately differentiate between different types of polyurethane foam using FTIR and NIR spectroscopy, but also, in the case of PU-PES foams, that it is potentially possible to follow the course of degradation using these techniques and to correlate spectrally derived signatures with the measured physical properties of these materials. This, therefore, potentially provides a non- or micro-sampling method of assessing the state of foams within objects and thus informing conservation display and storage conditions. Furthermore, it is possible to identify these materials within objects, even where the foam is obscured by layers of other fabrics, and these analyses can readily be carried out *in situ*.

The authors hope to extend and develop this work beyond the simple spectral analyses presented here in order to find appropriate spectral markers that would allow the deterioration of PU-PEt foams to be monitored in a similar way to the PU-PES materials; initial investigations suggest that the use of

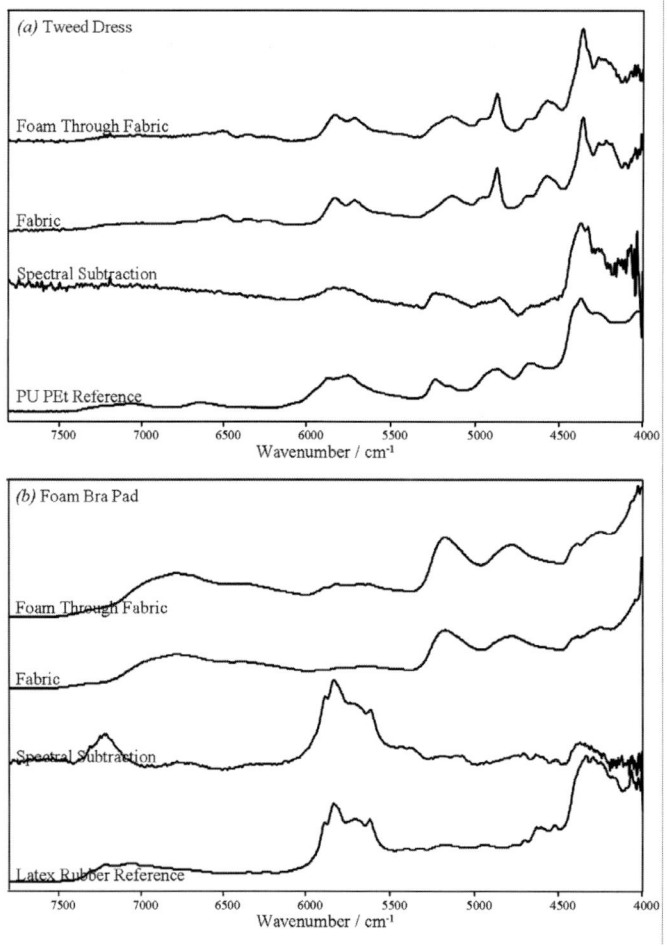

Figure 6 Results of *in-situ* NIR analyses of foam from (a) a tweed dress and (b) a bra pad, with spectral subtraction and appropriate reference material.

first or second derivatives of the spectra may provide a useful approach. This would hopefully allow a technique to be developed that would not only allow the materials to be identified by non-invasive, non-destructive methods, but also adequately characterised in terms of physical and chemical state. Ideally the technique would be able to be used *in situ*, which suggests that a NIR approach would be the most suitable.

Acknowledgements

The authors would like to acknowledge the support of the AHRC in funding the work via the Research Centre, Dr David King (Recticel Manufacturing) for providing the foam samples, the Museum of London and Sarah Howard (Hampshire County Council Museum Services) for access to collections. They would also like to thank colleagues at the Textile Conservation Centre, in particular Nell Hoare, Paul Wyeth, Maria Hayward and Dinah Eastop.

Notes

1. A variety of nomenclature exists to denote the different forms of polyurethane foams, and their use depends on both discipline and location; for the purposes of these experiments, polyurethane polyester is referred to as PU-PES, and polyurethane polyether as PU-PEt.
2. Personal communication with B. Keneghan, 2005.

References

Horie, C. (1987) *Materials for Conservation*. London: Butterworth.

Keneghan, B. (1999) 'Plastics research in the Victoria and Albert Museum', in *Modern Art: Who Cares?*, I. Hummelen and D. Sille (eds), 356–61. Amsterdam: Foundation for the Conservation of Modern Art and the Netherlands Institute for Cultural Heritage.

Kischel, M., Kisters, D., Strohe, G. and Veeman, W.S. (1998) 'Dynamic infrared spectroscopy: a tool to detect hydrogen bonds in polymers?' *European Polymer Journal* 34(11): 1571–7.

Lovett, D. (2003) *The Deterioration of Polyurethane Foam, with Reference to Foam-Laminated 1960s Dresses*. MA dissertation, Textile Conservation Centre, Winchester.

Lovett, D. and Eastop, D. (2004) 'The degradation of polyester polyurethane: preliminary study of 1960s foam laminated dresses', in *Modern Art, New Museums: Contributions to the IIC Bilbao Congress*, A. Roy and P. Smith (eds), 100–104. London: International Institute for Conservation of Historic and Artistic Works.

Pullen, D. and Heuman, J. (1988) 'Cellulose acetate degradation in the sculptures of Naum Gabo', in SCCR *Modern Organic Materials Meeting Preprints*, 57–66. Edinburgh: SCCR Publications.

Uhlig, K. (1999) *Discovering Polyurethanes*. Munich: Hanser.

Williams, S. (1998) 'Plastics in conservation materials and museum objects', in *The Costume Institute Colloquium Program: Care and Preservation of Modern Materials in Costume Collections*. New York: Costume Institute and the Metropolitan Museum of Art.

The authors

- Paul Garside is a postdoctoral Research Fellow at the AHRC Research Centre for Textile Conservation and Textile Studies. He has developed significant expertise in applying analytical methodology to conservation science problems.
- Doon Lovett works for Conservation Services at the Textile Conservation Centre as a general conservator, maintaining her interest in modern material with AHRC research conservator funding. The work in this paper follows on from her MATC dissertation on foam laminated dresses from the 1960s.

Addresses

- Paul Garside, AHRC Research Centre for Textile Conservation and Textile Studies, Textile Conservation Centre, University of Southampton, Park Avenue, Winchester, Hants SO23 8DL, UK (pg1@soton.ac.uk).
- Doon Lovett, Textile Conservation Centre, University of Southampton, Park Avenue, Winchester, Hants SO23 8DL, UK (dll1@soton.ac.uk).

Sticky oilskins and stiffened rubber: new challenges for textile conservation

Irene Skals and Yvonne R. Shashoua

ABSTRACT Twenty-five examples of military protective clothing in the collection of the Royal Danish Arsenal Museum, Copenhagen, dating from 1880 to 1980, exhibited extensive deterioration. Many of the uniforms had never been worn as they were samples to demonstrate the newest styles and materials of their period. The purpose of the project was to identify the materials present, causes of deterioration, and to develop techniques to prolong the longevity of this historically and technologically important collection. Fourier transform infrared–attenuated total reflectance (FTIR–ATR) spectroscopy was used to identify the waterproofing agents, which included bitumen, oils, rubber, polyvinyl butyral (PVB), polyvinyl chloride (PVC) and nylon. The production techniques of these waterproof garments included surface treatment of an underlying textile, lamination of a waterproofing agent between two layers of textile and construction from polymer sheets. The requirements for improved storage of the clothes varied according to the material. Oil- and bitumen-treated uniforms proved to be the most challenging because materials traditionally used in conservation, such as acid-free tissue paper, adhered to their tacky surfaces. Accelerated ageing was used to identify chemically and physically stable materials, which would be suitable as long-term packing, barriers and support. Rainwear containing rubber was stored in oxygen-free microclimates and polyethylene was used for supports and dust covers.

Keywords: uniform, waterproof, historical development, Fourier transform infrared–attenuated total reflectance (FTIR–ATR) spectroscopy, preventive conservation

Introduction

The Royal Danish Arsenal Museum in Copenhagen has a large collection pertaining to the Danish history of defence and warfare, including uniforms, dating from *c.*1600 to the present. The uniforms are stored in the attic above three floors of exhibition halls in an old building with no climate control. The Museum has an agreement with the National Museum of Denmark to provide conservation services and as part of this agreement, textile conservators inspect the uniform collection annually for insects.

Entering the large attic rooms and seeing endless rows of cupboards where the uniforms are stored is overwhelming. It is hard physical work as the uniforms were designed to protect the wearer and were often made of heavy wool. The cupboards are high and the space tight, so the uniforms hang closely. Many of the uniforms were samples approved by and marked with the date and wax seal of the respective ministries. They were made of the best quality materials of their respective periods and have never been worn. While doing this work, the critical condition of many of the examples of rainwear was noticed. Some of the garments were extremely stiff; other examples were sticky and had adhered to the walls of the cupboards and to neighbouring objects. Many raincoats were folded and deformed due to inadequate support and cramped storage on coat hangers (Fig. 1).

The aim of the study was to improve the storage of the items, thereby slowing their degradation. Twenty-five examples of

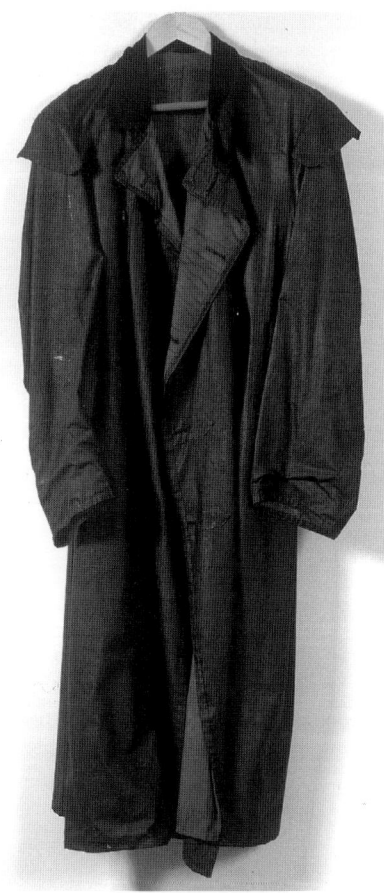

Figure 1 Sample uniform (1918): surface treated with bitumen. Stiffened, deformed and tacky (Plate 39 in the colour plate section).

rainwear representing the range of styles, materials and historical periods in the Royal Danish Arsenal Museum's collection were selected for the investigation. An integrated approach was employed involving the analysis of the waterproofing agents and their deterioration, and the evaluation of support and packing materials. The methods of waterproofing and the cut and design of the uniforms were also documented.

The history of rainwear

The history of modern rainwear starts at the beginning of the 19th century and a connection can be drawn between the manufacture of these garments and the development of chemical, technological and industrial innovations. Before this time, Europeans are believed to have used wool and hides for protective clothing, which they waterproofed as needed with materials at hand (Houston 2002). We must imagine that sailors from the many seafaring European nations and the farmers and fishermen in the rainy climate of northern Europe found solutions in their immediate surroundings to avoid getting soaked, solutions which evolved and were improved by the industrial revolution. As none of the rainwear in the collection of the Royal Danish Arsenal Museum dates from before the 1850s, it does not help us in answering questions about the earlier history of this type of clothing, but it is nonetheless interesting to discuss why and how certain types of materials were used.

The use of bituminous materials for waterproofing buildings and boats has been known since biblical times. A solution of bitumen in oil was also used to coat the cloth for mummy wrappings in ancient Egypt (Brydson 1999). Bituminous materials are formed over time from deposited biological remains. Their composition may partly reflect the nature of the original material; in general, bitumens are complex mixtures of paraffinic, aromatic and cyclic hydrocarbons in liquid and solid form. They may be dissolved or thinned in chlorinated organic solvents or emulsified to produce water-based surface coatings, which can be applied to textiles. On drying, bitumen forms a thermoplastic, hydrophobic film. The cloth to which it is applied becomes stiffer and is probably slightly uncomfortable to wear.

'Oilskins', as the name indicates, were produced by improving the natural waterproofing ability of an animal skin by oiling it with, for example, tallow. It is likely that common sailors used canvas sailcloth in place of the more expensive skin, which they waterproofed with tar or a mixture of tar and oil; such materials were readily available on a ship (Houston 2002). From c.1900, industrial production of cellulosic textiles is documented, including cotton, linen and calico, which were then treated with drying oils prior to being used to construct oilskins or raincoats. Earlier, garments had been sewn first and multiple coats of drying oils applied afterwards (Allington 1987). Such oils frequently included linseed, obtained by pressing flax seeds, which is rich in triglycerides of fatty acids, primarily linolenic and linoleic acids (Mills and White 1994). It was necessary to heat these oils together with a catalyst such as lead, manganese or cobalt salts, to oxidise the double bonds in the structure or 'body' them, resulting in a highly dense, dry film. The resulting garments were often stiff but also highly weather-resistant.

Rubber became popular in the 19th century when methods of applying the material to textiles were improved (Jentzsch 1994). The Mexican Indians were first recorded as applying natural rubber latex to the surfaces of cellulosic textiles to waterproof them in 1615 (Loadman 1993). Macintosh further developed the process in 1823, when he discovered that if a fabric was coated with rubber paste and a second layer of fabric applied under pressure, the resulting sandwich was waterproof. The paste was usually made by masticating rubber, sulphur and mineral filling agents and converting the mass into sheets. The sheets were mixed with a solvent, naphtha, until a sticky paste was formed. Examples of both surface-applied latex and rubber-sandwich technology were found in the uniform collections of the Royal Danish Arsenal Museum. In 1935, nylon 66 was invented in the USA. From the start it was used as a textile fibre, replacing silk in stockings (1940) and woven or knitted to form a fabric (Arbeitsblätter 1993). It is a hydrophobic material and if produced as a sheet it is impermeable to air.

The synthetic resin polyvinyl butyral (PVB) was invented in the US in 1936 and is still widely used as a coating for metal and wood, as a waterproofing treatment and as an adhesive. Experiments to strengthen archaeological textiles with PVB were performed but quickly abandoned, due to the irreversibility of the process.

Because of the shortage of natural rubber during the Second World War, polyvinyl chloride (PVC) was used as a substitute in the US, Germany and the UK from 1942, and was applied to the surfaces of textiles in paste form with heat to produce leather cloth (Brydson 1999). PVC's great advantage was its low cost, ease of production and handling, high mechanical resistance and potential for pale and mixed colours.

Identification of waterproofing materials

In order to determine the degradation reactions and therefore the factors most likely to accelerate their rates, knowledge of the waterproofing materials present was essential. Non-destructive Fourier transform infrared–attenuated total reflectance (FTIR–ATR) spectroscopy was used to identify the organic waterproofing agents present on the surfaces of the uniforms, including natural materials such as bitumen, rubber, drying oils and synthetic materials such as PVB, PVC and nylon (Shashoua and Skals 2004). In addition to its ability to analyze the surfaces of materials non-destructively, a great advantage of FTIR-ATR spectroscopy over other infrared (IR) techniques is its ability to provide well-resolved spectra for dark and uneven surfaces. Spectra were collected over 30 scans, at a resolution of 4 cm^{-1}, over the range 4,000–600 cm^{-1} (the lower limit of sensitivity for ATR), using an ASI DurasamplIR single reflection accessory (with an angle of incidence of 45° and fitted with a diamond internal reflection element) in a PerkinElmer Spectrum 1000 FTIR spectrometer.

The high refractive index of diamond compared with that of organic materials (approximately 2.5 and 1.5 respectively) means that the absorbance data are collected from a depth approximately equal to that of the wavelength of the IR radiation (a maximum depth of a few microns); this technique allows analysis of the surface alone, since the IR beam does not penetrate more deeply into the supporting textile. The quality of spectra depends on intimate contact between the DurasamplIR reflection element (the diamond crystal that forms this element has an active area of 1 mm diameter) and the surface of the sample. Uniform pressure distribution was achieved for all samples using the flat, circular tip of the pressure device (3 mm in diameter) supplied with the accessory in combination with a torque limiter which allowed the press to be tightened to a consistent pressure each time it was used.

The uniforms were analyzed without taking samples. Corners and edges of interest were positioned over the

Table 1 Date, materials and methods of waterproofing military rainwear.

Date	Object	Waterproofing material	Method of waterproofing	Textile base
1883	Sample. Long coat	Rubber	Surface treatment	Heavy black cotton
1890	Sample. Long coat	Bitumen	Surface treatment	Heavy black cotton. Lined upper part
1890	Sample. Long coat	Linseed oil. Possibly dyed	Surface treatment	Heavy black cotton. Fully lined with black cotton
1910?	Long general's coat. From the wardrobe of King Frederik 8th (reigned 1906–1923)	Rubber	Sandwich	Shell – black wool tabby. Lining – black cotton/wool twill
1911	Sample. Long coat	Rubber	Surface treatment	Heavy black cotton. Lined at shoulders on back
1918	Sample. Long black coat	Bitumen	Surface treatment	Heavy black cotton. Fully lined with brown cotton
Undated	Cape	Bitumen	Surface treatment	Black cotton
1922	Sample. Long coat	Rubber	Surface treatment	Medium thickness black cotton
1945	Long jacket	Chinese tung oil	Surface treatment	Thick cotton. Khaki. Lined upper part
1945	Motorcycle dispatch rider's coat	Rubber	Sandwich	Heavy cotton twill. Khaki
1945	Motorcycle dispatch rider's trousers	Rubber	Sandwich	Heavy cotton twill. Khaki
Undated	Motorcycle dispatch rider's coat	Rubber	Sandwich	Heavy cotton twill. Khaki
1945–46	Long coat. English. Used by Danish army. Tailored with space for rucksack	Unidentified oil	Surface treatment	Very thin cotton. Greenish/brown
1946	Long coat. Tailored with space for rucksack	Unidentified oil	Surface treatment	Very thin cotton. Army green
Undated	Short jacket	Unidentified oil	Surface treatment	Shell – medium thickness cotton. Lining – wool. Dark khaki
Undated	Long coat	Fish oil	Surface treatment	Heavy cotton. Khaki
Undated	Asymmetric cape. American. With a secondary function as groundsheet	Polyvinyl butyral	Surface treatment	Heavy cotton. Khaki
Undated	Cape. With push buttons and rivet holes along edges	PVC	Surface treatment. Very thin. Talcum adhering to surface	Medium thickness green cotton. Shiny surface inside. Matt outside
1952	Motorcycle dispatch rider's coat	Rubber	Sandwich	Heavy cotton twill. Khaki
1950s	Long coat. Norwegian	PVC with 50% phthalates. Labelled Plavex, which was developed in the 1950s	Surface treatment	Thin cotton. Green
1959	Sample. Long coat	PVC softened with phthalates	Surface treatment	Medium thickness cotton. Black outside. Green on inside
Undated	Jacket and trousers. Swedish. Waterproofed elbow patches	PVC. Waxed	Surface treatment	Unidentified material
Undated	Long coat	PVC softened with DEHP (di2ethylhexylphthalat)	Sheet PVC. Dark grey	
Undated	Short coat	PVC softened with phthalates	Sheet PVC. Blue	
1980	Sample. Coat	Nylon. Labelled 'Specially coated nylon'	Sheet nylon. Khaki	

DurasampleIIR reflection element, while the rest of the garment was supported on tables adjacent to the FTIR spectrometer. A silicone septum was placed between the uniform and the tip of the pressure device to prevent the marking of objects during the one minute period of analysis. The materials were characterised by matching their spectra with those from either the *Infrared and Raman Users Group Spectral Database* (Price and Pretzel 2000) or with spectra of suitable reference materials. The waterproofing materials were identified as drying oils, namely linseed, tung and fish oils, natural rubber, bitumen and plasticised PVC. Nylon textiles were also identified by this technique.

Three waterproofing techniques were represented in the rainwear selected for examination: surface treatment of a textile base with the waterproofing agent, sandwiching a layer of waterproofing material between two layers of textile and, most recently, using sheets of waterproofing material (Table 1). In the earliest garments, bitumen, rubber and linseed oil were applied to a heavy cotton textile. Three examples of rainwear selected for examination in this project were surface treated with bitumen, which made them heavy, particularly those with linings. Bitumen was not used as a waterproofing agent after about 1920, whereas rubber was used until the early 1950s. Three uniforms contained rubber as a surface treatment and five had rubber sandwiched between two layers of textile, usually heavy cotton twill, again imparting considerable weight to the garments. Because they were uniforms for motorcycle dispatch riders, they were designed to protect against both strong wind and wet weather. Although there was a great shortage of natural rubber around the time of the Second World War, it was still used in military clothing. There are six examples of oilskins in the sample garments dating from before the 1940s, after which rainwear was constructed from lighter weight base textiles, coatings of synthetic polymers and finally sheets of synthetic materials. After about 1950, different formulations of plasticised PVC were employed for waterproofing, of which six examples from the collection were selected for this study. There is also one example of nylon.

Degradation of waterproofing materials and means to improve their storage

Correlating the results obtained from FTIR–ATR spectroscopy with the visible condition of the materials used to construct the uniforms suggested that those treated with bitumen, natural rubber or oils exhibited more deterioration than the others and, therefore, required conservation as a matter of urgency (Table 2). It was evident that all the uniforms required internal support to hold their original forms and that the tacky uniforms needed external covers to prevent damage to other objects in the vicinity.

Bitumen-based waterproofing materials contain a small proportion of unsaturated (double) bonds, which react slowly with oxygen, usually in the presence of UV light. The degradation products have lower molecular weights and lower softening points than the original bituminous materials. This results in the formation of a tacky surface, which was the most frequent manifestation of degradation of these uniforms. This surface tackiness was accompanied by stiffening of the underlying substrate. The degraded bitumen had adhered to itself and to everything it touched. The textile base was fragile and tore easily when attempts were made to open the crease folds, which had stuck together. It was difficult to remove the dirt and pieces of paper that had become embedded in the bitumen. To improve the longevity of these objects, it was necessary to isolate the tacky surfaces and prevent them from sticking to themselves and to other objects in their vicinity, using a protective covering to which they could not adhere. In addition, the uniforms required support, particularly at the shoulders, to limit the possibility of further damage. Most uniforms treated with natural rubber exhibited cracking and crazing, accompanied by stiffening of the whole (Fig. 2).

Of the three raincoats surface treated with rubber, only one retained its original suppleness and flexibility, as did one of the five made according to the sandwich method. In several cases there were signs of fungal growth. The degradation reactions of all chemically unsaturated rubbers, synthetic as well as natural, involve atmospheric oxygen, reactions which are greatly accelerated by a rise in temperature, the presence of metallic ions (which may be found due to metal fittings, such as fasteners) and by absorption of light. It is not thought that

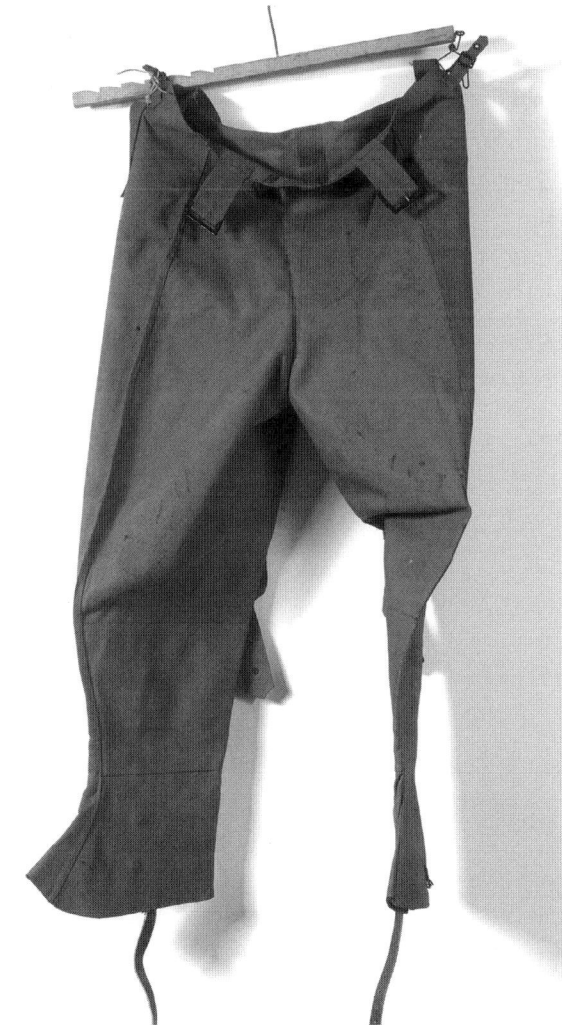

Figure 2 Motorcycle dispatch rider's trousers: waterproofed with rubber sandwich method. Stiff and deformed from poor storage (Plate 40 in the colour plate section).

Table 2 Date and degradation of materials.

Waterproofing material	Date	Method of waterproofing	Degradation of waterproofing material	Degradation of textile base
Bitumen	1890	Surface treatment	Extremely sticky, adhering to itself, also inside and in pockets. Bitumen adheres to surface of worktable and to hands. Dirt, bits of paper and paint sticking to surface	Deformed and flat. Several tears probably from forcing material apart
Bitumen	1918	Surface treatment. Possibly of both shell and lining	Sticky and greasy surface. Stiff	Deformed and flat
Bitumen	Undated 1920?	Surface treatment	Sticky surface also inside. Stiffened	Deformed. Dirty. Woollen collar damaged by moths
Rubber	1883	Surface treatment	Very stiff. Surface dry, cracked and powdery. Surface covered with brownish spots – possibly from fungus	Extremely deformed
Rubber	1910?	Sandwich	Very stiff	Deformed. Adhesion at seams has lost ability to stick
Rubber	1911	Surface treatment	Stiff. Surface dry, cracked and powdery. Small white crystals embedded in rubber – possibly from fungus. Spots of oil from oilskin hanging close to coat	Wrinkled and deformed
Rubber	1922	Surface treatment	Soft and flexible	A little bit wrinkled
Rubber	1945	Sandwich	Soft and flexible	A bit flat and deformed. Dirty and covered by whitish spots – possibly from fungus. Corrosion of metal push buttons
Rubber	1945	Sandwich	Stiff	Deformed. Dirty and discoloured
Rubber	Undated	Sandwich	Very stiff	Deformed. Dirty and discoloured
Rubber	1952	Sandwich	Very stiff	Deformed, flat and wrinkled. Dirty
Linseed oil. Possibly dyed	1890	Surface treatment	Extremely sticky surface also inside. Damaged by fungus	Deformed and flat
Chinese tung oil	1945	Surface treatment	Extremely sticky also inside. Lining adhering to shell. Possibly damaged by fungus. Discolorations. Dirt and paint sticking to surface	Deformed and flat. Corrosion of push buttons
Unidentified oil	1945–46	Surface treatment	Stiff and extremely sticky. The coat had been wrapped in Melinex which had adhered to surface and given a shiny surface	Extremely deformed and flat. Corrosion of push buttons
Unidentified oil	1946	Surface treatment	Extremely sticky. Dirty and stained	Extremely deformed and flat. Corrosion of push buttons
Unidentified oil	Undated	Surface treatment	Only slightly sticky. Discolorations. Dirty	Textile fragile and torn in places
Fish oil	Undated	Surface treatment	Sticky surface also inside. Small drops of oil have gathered dirt. Discolorations. Dirty and possibly damaged by fungus also inside	Deformed and flat
Polyvinyl butyral	Undated	Surface treatment	Extremely stiff. Surface cracked, powdery and disappeared in places. Dirty and discoloured. Brownish spots possibly from fungus	Extremely deformed. Material at shoulder without seam is stretched and bulged from hanger
PVC	Undated	Surface treatment. Very thin	A little bit stiff	Sharp folds have formed. Corrosion of push buttons
PVC with 50% phthalates. Labelled Plavex, which was developed in the 1950s	1950s	Surface treatment	Soft and flexible. Greasy surface. Brownish discolorations	Condition fine
PVC softened with phthalates	1959	Surface treatment	Soft and flexible. A little dirty	Condition fine
PVC. Waxed	Undated	Surface treatment	Surface of elbow patches a little greasy	Condition fine
PVC softened with DEHP (di2ethylhexylphthalat)	Undated	Sheets of waterproofing material. Dark grey	Wrinkled but soft and flexible	
PVC softened with phthalates	Undated	Sheets of waterproofing material. Blue	Wrinkled but soft and flexible. A little dirty and a few holes	
Nylon. Labelled 'Specially coated nylon'	1980	Sheet nylon. Khaki	Wrinkled but soft and flexible. Few holes	

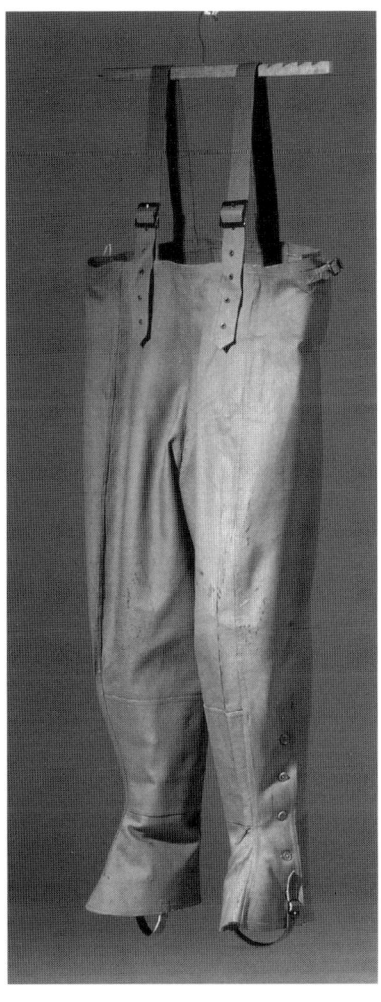

Figure 3 Motorcycle dispatch rider's trousers: original shape recreated (internally supported) (Plate 41 in the colour plate section).

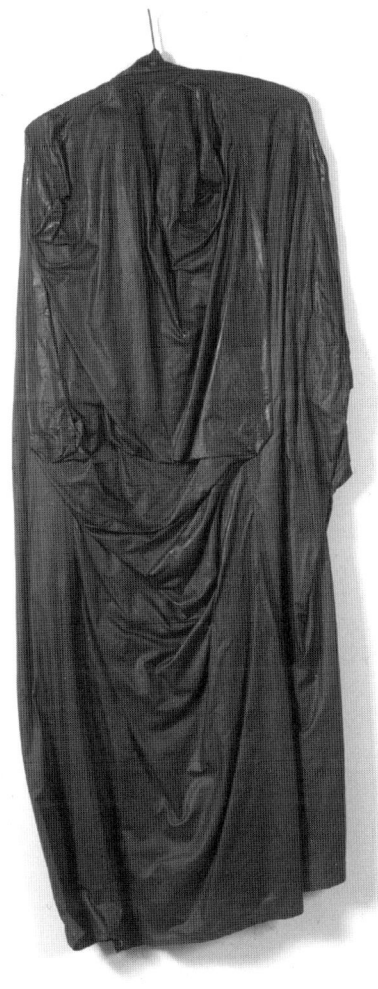

Figure 4 Oilskin uniform (1945): flat and sticky. The shape of the coat is unidentifiable (Plate 42 in the colour plate section).

relative humidity (RH) is an important factor in this process at room temperature; however, water does influence the rate of growth of micro-organisms (Michalski 1993). Rubber materials, stored at ambient temperatures in the dark, are susceptible to shelf ageing, a term used to describe autoxidation. Shelf ageing can result in either the hardening or softening of rubbers depending on their formulation (Buist 1955). Delicately, with the aid of a little heat, the deformations of the stiffened rubber-treated uniforms were straightened and the uniforms supported internally with acid-free paper (Fig. 3).

It was concluded that these uniforms all needed oxygen-free storage to inhibit the progress of deterioration reactions. This was achieved by enclosing the uniforms in double-layer bags prepared by heat-sealing sheets of oxygen barrier film, Cryovac BDF-200 (laminate of polyethylene and polyester). As much air as possible was sucked out of the bags and an oxygen absorber, Ageless, was introduced just prior to heat-sealing the bags closed. The use of an oxygen absorber is an effective, low-cost alternative to the use of inert gas atmosphere for the storage of rubber-containing materials (Shashoua 1999).

Of the six coats treated with oil, five were extremely tacky. They were flattened due to having been stored too close together and their shape was impossible to determine (Fig. 4). The crease folds had to be separated very delicately in order to recreate their original shapes. It is likely that the waterproofing oil applied to the uniforms had never dried thoroughly, but instead only became dry to the touch. In order to produce a hard, dry film from raw oil, it is necessary to heat the oil together with a metal catalyst, so that the double bonds in the structure can react with oxygen; this is a complex and slow process. Diffusion of oxygen into oil is impeded if films are thickly applied to textiles, thus inhibiting drying. One sample was not sticky – its surface appeared worn as if cleaned and the oil partially washed off. Requirements regarding the future storage of the oil-treated uniforms were similar to those treated with bitumen. They needed support and a covering that would isolate them and prevent them from sticking to themselves and neighbouring uniforms.

A cape which was surface treated with polyvinyl butyral was too deteriorated for any treatment. The surface coating had cracked and fell in flakes on contact. The cape was packed in a box instead of hanging. The six raincoats waterproofed with PVC and the nylon coat were in good condition. They were slightly wrinkled from cramped storage.

Evaluation of suitability of storage, support and packaging materials for bitumen- and oil-treated uniforms

It was clear that coat hangers were ineffective at supporting the raincoats evenly; distortion of seams, arms and creasing of the garments demonstrated this. Since coat hangers provided a good, basic form, however, it was thought that they could be upholstered to better support the uniforms. The raincoats also required protective covers to prevent them adhering to themselves and other objects.

It was necessary to find materials for the internal supports and covers for the tacky bitumen- and oil-treated uniforms. These materials were required to be flexible and easily worked. Since they would be in close contact with deteriorated surfaces for at least ten years, it was necessary that they should be chemically inert, physically stable and poor absorbers of oily liquids. Materials suitable for constructing the covers were also required to be transparent in order to facilitate future examinations, have low permeability to air and water vapour, to contain the microclimate surrounding the uniform and to be physically and chemically stable.

Acid-free tissue, silicone paper, polyethylene sheet, bubble plastic, Vilene, Tyvek, Melinex (polyester) and Cryovac BDF-200 were evaluated for their suitability as construction materials for supports and protective covers. These materials were selected on the basis of the authors' previous positive experiences with them, their suitable physical properties and their availability. Since it was not possible to use original material from the uniforms as experimental substrates to evaluate the stability of potential support and covering materials, model substrates were constructed for this purpose. Accelerated thermal ageing of support and protective covering materials in contact with model substrates was used to simulate the long-term storage of the uniforms; it was considered to represent the natural ageing environment in a store more accurately than accelerated light ageing. Support materials were judged suitable if they poorly absorbed oils or bitumen from model substrates and retained their initial physical properties after ageing, particularly flexibility. In addition to the above requirements, materials suitable for the construction of covers were required to retain their original transparency (see Shashoua and Skals 2004 for full details and results of the ageing and evaluation procedures).

The results of accelerated ageing suggested that oil-treated textiles could be safely supported using silicone paper or polyethylene-containing films; suitable covering materials were polyethylene and Cryovac BDF-200. Bitumen-treated textiles could be safely supported by polyethylene film and covered by polyethylene, Melinex or Cryovac BDF-200 (Shashoua and Skals 2004) (Fig. 5).

Conclusion

Many examples of military rainwear in the collection of the Royal Danish Arsenal Museum exhibited extensive visible deterioration despite never having been worn. Deterioration was manifested as physical changes to surfaces, and as chemical and biological changes to both textiles and waterproofing materials. Although inappropriate and uncontrolled storage climates were likely to have contributed to biological deterioration reactions, the physical and chemical deterioration observed may have been due simply to reaction with atmospheric oxygen. The decision to improve the storage environments of the uniforms with the aim of slowing the most prevalent degradation reactions was considered to be the most resource-effective long-term solution for the collection as a whole, rather than repairing the mechanical damage of individual uniforms.

The inhibitive conservation strategy described in this paper was developed using knowledge of the materials used to construct the uniforms and that of the most influential factors in their degradation reactions. The first stage was to identify the waterproofing agents and textiles present at the surfaces of each costume using FTIR–ATR spectroscopy. Knowledge of the materials present proved essential to define the major degradation reactions and the factors most likely to accelerate their rates.

Physical barriers were needed to separate specific microclimates. All uniforms that had been treated with natural rubber exhibited advanced oxidation; the implementation of an oxygen-free microclimate was deemed necessary to inhibit the progress of the deterioration reactions after a reconstruction and support of their original shape. The uniforms treated with oils or bitumen required internal support to hold their original forms and external covers to prevent their tacky surfaces from damaging objects in their vicinity. The uniforms treated with synthetic materials mainly needed physical support. Since this study was carried out, the raincoats have been moved from the attic store to one located in an underground bunker. The

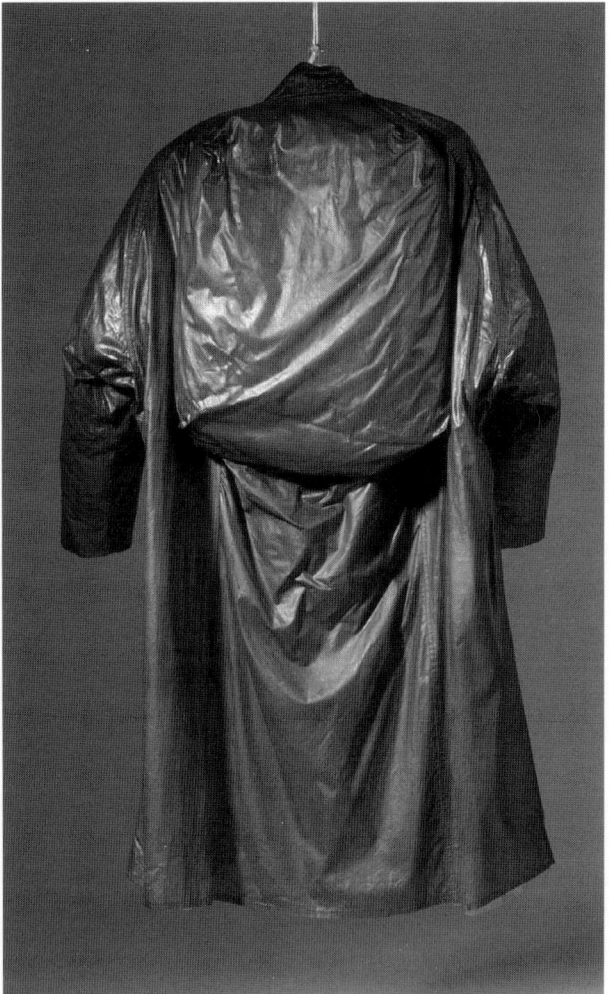

Figure 5 Oilskin uniform (1945): internally supported with bubble plastic and polyethylene (Plate 43 in the colour plate section).

climate in the bunker, although uncontrolled, buffers external seasonal changes better than the attic and is cooler with a lower RH. Reducing the RH below 60% is expected to inhibit biodeterioration processes.

Although the main goal of this project was to prolong the useful lifetime of this historically and technologically important collection of military rainwear, it also gave an ideal opportunity to investigate the development of this type of garment. The choices of the different types of waterproofing materials are rooted in both the knowledge of materials and historical tradition. The industrial developments of the 19th century improved the range of materials available and changed the fabrication of rainwear from garments that were home-made to those that were industrially produced.

The rainwear in this study all showed signs of degradation, which can be explained by the climatic and physical conditions to which they have been exposed. There are, however, examples in the collection at the Royal Danish Arsenal Museum that do not exhibit the expected degradation patterns although they have been exposed to exactly the same conditions, which still leaves questions to be answered. This study provided an opportunity to learn much about these relatively new types of materials, which will find their way to textile conservation workshops as they degrade. Solutions have been suggested to improve their storage thereby slowing their degradation. An examination of objects of this type is recommended before degradation becomes irreversible. Although some questions have been answered concerning these types of materials, their degradation and methods of prolonging their lifetime, further study is required.

References

Allington, C. (1987) 'The conservation of oilskins', in *Recent Advances in the Conservation and Analysis of Artifacts*, J. Black (ed.), 195–7. London: Summer Schools Press, Institute of Archaeology.

Arbeitsblätter (1993) *Zeittafel zur Kunststoffgeschichte*. Heft 2, 133–9. Gruppe 16: Materialen.

Brydson, J.A. (1999) *Plastics Materials*. 6th edn. Oxford: Butterworth-Heinemann.

Buist, J.M. (1955) *Ageing and Weathering of Rubber*. London: W. Heffer and Sons Ltd.

Houston, D. (2002) 'Wetter the better', *Classic Boat* July: 64–9.

Jentzsch, J. (1994) 'Gummi – elastische Materialien aus Natur – und Synthesekautschuk', *Restauro* 5: 314–19.

Loadman, M.J.R. (1993) 'Rubber: its history, composition and prospects for conservation', in *Saving the Twentieth Century: The Conservation of Modern Materials*, D.W Grattan (ed.), 59–80. Ottawa: Canadian Conservation Institute.

Michalski, S. (1993) 'Relative humidity: a discussion of correct/incorrect values', in ICOM-CC *10th Triennial Meeting Washington DC*, J. Bridgland (ed.), 624–9. London: James and James.

Mills, J.S. and White, R. (1994) *The Organic Chemistry of Museum Objects*, 2nd edn. Oxford: Butterworth-Heinemann.

Price, B. and Pretzel, B. (2000) *Infrared and Raman Users Group Spectral Database*. Philadelphia, PA: Infrared and Raman Users Group (www.irug.org).

Shashoua, Y. (1999) 'Ageless® oxygen absorber: from theory to practice', in ICOM-CC *12th Triennial Meeting, Lyon*, J. Bridgland (ed.), 881–7. London: James and James.

Shashoua, Y. and Skals, I. (2004) 'Development of a conservation strategy for a collection of waterproofed military uniforms', *The Conservator* 28: 46–54.

The authors

- Irene Skals is a Textile Conservator at the National Museum of Denmark. The collection of textiles, which spans from *c*.2000 BC to the present, offers a good opportunity for experience in all facets of textile conservation, including analytical work such as identification of materials and analysis of techniques and inhibitive conservation measures in exhibition and storage. Lately the role of the conservator in the non-verbal communication through exhibition of textiles has been the focus of her studies as well as the determination of the degradation of silk by scientific instrumental methods of analysis.
- Yvonne Shashoua is a Senior Researcher at the National Museum of Denmark investigating the degradation mechanisms and inhibitive conservation processes for synthetic materials. She joined the British Museum as a conservation scientist in 1988, specialising in the deterioration reactions and conservation of cellulose nitrate, PVC and rubber objects. In 1998, she was offered a PhD stipendium to research into the deterioration reactions associated with plasticised PVC at the National Museum of Denmark and the Technical University of Denmark. The thesis (Inhibiting the deterioration of plasticised PVC) was completed in 2001.

Addresses

- Author for correspondence: Irene Skals, Department of Conservation, National Museum of Denmark, PO Box 260 Brede, DK-2800 Kongens Lyngby, Denmark (irene.skals@natmus.dk).
- Yvonne R. Shashoua, Department of Conservation, National Museum of Denmark, PO Box 260 Brede, DK-2800 Kongens Lyngby, Denmark (yvonne.shashoua@natmus.dk).

The effect of acid dyes on the photodegradation of knitted nylon conservation support net

M.K. Sinha, R.M. Christie and R. Shamey

ABSTRACT The properties of knitted nylon 66 net, used as a conservation support material, were investigated. This net, constructed from 20-denier monofilaments, was supplied by the National Museum of Scotland (Edinburgh). Undyed and acid-dyed net were exposed to a xenon arc light for 350 and 750 hours. The morphological changes to the dyed and undyed samples, induced by irradiation, were characterised by scanning electron microscopy (SEM), ultraviolet-visible (UV-vis) spectrophotometry, differential scanning calorimetry (DSC) and X-ray diffraction (XRD) techniques, along with measurements of mechanical properties. The tensile strength and elongation at break of the net were measured using a constant-rate-of-extension machine, and the work of rupture of the sample was obtained.

Characterisation of the structural properties of the net revealed that its mechanical properties are direction-dependent, with one direction being rigid and the other stretchable. Exposure was found to have significantly increased the rate of degradation of dyed net as compared to undyed net. The mechanical properties in both the rigid and stretch directions were found to be sensitive to exposure. The SEM micrographs revealed both elongated and irregularly shaped cavities, thus indicating damage arising from the combination of dyeing and exposure to light. A marginal increase in the degree of crystallinity observed after exposure to light was probably due to the formation of crystals in the low-ordered regions arising from the breakdown of the polymer chains. Of the dyes used, Tectilon Red 2B had the most pronounced degradative effect while Tectilon Blue 6G exhibited the least. Statistical analysis was performed to ensure the validity of results. The results of the investigation will assist in the selection of appropriate dyes for the coloration of museum conservation net and are also relevant for outdoor performance nylon products.

Keywords: museum conservation net, acid dye, photodegradation, ANOVA, xenon arc light, mechanical properties

Introduction

The preservation of material heritage is a significant issue in the field of cultural history, so that we can continue to experience the past and understand how it shapes our world. It is important to ensure that historic textiles are preserved for future generations by ensuring minimal damage to these artefacts in storage and on display. The stability of an artefact and the need for it to be handled, treated and stored with care are major concerns, and therefore the mechanical properties of dyed nets used as support materials are of particular importance. This study of a conservation net, with due regard to the dyes commonly used with it, presents an example of how an understanding of materials and their behaviour can be used efficiently in order to maximise their photostability.

The phenomenon of photodegradation of polymers in the presence of dyes and pigments is of particular interest in a variety of applications, including plastics and textiles. A variety of dye types has been observed to sensitise the host polymer to photodegradation (Allen *et al.* 1990; McKellar and Allen 1978; McKellar 1971; Milligan 1986). The extent of dye-sensitised degradation is dependent on the chemical nature of the dye, and even within specific dye classes the degree of photodegradation may vary (Allen and McKellar 1980). The presence of oxygen, oligomers and other additives such as delustrants and polymerisation catalysts can also influence these photochemical and photophysical processes. Some research into oxidative degradation has already been carried out (Stowe *et al.* 1974; Vachon *et al.* 1968).

This paper describes results from an investigation into the factors responsible for the deterioration of nylon 66 net, in particular the influence of acid dyes on the light-induced degradation. This research is aimed at providing an understanding of the behaviour of the material in order to prevent its deterioration over time, and hence to assist in the conservation of those historic textiles with which it is used. This project was initially instigated by the National Museum of Scotland (Edinburgh), who suspected that there may be photodegradation problems with this kind of dyed net.

The conservation net was knitted from 20-denier monofilaments. Its method of knit construction means that it is rigid in one direction while being stretchable in the other. In order to establish quality guidelines for this net, samples were evaluated in both the rigid and stretch directions using various parameters. The effect of dyeing the net with acid dyes on the structural and physical characteristics of the fibres was assessed after light exposure in a Xenotest 150s instrument. The xenon arc light used in this machine has a spectral profile broadly similar to that of daylight, with a particularly good match in the damaging UV region, but with much higher intensity, and is commonly used for the accelerated simulation of exposure to sunlight. Dyed and undyed nylon 66 net samples were exposed to light in the instrument for up to 750 hours. Changes in mechanical properties (tensile strength, elongation and work of rupture) of the samples were measured. Morphological changes in the dyed samples were characterised by SEM, and similarly the effects on the structural properties of fibres were investigated by determining the degree of crystallinity using differential scanning calorimetry (DSC) techniques. An initial attempt was made to study the chemical changes in the net due to dyeing and exposure, using UV-visible (UV-vis) spectroscopy; this aspect of the experiment was limited to the Tectilon Red 2B dyed samples.

The information obtained in this study provides a basis not only for understanding the degradation of the dyed museum support net, but also the behaviour of the materials in other applications, such as the weathering of industrial fabrics as used, for example, in outdoor sportswear.

Experimental

Materials

The knitted nylon 66 net, supplied by the National Museum of Scotland, was manufactured by Dukeries Textiles and Fancy Goods and was constructed from yarns made of 20-denier monofilaments. The structural properties of the net differed in the directions of the length and width. One direction was rigid while the other was stretchable (described throughout this paper as the rigid and stretch directions respectively). The direction in which the loops are more or less arranged in one straight line is characterised as the rigid direction of the net. When the rigid direction is tensioned, the linear arrangement of the loops ensures that it undergoes little deformation. In the stretch direction, the loops are arranged in a diagonal fashion, in a similar manner to a twill fabric. When the stretch direction is tensioned, there will be a large deformation because of this zig-zag arrangement. Thus the overall effect on the mechanical properties is that the response to tension is different in stretch and rigid directions.

The acid dyes used in this study were Tectilon Yellow 3R 200% (Colour Index Acid Yellow 246), Tectilon Red 2B 200% (Colour Index Acid Red 361) and Tectilon Blue 6G 200% (Colour Index Acid Blue 258) supplied by Ciba Speciality Chemicals. These materials were chosen for this study because they are the dyes most frequently used by the National Museum of Scotland for nylon 66 net, due to their desirable properties: ease of dyeing, high affinity toward the net, overall good colourfastness, better exhaustion and fixation, and good levelling properties.

Dyeing procedure

The nylon net samples were dyed in a Zeltex Polycolor dyeing machine in stainless steel tubes. The dyebath contained dye (2.0%), acetic acid (1.0%) and Univadine PA (levelling agent 1%). The quantities were calculated as percentages on the basis of the total weight of the net. Each specimen (2–3 g) was dyed using a liquor to goods ratio of 100:1. The specimens were introduced into the dyebath at an initial temperature of 40 °C, which was maintained for 20 minutes. The temperature was then increased to 100 °C at a rate of 2 °C per minute (microprocessor controlled system) and maintained for a further 45 minutes. The dyebath was finally cooled to 40 °C. After dyeing, each specimen was rinsed thoroughly with cold distilled water and air-dried.

Light exposure studies

In order to determine the effects of simulated sunlight and the synergistic effects of sunlight and acid dye on nylon net, the undyed and dyed net were exposed under the same temperature and humidity conditions. A lightfastness tester, Xenotest 150s, was used to expose the dyed and undyed net for specific periods of time (350 and 750 hours). The instrument uses a xenon arc light, with borosilicate inner and outer filters; settings of 65±5% relative humidity (RH) and 25±3 °C were employed. The irradiance level measured for the xenon bulb from 300 to 800 nm was typically 854 W/m^2; the power rating was 1500 W.

Measurement of tensile properties at break

An M5 Nene Instruments tester was used to measure the tensile strength of the undyed and dyed net, before and after exposure. The tensile strength of the net was calculated using a constant rate of extension method (ISO 1999), with a speed of 100 mm/min^{-1}. The breaking strength and percentage elongation measurements were carried out under controlled standard atmospheric conditions of 20±2 °C and 65±2% RH. Five samples (104 × 40 mm) were used for each measurement, cut both in rigid and stretch directions. A graph of load vs. displacement was used to provide load at peak (N), elongation at break (mm) and work of rupture (J) (ISO 1999).

Degree of crystallinity

DSC was carried out using a TA Instruments Ltd DSC 2010 v4.4E. Samples were cut from both rigid and stretch test specimens, and accurately weighed. The measurements were made in aluminium sample pans, over the range 25–300 °C, with a heating rate of 10 °C min^{-1} and with flowing nitrogen as the purge gas. The enthalpies of melting (H_m, J.g^{-1}) were determined by integration. The percentage crystallinity was determined using the following equation (PerkinElmer Instruments 2000):

$$\% \text{ Crystallinity} = [H_m - H_c] \times 100 / H_{rm}$$

where H_c is the enthalpy of cold crystallisation, set to zero as in our case no cold crystallisation exothermic peak was observed. The term H_{rm} is the enthalpy of melting if the polymer were 100% crystalline, which has been established for nylon 66 as 255.8 J.g^{-1} (PerkinElmer Instruments 2000).

x-ray diffraction images of nylon samples

The images were recorded on a Bruker Nonius x8 Apex2 CCD diffractometer. The samples were mounted centrally such that the beam would pass through them. The x-radiation was generated at 50 kv and 40 mA from a molybdenum metal target in a sealed x-ray tube. A Land diffraction unit containing Polaroid 4×5 Land type 57 films was mounted in front of the standard scintillation counter detector and the sample exposed to x-rays for 30 minutes.

Scanning electron microscopic (SEM) analysis

A Hitachi Stereoscan-530 SEM was used to study surface changes on the fibre. Specimens were coated with gold/palladium (thickness $c.$120 Å) using an automatic sputter coater (Polaron Equipment Ltd., scanning electron microscope coating unit SC 7620). The accelerating voltage was 15 kv.

UV-visible absorption analysis

UV absorption measurements for solutions of the net were carried out using a PerkinElmer Lambda 2 UV-vis spectrophotometer. 1,1,1,3,3,3-hexafluoro-2-propanol (HFIP), from Alfa Aesar, was used to dissolve the nylon 66 net (5 mg of nylon 66 net in 1 cm^3 of HFIP solution). Spectra were obtained over the range of 200–600 nm, using pure solvent as a reference. The same procedure was followed for each of the samples: undyed net, undyed exposed net (350 and 750 hours), Tectilon Red 2B dyed net and Tectilon Red 2B dyed exposed net (350 and 750 hours).

Statistical analysis of data

In order to evaluate the single and interactive influence of the dyeing and exposure parameters on the dyed and undyed conservation net, an ANOVA test was carried out. The purpose of the analysis is to test for significant differences between two or more groups of data. The ANOVA analysis tools provide different types of variance analysis. The test is used in this study to analyze the significance of inter- and intra-sample variations in various experiments, based around the null hypothesis that exposure to light makes no significant difference to the mechanical properties of the net, i.e. the likelihood that any correlation observed is mere coincidence. In general terms, the null hypothesis (H_0) is a hypothetical assumption that the standard deviations are equal, made before calculating F_o values. This hypothesis is rejected if $F_o > F$, where the F_o values are calculated from the experimental data and tabulated F values are given in the literature. The significance level for the test is the confidence limit for the occurrence of a particular event; for example, if this level is set at 0.01 then if the null hypothesis is true there is only a 1% chance that the observed effect would have occurred through random variability. The significance was tested by comparing the calculated ratio (F_o) with the tabulated values (F) at the same degree of freedom and level of significance, taken from the literature. Thus, for each hypothesis tested, if the F_o value was greater than the tabulated F values at a significance level of 0.01, it was concluded that the correlation between the factors was significant, and not a matter of chance.

In these experiments, a two-way ANOVA table was employed for statistical analysis to calculate a F value, which has a distribution related to the sample size and number of conditions (degrees of freedom), and the hypothesis was tested by assessing these F values (Leaf 1987).

Results and discussion

Conservation net characterisation

This study was carried out to develop a quantitative understanding of the properties of the conservation net and to provide quality guidelines for its construction. The characteristics evaluated for the undyed net were: melting point, bursting strength, tearing strength and mechanical properties. The results are given in Table 1.

Table 1 Characteristics of the undyed net.

Quality parameters		Mean values
Melting temperature °C		248.1
Bursting strength (KPa)		116.6
Tearing strength (N)	Net rigid	14.7
	Net stretch	7.4
Mechanical properties		
Tensile strength (N)	Net rigid	55.7
	Net stretch	18.3
% Elongation at break	Net rigid	24.9
	Net stretch	69.5
Work of rupture (J)	Net rigid	1.4
	Net stretch	1.3

The melting point is consistent with the results of Corbman (1983). In this case, bursting strength (the combined tensile strength and stretch of a material, measured by the ability of the material to resist rupture) is not considered an important factor for the net. Nylon 66 fibres, however, have numerous technical end uses where bursting strength is of importance. The average bursting strength of the undyed conservation net was found to be 116.6 KPa. The mean value for the tearing strength of the undyed net in the rigid direction was found to be roughly twice that of the stretch direction. Similarly, large differences were observed between the tensile strength in the rigid and stretch directions (mean 55.7 and 18.3 N, respectively). Table 1 also shows sizeable differences in percentage elongation at break between the rigid and stretch directions (mean 24.9 and 69.4%) for the undyed net. Thus an opposite trend is observed in elongation at break compared with tensile strength. Work of rupture is a measure of the amount of energy exerted to destroy the fabric. The data gathered (Table 1) indicate a similar trend (tensile strength) to the tear strength results, where marginally more energy (work of rupture) is required to destroy the net in the rigid direction compared to

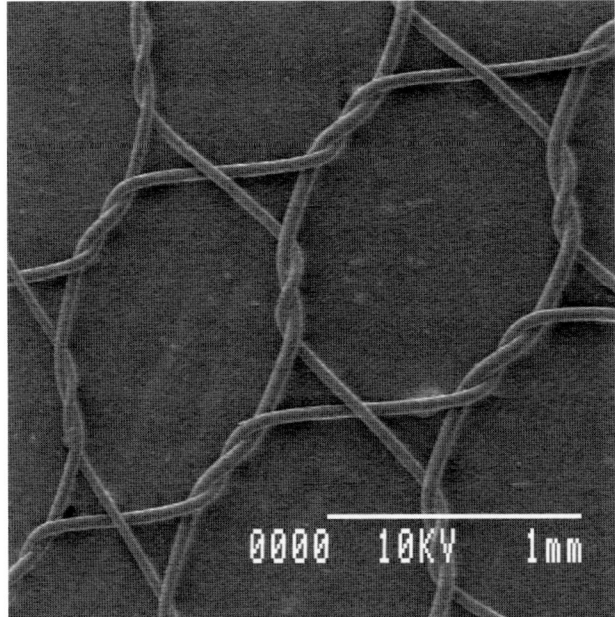

Figure 1 SEM image of two consecutive loops of the conservation net.

the stretch direction. The directional effect observed is due to the way the netting is manufactured with a special type of bobbin net machine dating from the 19th century. The number of loops and their geometry in each direction are the important criteria for the directional properties.

Thus the net is characterised by its anisotropic behaviour. An SEM micrograph (Fig. 1) of nylon 66 net highlights the approximately hexagonal shape of the loop and the form of the twist that ensures the net has the firm structure, which is important to its role as a support material.

Effect of exposure on mechanical properties

The combined effect of dyeing and exposure to light on the physical properties of the net (tensile strength, elongation at break and work of rupture) was investigated. The mean changes in the mechanical properties of undyed and acid-dyed net with different exposure times are shown in Table 2.

It is observed that a significantly higher loss in the mechanical properties for both undyed and acid-dyed rigid net was found after exposure for longer times. The losses were possibly due to combined physical and chemical damage to the polymer, conceivably due to chain cleavage. Table 2 shows the high degree of photodegradation suffered by nylon 66 rigid nets dyed with all three acid dyes when compared to the undyed materials. Dye-induced photodegradation for nylon 66 has been clearly established by Egerton and Morgan (1971). In this study, Tectilon Red 2B and Tectilon Yellow 3R had the greatest degradative effect while Tectilon Blue 6G exhibited the least. The extent of the dye-induced photodegradation on the net thus depends on the nature of the dye. A possible explanation for this accelerated photodegradation of the dyed material is that it is caused by a free radical mechanism (Allen 1994; McKellar and Allen 1979). The dye molecules absorb radiation and then the photoactivated dye abstracts hydrogen from the nylon 66 polymer chains. This creates a free radical centre in the polymer, which is attacked by oxygen in the usual way. In this manner, progressive photodegradation of nylon 66 fibres dyed with the acid dye takes place in close agreement with the findings of earlier work (Allen 1994; McKellar and Allen 1979).

It was seen that the tensile strength increased in the stretch direction. This increase may be due to structural changes and possible cross-linking within the stretch net. Thus, light affects the fibre properties in many different ways (Bresee 1986). Elongation at break decreased in both directions. It is possible that the combination of exposure and dyeing caused some chain breaking and subsequent cross-linking; this is supported by the findings in previous photodegradation studies (Zimmerman 1960). There is always competition between the chain breaking and bond reformation mechanisms. The ratio of number of total chain breaks to the number of bonds formed is an important parameter for the photodegradation of the material (Stephenson et al. 1961), and will depend on the nature of the net and dye. These bond reformation arguments are strengthened by the corresponding data of the elongation at break, which was found to decrease in spite of the increase in the tensile strength for the undyed and dyed stretch net. The study revealed that the net structure is one of the major influencing factors that govern the photodegradation reaction of the nylon 66 fibres.

For the ANOVA analysis, the test statistic is defined by $F = s_1^2 / s_2^2$, where s_1^2 and s_2^2 are the standard deviations of the sample data. The more this ratio deviates from 1, the stronger the evidence for unequal population variances. The higher the F_o value the more significant the damage in the net on exposure. The comparison of the F_o (tensile strength) for undyed exposed rigid net ($F_o = 15.73$) with acid-dyed exposed rigid nets ($F_o = 43.90, 156.55,$ and 178.78) can be noted from Table 3. The increase in F_o after dyeing and exposure is a clear indication of the additional impact of dyestuffs upon the photodegradation of dyed net on exposure to light.

Table 2 Effect of exposure on the mechanical properties of undyed and dyed nylon 66 net.

Sample	Mean % change in tensile strength		Mean % change in elongation at break		Mean % change in work of rupture	
	350 hr exposure	750 hr exposure	350 hr exposure	750 hr exposure	350 hr exposure	750 hr exposure
Undyed net (rigid)	4.78	−11.97	−9.62	−25.62	−5.44	−34.85
Undyed net (stretch)	11.98	19.09	2.06	−24.47	13.77	−10.34
Blue-dyed net (rigid)	−13.61	−37.41	−8.90	−30.89	−21.26	−56.60
Blue-dyed net (stretch)	20.61	−3.04	−5.73	−8.44	13.62	−11.32
Red-dyed net (rigid)	−19.31	−40.81	−28.34	−43.12	−42.04	−66.35
Red-dyed net (stretch)	13.86	−9.38	−17.64	−25.52	−6.49	−32.59
Yellow-dyed net (rigid)	−28.68	−40.78	−19.16	−36.48	−42.44	−62.41
Yellow-dyed net (stretch)	25.38	−12.15	−7.80	−10.54	14.60	−19.84

Table 3 Two-way summarised ANOVA table (between samples exposed for different times and within the group of five samples measured) for the undyed and dyed net.[†]

Sample	FO Tensile strength		FO Elongation at break		FO Work of rupture	
	Between	Within	Between	Within	Between	Within
Undyed net (rigid)	15.73	0.66	24.17	0.34	26.72	0.316
Undyed net (stretch)	4.73	0.89	29.18	2.60	5.38	0.96
Blue-dyed net (rigid)	43.90	1.46	27.58	1.20	34.03	0.64
Blue-dyed net (stretch)	15.46	1.28	8.24	0.52	9.12	0.68
Red-dyed net (rigid)	156.55	0.57	186.59	1.13	210.34	1.15
Red-dyed net (stretch)	14.68	1.51	54.65	0.08	15.54	0.90
Yellow-dyed net (rigid)	178.78	0.93	68.72	2.31	111.07	1.73
Yellow-dyed net (stretch)	44.68	1.22	10.67	1.23	14.94	1.05

† F (2,8; 0.01)=8.70 and F (4,8; 0.01)=7.00 Tabulated F values from [10], where 2 and 8 and 4 and 8 are degrees of freedom for each way and 0.01 is level of significance.

Therefore, exposure had a statistically significant effect on the mechanical properties of the net. From the table, the relative stability of net treated with acid dyes may be arranged in the order of decreasing F_o (work of rupture) as follows: Tectilon Red 2B > Tectilon Yellow 3R > Tectilon Blue 6G, thus indicating the relative degrees of influence on photodegradation. This order of photodegradation may be due to the fact that the yellow and red dyes are azo-based dyes but the blue dye is anthraquinone based.

Effect of exposure on surface properties

SEM

Changes to the surface of the undyed and dyed nets were observed using the scanning electron microscope. SEM micrographs are shown in Figures 2a undyed, 2b undyed exposed, 2c dyed and 2d dyed exposed net samples respectively. As can be seen from Figures 2a and 2c, no surface damage was visible for the unexposed net. Figures 2b and 2d, however, revealed both elongated and irregularly shaped cavities and pits on the exposed net, thus indicating damage due to combined dyeing and exposure to light. Pit and cavity formation correlates well with the extent of degradation as determined by other measurements. It is possible that the pits and cavities formed on the net fibres surface as a result of exposure are due to chain cleavage produced in the photo-oxidative degradation.

X-ray diffraction (XRD) images of nylon samples

Diffractometer data were recorded in order to investigate the crystalline properties of the unexposed and exposed nylon 66 net fibres. XRD patterns are shown in Figures 3a for undyed unexposed and 3b undyed exposed net samples respectively. It is evident from the images that the equatorial arcs (rings) of exposed net become sharper/brighter (reflections) and more symmetrical after the light exposure. The sharpness of the pattern observed in XRD of the exposed net samples indicates

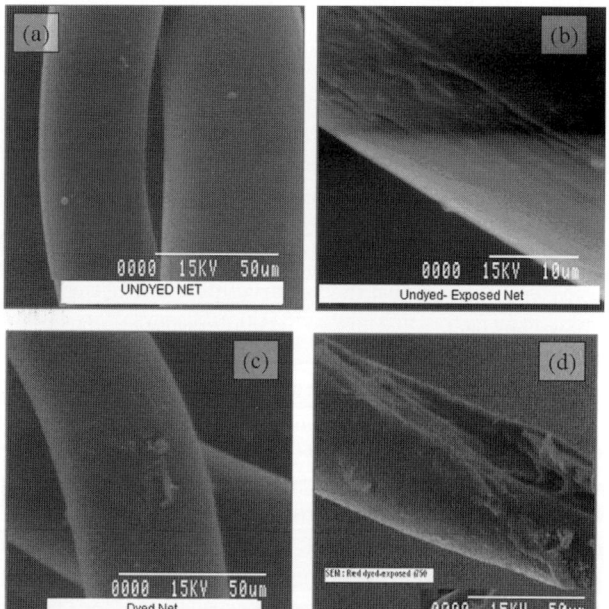

Figure 2 SEM micrographs of (a) undyed net; (b) undyed exposed net; (c) dyed net; (d) dyed exposed net.

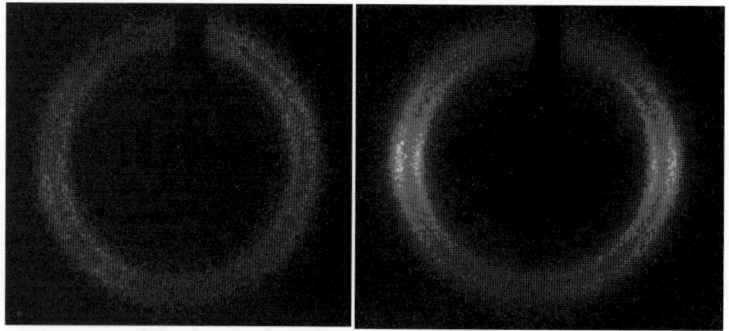

Figure 3 XRD pattern for the undyed net (a) before exposure and (b) after exposure (Plate 44 in the colour plate section).

higher crystallinity compared to undyed net samples. It was difficult, however, to determine visually the relative sharpness of the exposed samples as compared to the unexposed ones.

Differential scanning calorimetry (DSC)

In order to determine structural physical changes in the net fibres due to the exposure, the crystallinity was measured using the DSC technique (PerkinElmer Instruments 2000; Thanki *et al.* 2001). The DSC results are given in Table 4. The crystallinity changed marginally after exposure to xenon arc light. The DSC melting endothermographs are shown in Figures 4a undyed, 4b undyed exposed, 4c dyed and 4d dyed exposed net samples respectively.

It can be seen from Figures 4b and 4d that the melting peak consists of two peaks after exposure, but at the same time there is no such peak resolution for undyed or dyed unexposed samples (Figs 4a and 4c). These DSC results confirmed that there was formation of new crystalline morphology upon photo-irradiation, an effect which is more prominent in the case of dyed exposed (Fig. 4d) than the undyed unexposed net (Fig. 4b). This formation of new crystalline morphology has also been reported in earlier nylon 66 polymer studies (Thanki *et al.* 2001). These interpretations support the effects of degradation in terms of strength loss, elongation loss and work of rupture loss found in the dyed net samples with exposure.

Chemical characterisation of the net before and after exposure

Analysis of the undyed exposed net samples by UV-vis spectroscopy. Some of the important spectral changes of undyed nylon 66 net before and after exposure are illustrated in Plate 45 (in the colour plate section). Clearly, the band at 288 nm is significant. The unexposed net did not exhibit this absorption, it is not due to solvent absorption, and its intensity significantly increased with the longest exposure time (750 hours).

Table 4 Effect of exposure on crystallinity by DSC method for undyed and acid nylon 66 net.

Sample name	% Crystallinity
Undyed net	
Undyed	28.26
Rigid 350 hr exposure	31.59
Stretch 350 hr exposure	31.18
Rigid 750 hr exposure	32.94
Stretch 750 hr exposure	32.51
Net dyed with Tectilon Yellow 3R	
Rigid (before exposure)	26.28
Stretch (before exposure)	27.80
Rigid 350 hr exposure	29.51
Stretch 350 hr exposure	30.03
Rigid 750 hr exposure	31.11
Stretch 750 hr exposure	30.45
Net dyed with Tectilon Red 2B	
Rigid (before exposure)	26.08
Stretch (before exposure)	27.36
Rigid 350 hr exposure	28.84
Stretch 350 hr exposure	27.55
Rigid 750 hr exposure	30.23
Stretch 750 hr exposure	29.75
Net dyed with Tectilon Blue 6G	
Rigid (before exposure)	31.77
Stretch (before exposure)	30.33
Rigid 350 hr exposure	33.44
Stretch 350 hr exposure	31.96
Rigid 750 hr exposure	32.39
Stretch 750 hr exposure	32.10

There is little obvious change after the shorter exposure time (350 hours). This absorbance is due to chemical degradation products and a qualitative correlation between the mechanical properties and build-up of degradation products from the undyed nylon 66 net was seen. The low accumulation of degradation products is consistent with the small loss in mechanical

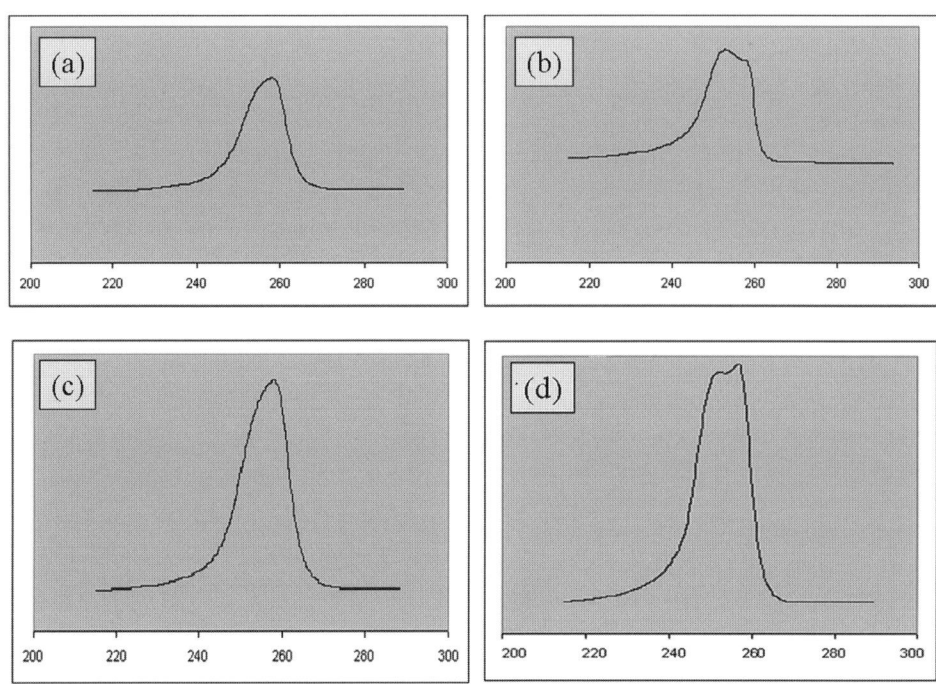

Figure 4 DSC endothermographs of (a) undyed net; (b) undyed exposed net; (c) dyed net; (d) dyed exposed net. Temperature (°C) is given on the horizontal axis.

properties after 350 hours of exposure. Similarly, the higher concentration of degradation products is consistent with the much greater loss in mechanical properties after 750 hours. This may be explained by a direct increase in carbonyl group concentration with exposure time, an indication of the formation of an imide compound via a free radical chain cleavage mechanism (Allen 1982; Stowe et al. 1974). Allen reported a strong absorption at 290 nm which is associated with carbonyl species in the photo-induced oxidation studies for nylon 66 film. Thus these results are consistent with previous studies of the photo-oxidation mechanism in the photodegradation process of these materials on exposure to UV light.

Analysis of the dyed exposed net samples by UV-vis spectroscopy. The changes in the UV-vis spectrum of dyed (Tectilon Red 2B) net samples are shown in Plate 46 (in the colour plate section). Two broad peaks were observed at 500 and 300 nm in the dyed and exposed samples. Plate 46 shows continuous decrease in these absorptions with photo-irradiation, dependent on the lengths of exposure. The results of this investigation show indirectly the photochemical degradation of net dyed with Tectilon Red 2B dye, in a manner that agrees with measurements and qualitative observations of other previous researchers (Allen and Ledward 1993; Allen 1994). Photodegradation of the dye accelerates the photochemical degradation of dyed net fibres, and the reduction in the concentration of the dye is reflected as a significant decrease in the observed band intensities. The decrease in absorbance can be regarded as an estimate of the amount of dye that has undergone degradation in the process, causing a significant loss in mechanical properties. Photo-excited dye molecules are reactive and can lead to abstraction of a hydrogen atom from a polymer substrate (in this case, the nylon 66 net). As a result of this abstraction, the azo dye forms a colourless hydrazo compound, which results in the formation of free radical polymer reactive species. Degradation continues as a result of reactions undergone by these reactive free radicals. The results of the spectroscopic study show a degree of correlation with the substantial loss in mechanical properties of red-dyed net.

Conclusion

The study of the undyed net provides an understanding of its structure and two directional properties, and thus its behaviour when used by the conservator; this will, therefore, facilitate the development and validation of analytical conservation techniques. A significant reduction in the mechanical properties of dyed and undyed net after xenon arc light exposure is attributed to possible chain cleavage. UV-vis spectroscopy revealed the apparent formation of carbonyl compounds and thereby indicated photo-oxidative degradation in the undyed nylon 66 net. It has also been shown that the application of certain dyes accelerates the rate of fibre degradation as a result of exposure to light. In this study, Tectilon Yellow 3R (azo dye) and Tectilon Red 2B (azo dye) had reasonably similar degradative effects while Tectilon Blue 6G (anthraquinone base) exhibited a lesser deleterious effect. It seems, therefore, that the structure of dyestuffs plays an important role in this regard. Further evidence in favour of a catalytic photodegradation effect by dyes was reflected in the significant loss in the mechanical properties of red-dyed net on exposure to light (750 hours). The increase in the tensile strength of the stretch net may be due to structural changes and possible cross-linking within the material. This is because of its inherently higher ability to undergo elongation as compared to the rigid net. Hence net structure also plays an important role in the photodegradation. The XRD and DSC studies show modification in the fibre morphology due to irradiation. A marginal increase in the percentage of crystallinity was observed after exposure to light. This increase may be attributed to the formation of new crystals in the low-ordered amorphous region on exposure to light. Surface damage to exposed net fibres, as observed by SEM, reveals a combination of types of physical damage. The exposed samples exhibit cracks and pits on the surface of the fibre.

The analysis of these data by ANOVA tests suggests that there is a statistically significant link between the exposure of the various samples to light and their subsequent degradation, as measured by changes in physical properties.

Acknowledgements

The authors would like to thank the National Museum of Scotland, Edinburgh, for providing the knitted net, along with technical information. They also acknowledge the help and continuous technical support given by Margaret Robson and Ann Hardie (Heriot-Watt University).

References

Allen, N.S. (1982) 'Ultraviolet derivate absorption spectra of nylon 6.6: effect of photolysis versus photo-induced oxidation', *Polymer Degradation and Stability* 239(4): 239–44.

Allen, N.S. (1994) 'Photofading and light stability of dyed and pigmented polymers', *Polymer Degradation and Stability* 357(44): 357–74.

Allen, N.S. and McKellar, J.F. (1980) 'Photosensitised degradation of polymers by dyes and pigments', in *Photochemistry of Dyed and Pigmented Polymers*, J.F. McKellar and N.S. Allen (eds), 161–245. London: Applied Science Publishers.

Allen, N.S. Ledward, M. and Follows, G.W. (1990) 'Influence of acid dyes on the photooxidation of nylon 6.6 film: relationship with rate of photofading and luminescent species', *Journal of Photochemistry and Photobiology A: Chemistry* 373(53): 373–86.

Allen, N.S., Ledward, M. and Follows G.W. (1992) 'Photooxidation and photofading of dyed nylon 6.6 film: influence of acid dye concentration and relationship with luminescent species', *European Polymer Journal* 23(28): 23–7.

Bresee, R.R. (1986) 'General effects of ageing on textiles', *Journal of the American Institute for Conservation* 39(25): 39–48.

Corbman, B.P. (1983) 'Nylon the polyamide fibre', in *Textiles Fibre to Fabric*, E. Kennedy (ed.), 6th edn, 362–3. Singapore: McGraw-Hill Company.

Egerton, G.S. and Morgan, A.G. (1971) 'The photochemistry of dyes IV. The role of singlet oxygen and peroxide in photo-sensitised degradation of polymers', *Journal of the Society of Dyers and Colourists* 268(87): 268–77.

ISO (1999) EN ISO 13934-1 (1999) 4: *Standard Test Method*.

Leaf, G.A.V. (1987) 'Analysis of variance', in *Practical Statistics for the Textile Industry Part II*, K. Doghlas (ed.), 50–68. Manchester: Textile Institute.

McKellar, J.F. (1971) 'Photodegradation of the anthraquinone vat dyes: a review', *Radiation Research Review* 141(3): 141–65.

McKellar, J.F. and Allen, N.S. (1978) 'The photochemistry of commercial polyamides', *Macromolecular Reviews* 241(13): 241–81.

McKellar, J.F. and Allen, N.S. (1979) 'Photosensitised process involving dyes and pigments', in *Photochemistry of Man-Made Polymers*, J.F. McKellar and N.S. Allen (eds), 199–205. London: Applied Science Publishers.

Milligan, B. (1986) 'The degradation of automotive upholstery fabrics by light and heat', *Review of Progress in Coloration* 1(16): 1–7.

PerkinElmer Instruments (2000) *Thermal Analysis Application Note*. PETech-40.

Stephenson, C.V., Moses, B.C., Burks Jr., R.E., Coburn Jr., W.C. and Wilcox, W.S. (1961) 'Ultraviolet irradiation of plastics II. Crosslinking and scission', *Journal of Polymer Science* 465(55): 465–75.

Stowe, B.S., Fornes, R.E. and Gilbert, R.D. (1974) 'UV degradation of nylon 6.6', *Polymer Plastic Technology Engineering* 159(2): 159–96.

Thanki, P.N., Ramesh, C. and Singh, R.P. (2001) 'Photo-irradiation induced morphological changes in nylon 6.6', *Polymer* 535(42): 535–8.

Vachon, R.N., Rebenfeld, L. and Taylor, H.S. (1968) 'Oxidative degradation of nylon 6.6 filaments', *Textile Research Journal* 716(38):716–34.

Zimmerman, J. (1960) 'Degradation and cross-linking in irradiated polyamides and the effect of oxygen diffusion', *Journal of Polymer Science* 151(46): 151–62.

Suppliers

- Dukeries Textiles and Fancy Goods, Spenica House, 15A Melbourne Rd, West Bridgford, Nottingham NG2 5DJ, UK.
- Ciba Speciality Chemicals, Hulley Road, Macclesfield, Cheshire, SK10 2NX, UK.
- Alfa Aesar, Shore Road, Port of Heysham Industrial Park, Heysham, Lancs LA3 2XY UK.
- Xenotest, W.C. Heraeus GmbH, Postfach 15 53, D-6450 Hanau 1, Germany.

The authors

- Mukesh Kumar Sinha worked in the textile industry for eight years in the capacity of Fabric Technologist and Manager in the textile processing field in different industries in India and Bangladesh. In 2002 he joined the PhD course in Textile Coloration at Heriot-Watt University, Scotland, which he is about to complete. His area of interest is textile processing technology. In 2003 he was awarded Chartered Colourist status by the Society of Dyers and Colourists in the UK.
- Dr R.M. Christie is currently Director of Research and Reader in Colour Chemistry in the School of Textiles and Design at Heriot-Watt University. His career has included periods in industrial research with two international colour manufacturers, Ciba Pigments (Paisley, Scotland) and Dominion Colour Corporation, Toronto, Canada. His principal research interests are in the chemistry and technology of organic pigments, especially in computer-aided molecular design, synthesis, characterisation and application performance.
- Dr Renzo Shamey currently works in the Polymer and Color Chemistry (PCC) program in TECS, COT at North Carolina State University. Following his PhD from Leeds University, he worked as a post-doctoral Research Associate and contributed to the teaching programme of the Colour Chemistry Department at the university until 1998 when he joined Heriot-Watt University as a lecturer in dyeing, printing and finishing. He established (and was head of) the **Automation and Novel Coloration** Research Group. In 2000, he was awarded the Chartered Colourist status by the Society of Dyers and Colourists in the UK.

Addresses

- Corresponding author: M.K. Sinha, School of Textiles and Design, Heriot-Watt University, Scottish Borders Campus, Galashiels TD1 3HF, Scotland (M.K.Sinha@hw.ac.uk).
- R.M. Christie, School of Textiles and Design, Heriot-Watt University, Scottish Borders Campus, Galashiels TD1 3HF, Scotland.
- R. Shamey, North Carolina State University, Raleigh, NC, 27695, USA.

Freezing the present to preserve the future

Yvonne R. Shashoua

ABSTRACT The storage of plastic objects in domestic freezers (−20 °C) is theoretically a long-term, low-cost, low-maintenance technique for prolonging their useful lifetime, but it is not used in practice. A 5–10 °C reduction in temperature both halves the rate of chemical reactions and inhibits some physical degradation processes. The research described here quantified the changes in dimension caused by cooling from ambient to freezer temperature and the influence of the rate of cooling on the risk of formation of condensation for degraded plastics materials representative of those found in museum collections. Results suggested that thin, degraded cellulose nitrate (CN), polystyrene (PS), polyesters and acrylonitrile-butadiene-styrene copolymers (ABS) may be safely stored in a freezer protected only by a closed polyethylene bag to prevent contact with any condensation formed on return to ambient conditions. Manipulation at freezer temperatures should be limited, particularly for degraded CN and plasticised polyvinyl chloride (PVC), which lose flexibility in the cold. The dimensional changes and risk of formation of condensation for degraded plasticised PVC and composite materials, however, suggest storage in a freezer to be less suitable for these than for other materials. Use of insulation to slow the rate of cooling does not reduce the risks of formation of condensation on plastics.

Keywords: cold storage, freezer, plastics, shrinkage, condensation, cooling rate

Introduction

Although the modern frozen food industry, with its rapid freezing techniques, was launched in the USA in the 1930s, cold storage as a means of preservation has long historical roots. The Chinese were the first to prolong the longevity of food by freezing it after the winter cold, using ice cellars as early as 1000 BC. The Greeks and Romans stored compressed snow in insulated cellars, while the Egyptians and Indians discovered that rapid evaporation through the porous walls of clay vessels produced ice crystals in water contained inside. Freezing prevents food spoilage by inhibiting micro-organic and enzymic action.

Storing plastic-containing objects at low temperatures (between 10 °C and −20 °C) has been proposed as a relatively low-cost, low-maintenance technique for slowing the rate of the most common chemical degradation reactions, namely hydrolysis and oxidation (Michalski 2002). Many plastics exhibit degradation 1–50 years after manufacture, so all opportunities to extend their short lifetimes are of interest to museum professionals (European Union Commission 2000). The justification for cold storage is that reducing the storage temperature by between 5 and 10 °C halves the rate of chemical reactions. Some physical degradation processes are also inhibited by cold storage: for example, reducing the storage temperature from ambient to that of a domestic freezer reduces the rate of migration of plasticisers from polyvinyl chloride (PVC) by as much as 15 times (Shashoua 2004).

The implementation of low-temperature storage for photographic collections was first proposed in 1970 and is now an established conservation technique for negatives and rolls of movie film (Adelstein *et al.* 1970). Low-temperature conservation treatments for three-dimensional objects, however, have been limited to pest eradication. One of the few published trials of cold storage applied to three-dimensional objects suggested that the increase in crystallisation of rubber at low temperatures had caused irreversible stiffening of rubber shoe soles and knee convolutes in the Apollo spacesuit collection in the National Air and Space Museum, Washington DC (Kent 2000). Since the loss of elasticity of freshly vulcanised rubber is documented to take place below −55 °C (the storage temperature was 5 °C in this case), it seems likely that degradation had changed the material's response to cold (Hvostovskaya and Margaritov 1933). The reluctance by conservators to apply cold storage to plastic objects may be attributed in part to uncertainty of the resulting changes in the physical properties of the materials.

Plastics exhibit both reversible changes (including those of dimensional and tensile properties) and irreversible changes (including extent of crystallisation and mechanical failure) on cooling. They tend to contract or shrink considerably more than other materials found in museum collections such as metals, ceramics and glass. The linear coefficient of expansion for thermoplastics (4.0–20.0 × 10^{-5} °C^{-1}) is 5–10 times greater than those of most metals. For example, a copper pipe will shrink by 0.01% if the temperature is reduced by 10 °C. Under the same conditions, a high density polyethylene pipe would shrink by 0.07%, and polypropylene and hard polyvinyl chloride (PVC) pipes by 0.04% (Nason *et al.* 1951). Although the shrinkage of plastics is an unavoidable process on cooling, it is reversible and, in the absence of deterioration, a plastic object will assume its original dimensions on return to ambient conditions. When two different materials are in close contact, differential shrinkage on cooling may cause failure of one of them unless the stress introduced by the temperature change is absorbed by elastic or inelastic deformation.

In addition to the effect of temperature on plastics, the influence of the accompanying reduction in moisture content of the cold air surrounding the material must be considered. Recent research into cold storage of rolls of cellulose acetate-based movie film concluded that a maximum of six to ten degrees temperature variation between film and its storage container or anywhere in the mass of materials should be maintained to avoid formation of condensation (Padfield 2002). Air close to warmer areas of film can hold more water vapour than air close to colder areas, resulting in movement of moisture between the object undergoing cooling and its impermeable container, with the resultant possibility of formation of condensation and physical damage to the film.

It is important to limit contact between condensation and plastics because many plastics absorb moisture, swell and fail. Water acts as a plasticiser for many of the early plastics, notably casein, and can itself displace plasticiser from PVC (Shashoua 2001). Polyamides, such as nylon, are the most hygroscopic polymers in common use, containing up to 3% moisture by weight under ambient conditions. Cellulose acetate and poly (methyl methacrylate) contain 0.8%, PVC 0.4% and polystyrene (PS) 0.1% moisture by weight at ambient. Padfield (2002) suggests that slowing the rate of both cooling and return to ambient conditions would ensure that there is never more than six degrees temperature variation in the mass of materials, thereby minimising the risk of formation of condensation.

Transfer of objects from ambient to a domestic freezer is a readily available, relatively low-cost and convenient technique to achieve cold storage in museums. Chest freezers are designed to hold large volumes of air at a mean temperature of around −20–25 °C and at a stable relative humidity (RH) for long periods of time, if allowed to remain closed. The opportunity for development of structural damage including stress cracking on cooling to freezer temperature has been investigated for materials in good condition but not for those exhibiting deterioration, while the possibility for formation of condensation has only been studied in detail for photographic film (Daniels and Kibrya 1998). The research described here quantified the changes in dimension caused by cooling from ambient to freezer temperature, and the influence of the rate of cooling on the risk of formation of condensation for degraded plastics materials representative of those found in museum collections.

Sample materials

All the sample plastics selected were amorphous thermoplastics; such materials are frequently found in collections concerned with both the life of high society at the start of the 20th century and with everyday life after the Second World War. Furniture and flooring are usually constructed from thermoplastic vinyls or polyesters. Telephones, computers, television housings and electronic devices are commonly made of acrylonitrile-butadiene-styrene (ABS). The majority of flexible toys contain thermoplastic PVC. Selected experimental materials included both composite and solid plastics, thin- and thick-walled and semi-synthetic (regenerated) and synthetic materials.

Samples exhibited degradation and comprised a cellulose nitrate (CN) negative film, a polyester cassette tape in a polystyrene case, a pile of ten LP vinyl records in their original packaging, PVC photo album and a toy ship constructed from LEGO bricks (Fig. 1). The polymer types present in the sample objects were identified non-destructively using Fourier transform infrared–attenuated total reflectance spectroscopy (FTIR–ATR).

According to the literature, the glass transition temperatures of the sample plastics range from −20 °C (plasticised PVC in record sleeve) through 30–80 °C (vinyl acetate/vinyl chloride copolymer in record) to 80–100 °C (ABS LEGO bricks, PS cassette case and CN negative film (Brydson 1999)). Glass transition temperature (T_g) is a property of amorphous polymers. Above the T_g, there is sufficient energy for molecular movement, resulting in polymers that are soft and flexible. When cooled below this temperature, polymers become hard and brittle due to a lack of molecular mobility, so attain a glassy state in which their response to stress from handling and moving allows the possibility to 'freeze in' molecular orientations. If the orientations vary from place to place in the structure, sufficient stresses may be established to cause the plastics material to distort. The plastic components of cassette tapes, LEGO bricks and CN film negatives remain below their T_g (in their glassy state) both at ambient and at freezer temperature while the outer sleeves (plasticised PVC) of the vinyl records and the photograph album are above their T_g (rubbery state) at ambient but probably below it at freezer temperature.

CN was used as a film support for motion film and still photographic negatives between 1910 and 1950. The film is 11–15 μm thick and comprises a layer of gelatine containing silver particles which provide the image, attached to a film of plasticised CN. When both materials are in good condition, the thermal coefficient of expansion of the CN support layer is around 10.0×10^{-5} °C^{-1} compared with 3.0×10^{-5} °C^{-1} for photographic gelatine, suggesting that on cooling CN will shrink more than the gelatine layer to which it is attached. The moisture coefficients of expansion of the two materials are also different (Mecklenburg et al. 1994). Differences

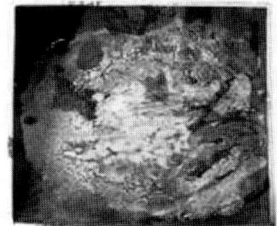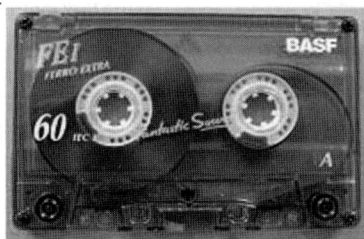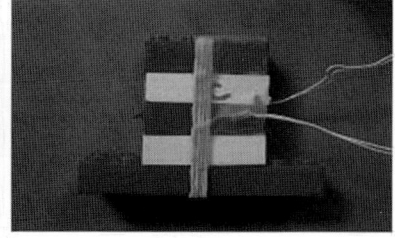

Figure 1 Sample plastics included a degraded CN negative film from the 1940s (left), polyester cassette tape from 1970s (centre) and a toy ship made from ABS LEGO bricks (right) (Plate 47 in the colour plate section).

in dimensions caused by change in temperature or RH give the potential for separation of the two layers unless the stress produced by the temperature change is absorbed by elastic or inelastic deformation.

The sample CN negative film used in this investigation dated from the 1940s and was degraded to such an extent that copying was necessary to preserve the image. The CN had lost plasticiser and was brittle, while the gelatine layer was tacky and much of the image had been lost. The recommended standard cold storage conditions for degraded CN film are −5±2 °C and 25±5% RH (ISO 1996).

A polyester cassette tape in a polystyrene case dating from the 1970s had been distorted by exposure to heat during playing, but otherwise appeared to be in good condition. The magnetic tape consisted of a polyester base layer and binder layer which contained magnetic particles. The linear coefficient of expansion for new polyester is 12.4×10^{-5} °C^{-1}. The current recommended storage environment for magnetic media is 11±2 °C and 50±2% RH (ISO 2000).

A pile of ten vinyl LP records dating from the 1980s in paper folders within card sleeves was protected by PVC pockets. The records were moulded in a vinyl chloride/vinyl acetate copolymer, with plasticisers, fillers, stabilisers and pigments. The vinyl records themselves were in good condition while the PVC pockets had shrunk and buckled due to loss of plasticiser. A photograph album from the 1980s also contained the polymer, with an opaque, cardboard-reinforced PVC cover and transparent plasticised PVC pockets. The first and last pockets (those in contact with the covers) were visibly buckled; this degradation was thought to be caused by shrinkage of the plasticised PVC due to loss of phthalate plasticiser. Beads of plasticiser were also clearly visible between the walls of pockets.

A toy ship constructed from six ABS LEGO bricks and dating from the 1960s showed some mechanical damage including scratches and missing areas. Plastics based on styrene have a very low thermal coefficient of expansion (around 7.2×10^{-5} °C^{-1}) and high dimensional stability which gives them resistance to physical damage when used or stored at low temperatures (Nason *et al.* 1951). ABS is hygroscopic, however, absorbing up to 0.3% moisture in 24 hours at 20 °C.

Cooling process

Before museum objects in the National Museum of Denmark are exposed to low temperatures to eradicate pests, they are first packed in close-fitting polyethylene bags under ambient conditions to prevent contact with condensation on returning to ambient after treatment. The same procedure was followed during this study. Plastic samples were conditioned at ambient temperature and RH (20±2 °C and 40±2% respectively) for 48 hours prior to enclosing in a close-fitting zip-lock low density polyethylene (LDPE) bag with as much air as possible removed. Thereafter, samples were placed in acid-free cardboard boxes with fitted lids both to reduce the local variations in temperature and RH, and to protect objects against physical damage during handling. Boxes had volumes approximately double those of the samples they contained.

To control the rate of cooling of the samples, the air space between the enclosed plastics and the box was filled either with closely packed polyethylene foam or with loosely packed expanded polystyrene insulating chips (Fig. 2). The boxes were then placed in a frost-free domestic chest freezer set to maintain a temperature between −25 and −30 °C.

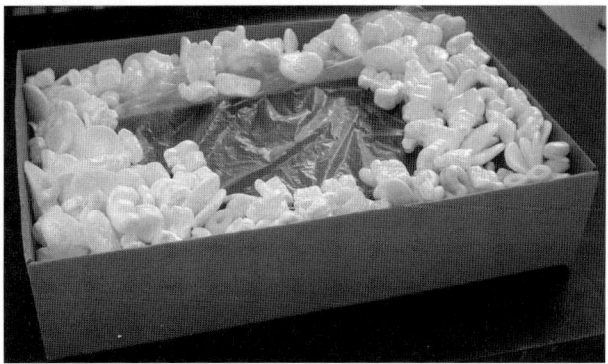

Figure 2 Sample plastics were insulated by packing them with expanded polystyrene chips prior to cooling (Plate 48 in the colour plate section).

Examination of sample materials

Rate of cooling of sample plastics

Tinytalk temperature loggers, fitted with 10k NTC Thermistor (encapsulated) sensors and sensitive within the range +50 °C to −30 °C were used to record temperature profiles at the surfaces (temperature sensor 1) and centres (temperature sensor 2) of the samples during cooling from ambient to freezer temperature. The sensor accuracy was ±0.2 °C and resolution 0.25 °C across the loggers' temperature range. The thermistors of the sensors were held in position with adhesive tape so that they made close contact with samples (Fig. 3). In addition, a Tinytalk temperature logger was placed in the freezer beside the boxed sample plastics to monitor the air temperature external to the experiment (temperature sensor 3).

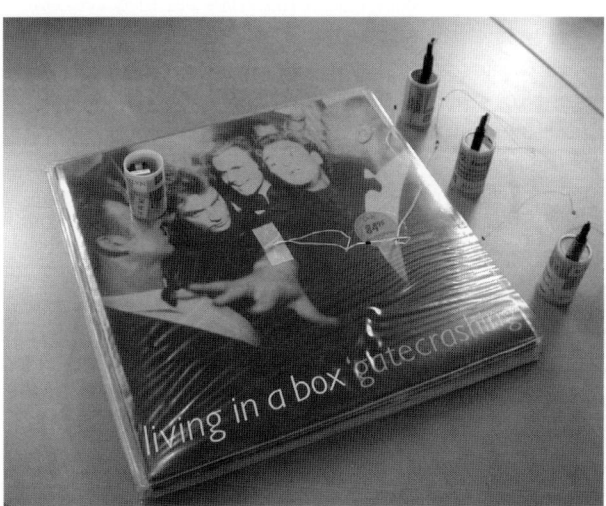

Figure 3 Thermistors of Tinytalk data loggers were held in good contact with surfaces of sample plastics using tape.

Temperature was recorded once per minute; values were downloaded via an interface cable and plotted against time using Gemini Data Logger software. Measurement of temperature profiles on cooling was repeated three times for each sample plastic with and without insulation. The RH around the enclosed sample plastics was not measured; past experience has shown that it is difficult to measure RH accurately under steadily falling temperatures and over a large temperature range.

In the absence of insulation, the CN negative film, polyester cassette and ABS LEGO ship (all thin-walled objects), cooled from ambient to freezer temperature in 45 minutes or less, compared with a period of 90 minutes for the PVC photograph album and pile of ten LP records (thick-walled objects). Insulating the objects prolonged the period taken by them to attain freezer temperature regardless of type or packing arrangement of insulating material (Table 1). The mean maximum temperature difference between surfaces and central areas of thin-walled materials was never greater than 10 °C, while it was as high as 31 °C for 5–10 minutes for thick-walled materials. The greatest temperature difference always occurred at the start of the cooling process (Figs 4 and 5).

Table 1 Rate of cooling sample plastics from ambient and detection of condensation.

Sample plastic/insulation	Mean time for all areas of sample plastics to attain freezer temperature (min)	Mean maximum temperature difference between temperature sensors 1 and 2 (°C)	Formation of condensation?
CN negative film			
– no insulation	15	2	no
– close PE foam	20	5	no
– loose PS chips	20	5	no
Cassette tape			
– no insulation	45	5	no
– close PE foam	90	6	no
– loose PS chips	90	10	no
LEGO ship			
– no insulation	30	6	no
– close PE foam	45	2	no
– loose PS chips	45	6.5	no
Photograph album			
– no insulation	90	26	yes
– close PE foam	105	30	yes
– loose PS chips	105	31	yes
Pile LP records			
– no insulation	90	6	no
– close PE foam	105	14	no
– loose PS chips	105	15	no

Insulating the samples slowed the initial rate of cooling of those surfaces closest to the insulating foam, thereby increasing the maximum temperature difference between surfaces and central areas.

Dimensional changes due to cooling regime

Although the expansion or contraction of materials with changing temperature is a three-dimensional phenomenon, the change in height was measured in this study to facilitate comparison between those samples which were composites in construction, those which were comprised of layers and those which were in block form. The heights at three predetermined positions of each sample plastic were measured using either callipers or a micrometer screw gauge (in the case of CN film) under ambient conditions (20±2 °C and 40±2% RH), at freezer temperature and again after 48 hours at ambient. Determination of change in dimensions on cooling and on achieving ambient temperature again was repeated three times for each sample plastic and the mean values calculated.

Percentage linear shrinkage on cooling at each of the three positions was calculated from the following formula and the mean shrinkage for the sample plastic was obtained.

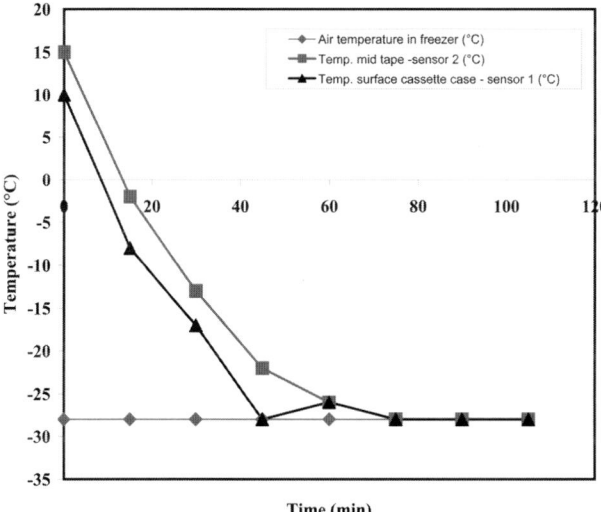

Figure 4 Rate of cooling of thin-walled cassette tape from ambient to freezer temperature (Plate 49 in the colour plate section).

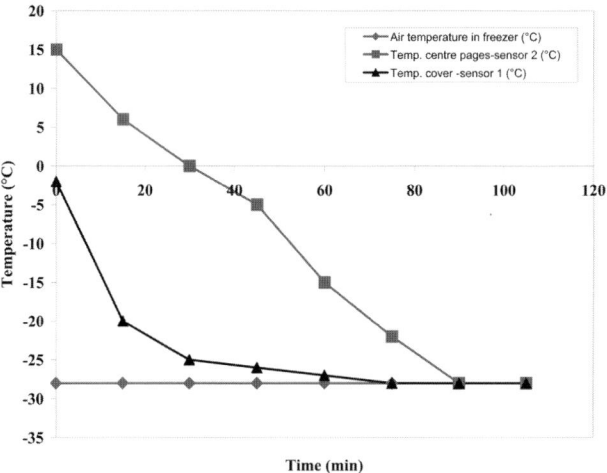

Figure 5 Rate of cooling of thick-walled photograph album from ambient to freezer temperature (Plate 50 in the colour plate section).

Percentage linear shrinkage on cooling (%) = $\dfrac{H_{RT} - H_F}{H_{RT}} \times 100$

Where: H_{RT} = height under ambient conditions (mm)
H_F = height at freezer temperature (mm)

A similar formula was used to calculate the percentage linear

Table 2 Dimensional changes of sample plastics on cooling and returning to ambient.

Plastics materials	Mean percentage shrinkage on cooling in freezer from ambient (%)	Mean percentage shrinkage on warming to ambient (%)	Comments
CN negative film	0.38	0.00	Very stiff at freezer temperature and failed on handling
Cassette tape case	0.40	0.00	
LEGO ship	0.32	0.01	
Photograph album	−97.00	0.01	Deformation and stiffening of PVC pages on cooling caused album to occupy more space than at ambient
Pile LP records	−48.30	0.02	Deformation and stiffening of PVC covers on cooling caused records to occupy more space than at ambient

shrinkage on return to ambient. All changes in dimensions induced by cooling were fully reversed within experimental error (±0.02 mm) when the samples attained ambient temperature again (Table 2). A percentage shrinkage of less than 1% was measured for the CN negative film, polyester cassette and ABS LEGO ship, all considered to be thin-walled.

Although linear shrinkage of CN negative film was low, the material became notably stiffer at freezer temperatures and failed despite careful handling while cold. The film regained its original flexibility on returning to ambient. The flexibilities of the two major components of the film, CN and gelatine, are directly related to temperature. For a CN sheet in good condition, the flexibility under flexural stress shows an increase in Young's modulus by 21% from 1378 MPa to 1654 MPa on cooling from 25 °C to −5 °C (Nason *et al.* 1951). In addition, the flexibility of gelatine varies with its moisture content. Young's modulus for commercial gelatine decreases when the surrounding RH (Calhoun and Leister 1959). A change in moisture content of the CN film was unlikely to be a contributing factor in this study. Samples were kept in closed polyethylene bags so that transfer of moisture between sample plastics and air was minimised.

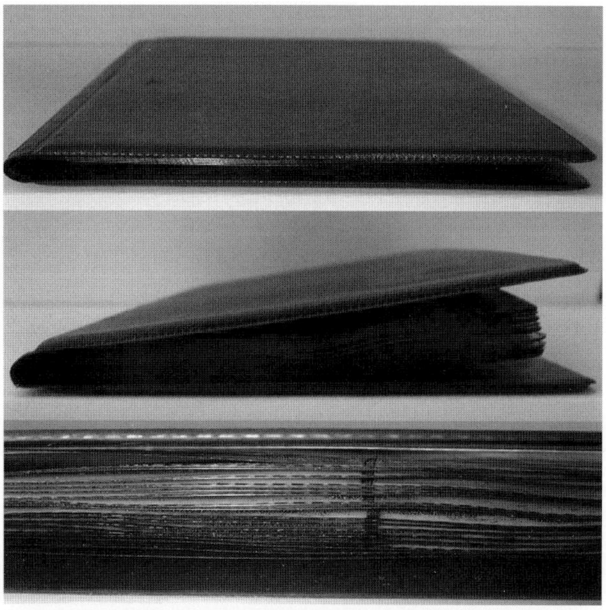

The mean linear shrinkage of the cassette tape case on cooling was greater than that of the CN negative film and of the LEGO ship, as expected from their respective linear thermal coefficients of expansion for materials in good condition (12.4 × 10^{-5} °C^{-1} for PS, 8.0 × 10^{-5} °C^{-1} for CN and 7.2 × 10^{-5} °C^{-1} for ABS). By contrast, the results from measurement of shrinkage of the PVC photograph album and pile of ten LP records were more complex to interpret. The linear thermal coefficient of expansion for plasticised PVC in good condition (25 × 10^{-5} °C^{-1}) suggested that the photograph album and pile of LPs should exhibit greater linear shrinkage than that shown by the cassette tape and LEGO ship. This, however, was not the case.

The pages of the photograph album, already slightly shrunken due to loss of plasticiser, became increasingly buckled and stiffened on cooling, reducing the ability of pages to pack together as closely as they did at ambient (Fig. 6). The T_g of plasticised PVC (−20 °C) indicated that the plastic components of the photograph album underwent a change in phase from rubbery at ambient temperature to glassy on attaining freezer temperature; this accounted for the increased stiffness (Wilson 1995). The combination of the buckling and stiffening of the pages on cooling resulted in a measurable linear expansion of the album, instead of the shrinkage expected. If the album had contained photographs or negatives, these would have been folded or distorted during cooling, probably resulting in irreversible damage. The PVC outer sleeves of the LP records exhibited identical behaviour to the pages of the photograph album, becoming distorted and stiff on cooling so that each record was pushed further from its neighbour. All changes were fully reversed 24 hours after removal from the freezer.

Formation of condensation

Methylene blue powder is highly sensitive to the presence of liquid water, changing colour from grey when dry to an intense blue colour when dissolved. It has been used to indicate the presence of condensation in similar investigations (Padfield 2002). Sheets of Whatman's filter paper were dusted lightly with methylene blue and the powder was distributed evenly using a soft brush to produce indicator sheets for condensa-

Figure 6 PVC photograph album at ambient (upper) and at freezer temperature (centre). Pages of the album became cockled and stiff on cooling (lower) (Plate 51 in the colour plate section).

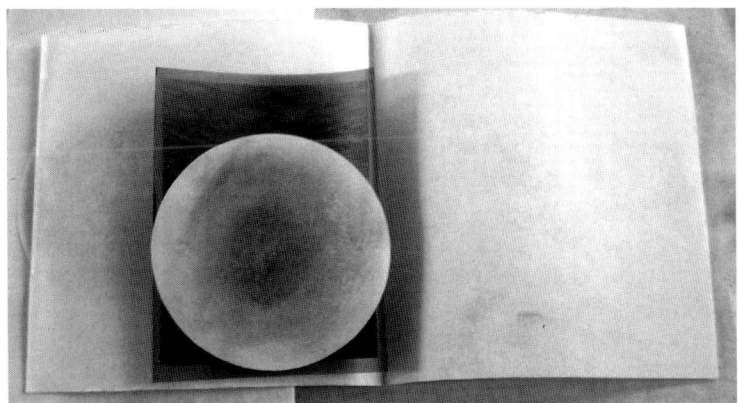

Figure 7 Methylene blue powder was applied to filter paper to detect the presence of moisture. No condensation developed on CN negative film in albums after removal from cold storage (Plate 52 in the colour plate section).

tion. To investigate whether there was a risk of condensation forming at the surfaces of CN negative film on cooling or during return to ambient temperatures, indicator sheets were placed in contact with the surfaces of films which were subsequently packed in paper folders (Fig. 7).

Indicator sheets, cut in thin strips of 10 mm × 2 mm, were placed in the middle of the centre wind, between the two outer winds of the cassette tape and in the air space between tape and case. A sheet of indicator paper was placed in contact with the external and internal surfaces of the LEGO ship. To investigate the possibility of formation of condensation in the photograph album, indicator sheets were placed in contact with front and back covers and their respective adjacent pages and between the centre pages. Indicator sheets were also placed between the outer PVC sleeve and card cover of the pile of LP records so that they made contact with a record close to the centre of the pile.

Condensation did not form during cooling or return to ambient for the CN negative film in its paper album, the polyester cassette or the ABS LEGO ship, a finding which agreed with Padfield's experience that a temperature difference lower than six to ten degrees did not support the formation of condensation in materials at equilibrium with 50% RH (Table 1). Condensation formed between the front cover (that uppermost in the cardboard box) and first, buckled page of the PVC photograph album, regardless of the material's rate of cooling. The mean maximum temperature differences between the cover and centre photo pockets (pages) were 26–31 °C, notably higher than for the other sample plastics. The use of insulation slowed the rate of cooling of the front cover of the album, while having no influence on the temperature between the centre pages. The result was an increase in the already high initial temperature gradient throughout the object and an opportunity for condensation to form.

Because the pile of LP records showed mean maximum temperature differences between external surfaces and the middle record higher than the six to ten degree 'limit' during cooling, the formation of condensation was expected. The fact that it was not detected may be attributed to the presence of layers of moisture-absorbing card, paper sleeves and covers which interrupted the free flow of water vapour between various areas of the record pile.

Conclusion

The risk of introducing either irreversible physical changes or damage due to the formation of condensation into degraded plastics materials during storage at freezer temperatures was insignificant for the plastics which contained material thinner than 1 cm (thin-walled), regardless of their polymer type or extent of degradation. The CN negative film, polyester cassette tapes and ABS LEGO bricks attained freezer temperatures rapidly, within 15–45 minutes. As a result, thin-walled plastics did not experience a temperature difference greater than ten degrees between surfaces and central areas, so did not support the formation of condensation. Insulating these plastics prior to placing in a freezer prolonged the period required to achieve −25 °C, increasing the maximum temperature difference within the materials and with it, the potential risk for formation of condensation.

Dimensional changes sustained on cooling thin-walled plastics were small (less than 0.4%) and dependent on the polymer type; PS shrank more than CN which, in turn, shrank more than ABS. All dimensional changes in thin-walled plastics, however, were fully reversed on attaining ambient temperature again. On a macroscopic scale, there was no evidence of delamination or distortion due to differential shrinkage for CN, polyester or ABS. The plasticised PVC underwent a change in phase from rubbery at ambient temperature to glassy at freezer temperature, however, which caused a stiffening of the photograph album's already crinkled pages. Although the phenomenon was reversed on warming to ambient, the distortion on cooling would have caused irreversible damage to any enclosed photographs or negatives. Degraded CN negative film lost flexibility and failed even on careful handling at freezer temperatures, which suggested that such materials should not be manipulated while cold. They should instead be warmed to ambient in their storage containers before handling.

Plastics that contained a bulk of material thicker than 1 cm (thick-walled) required at least 90 minutes for all areas to attain freezer temperature. The slower cooling resulted in temperature differences between peripheral and central areas of plastics higher than the ten degree 'limit' for the formation of condensation. Condensation was detected between the front cover and first page of the PVC photograph album.

If allowed to remain in contact, water readily displaces plasticiser from PVC, imparting an opaque appearance and loss in flexibility. Insulating objects prior to cooling them prolonged the period required for all areas of thick-walled plastics to attain freezer temperature, further increasing the temperature gradient throughout the plastic and the opportunity for condensation to form.

In conclusion, thin-walled CN, PS, polyesters and ABS may be safely stored in a freezer, protected only by a closed polyethylene bag, despite their degraded conditions. The use of insulation to slow the rate of cooling offers no advantages for storage of plastics. Manipulation at freezer temperatures should be limited, particularly for degraded CN film and plasticised PVC, which become brittle in the cold. By contrast, the high shrinkage and risk of formation of condensation associated with degraded, plasticised PVC and composite materials provides less certainty as to the suitability of low-temperature storage for these materials.

References

Adelstein, P.Z, Graham, C.L and West, L.E. (1970) 'Preservation of motion-picture color films having permanent value', *Journal of the Society of Motion Picture and Television Engineers* 79(11): 1011–18.

Brydson, J.A. (1999) *Plastics Materials*, 6th edn. Oxford: Butterworth-Heinemann.

Calhoun, J.M. and Leister, D.A. (1959) 'Effect of gelatin layers on the dimensional stability of photographic film', *Photographic Science and Engineering* 3(1): 1–29 (http://albumen.stanford.edu/library/c20/calhoun1959.html).

Daniels, V.D. and Kibrya, R. (1998) 'Effects of freezing on museum objects', *Conservation News* 66 (July): 20–21.

European Union Commission (2000) 'Green Paper on environmental issues of PVC (COM (2000)469FINAL' (http://www.europa.eu.int/comm/environment/pvc/index.htm).

Hvostovskaya, S. and Margaritov, V. (1933) 'Physical properties of rubber at low temperatures', *Journal of the Rubber Industry (USSR)* 10: 231–41.

ISO (1996) 10356: *Cinematography – Storage and Handling of Nitrate Base Motion Picture Films*.

ISO (2000) 18923: *Imaging Materials – Polyester-base Magnetic Tape – Storage Practices*.

Kent, E. (2000) 'Conservators race against time to delay space shuttle from decay', *Research Reports* 102 (Autumn) (www.si.edu/opa/insideresearch/01103/spacesuit.htm).

Mecklenburg, M.F., McCormick-Goodhart, M. and Tumosa, C.S. (1994) 'Investigation into the deterioration of paintings and photographs using computerized modeling of stress development', *Journal of the American Institute for Conservation* 33(2): 153–70 (http://aic.stanford.edu/jaic/articles/jaic33-02-007_3.html).

Michalski, S. (2002) 'Double the life for each five-degree drop, more than double the life for each halving of relative humidity'. *Paper presented at the 13th Triennial Meeting of the ICOM Committee for Conservation. Rio de Janeiro, September 2002*, 66–72. London: James and James.

Nason, K., Carswell, T.S and Adams, C.H. (1951) 'Low temperature behavior of plastics', *Modern Plastics* December: 127–203.

Padfield, T. (2002) 'Condensation in film containers during cooling and warming'. *Paper presented at Preserve, then Show at the Danish Film Institute, Copenhagen, October 2002*, 1–9.

Shashoua, Y. (2001) *Inhibiting the Deterioration of Plasticised Poly (Vinyl Chloride): A Museum Perspective*. PhD dissertation, Danish Polymer Centre, Technical University of Denmark.

Shashoua, Y. (2004) 'Modern plastics: do they suffer from the cold?' *Modern Art, New Museums: Contributions to the IIC Bilbao Congress*, A. Roy and P. Smith (eds), 91–5. London: International Institute for Conservation of Historic and Artistic Works.

Wilson, A.S. (1995) *Plasticisers: Principles and Practice*. London: Institute of Materials.

The author

Yvonne Shashoua is a Senior Researcher at the National Museum of Denmark investigating the degradation mechanisms and inhibitive conservation processes for synthetic materials. She joined the British Museum as a conservation scientist in 1988, specialising in the deterioration reactions and conservation of cellulose nitrate, PVC and rubber objects. In 1998, she was offered a PhD stipendium to research into the deterioration reactions associated with plasticised PVC at the National Museum of Denmark and the Technical University of Denmark. The thesis (Inhibiting the deterioration of plasticised PVC) was completed in 2001.

Address

Yvonne R. Shashoua, Department of Conservation, National Museum of Denmark, PO Box 260 Brede, DK-2800 Kongens Lyngby, Denmark (yvonne.shashoua@natmus.dk).

The pits of despair? A preliminary study of the occurrence and deterioration of rubber dress shields

Anna Hodson

ABSTRACT The presence and deterioration of rubber-proofed dress shields in museum collections are researched and the conservation problems they pose are considered. Two collection surveys, conducted at the Gallery of Costume, Manchester, and Hampshire County Council Museums and Archives Service (HCCMAS), reveal the extent, types and condition of dress shields in museum collections. Instrumental analytical testing is used to identify rubber and starch as the problematic proofing materials present in the dress shields; the observable deterioration processes of the dress shields are described and the probable causes suggested. The detrimental effect of degradation of the dress shields on the cotton textiles of the host garments is assessed through accelerated ageing experiments on selected samples. Accelerated thermal ageing experiments undertaken suggest that the cotton fibres undergo acid-catalysed hydrolysis as a result of being in contact with deterioration products of the vulcanised natural rubber in the dress shields. Possible preventive treatment strategies are highlighted and briefly considered for the storage of dress shields in museum collections.

Keywords: dress shields, rubber, perspiration, thermal ageing, acid-catalysed hydrolysis, cotton

Introduction

Dress shields, also known as dress preservers and dress protectors, are small crescent-shaped items that are composed of a textile layer secured to a waterproof layer. They are most often stitched into the underarm area of garments to prevent the transferral of perspiration from the wearer to their clothing. These protective and practical components of Western female dress were at their most popular during the second half of the 19th century until c.1910. Several variations of design exist but all have at least one layer of textile, either silk or more commonly cotton, and a proof layer of some form. The waterproofing layer in the dress shields examined in the course of this study was found to be predominantly natural rubber. Two major designs of dress shield are considered in this paper. The terms 'proof' and 'proofed' have been adopted in this paper to describe the waterproofing layer in dress shields as they were used in contemporary advertisements for these items (Fig. 1).

Another major class of dress shield – a separate garment incorporating underarm shields worn over underwear – are not considered in the study as these are not permanently attached to the inside of garments and thereby do not pose the same potential problems as a result of their deterioration.

Due to their past function, use and composition, dress shields found in costume collections today are often unsavoury looking. They are stained and discoloured, and show signs of deterioration. Their displeasing countenance has attracted the attention of costume curators and conservators,

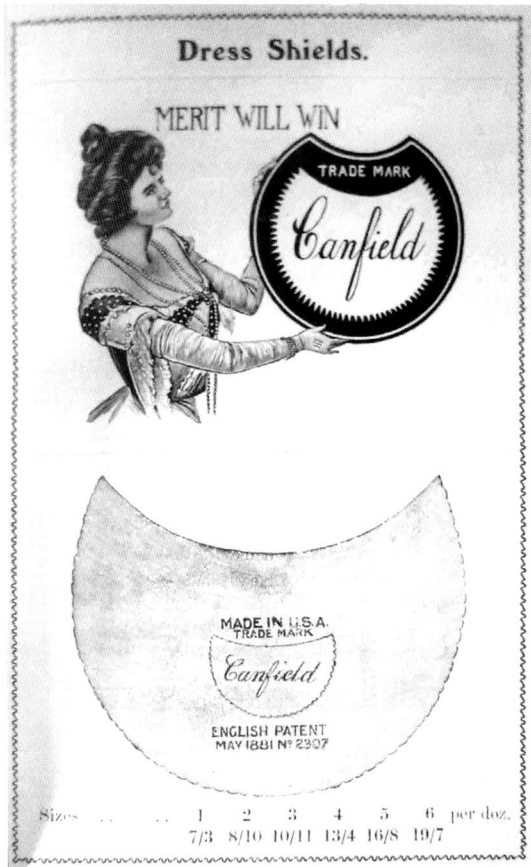

Figure 1 A dress shield advertisement from the Richard Evans and Company haberdashery catalogue.

who encounter them in the course of their work, since their appearance suggests that they may now be causing damage to the garments they were initially designed to protect.[1] Furthermore, it became apparent in the preliminary stages of the research that dress shields are, on occasion, being removed from costume items in response to the perceived damage they are causing to the surrounding garment.[2] Dress shields are of interest and value in the study of Western dress; removing them from garments eradicates not only their physical presence but potentially their place in dress history. Such action is therefore felt inadvisable. Patently there is concern, where they are present in costume collections, both for the long-term stability of the garments and the dress shields themselves. Despite this apprehension and occasional radical intervention, no formal study has been made to determine whether dress shields do indeed pose a potential deterioration problem for costume items. This paper, therefore, aims to address the following:

- to identify whether dress shields are actively deteriorating and to suggest the causes;
- to establish whether this poses a threat to the host garment;
- to consider the implications of results found.

In order to answer these questions, two dress collections were surveyed. The results identify different types of dress shield in collections and establish the extent of their occurrence. The types of deterioration present are highlighted and provide evidence of the causes of deterioration. Accelerated ageing tests were devised and constructed to further investigate the deterioration processes observed and to suggest future deterioration pathways of dress shields.

Collection surveys

Manchester City Galleries' Gallery of Costume (Platt Hall, Manchester) was chosen for the research because the collection contains a large number of dresses from the early Victorian period. The second collection selected was Hampshire County Council Museums and Archives Service (HCCMAS) (Winchester) because it holds a large number of costume items from the period when dress shields were most popular. All female garments that cover the armpit area, dated from 1840 to 1910, were examined in both collections. In total, 1,144 garments were surveyed, revealing 134 examples of garments with dress shields – 12% of the total garments studied. Almost all of the dress shields surveyed showed signs of damage and deterioration of some description. The trends in deterioration observed will be outlined later.

Types of dress shield encountered

For the purposes of this paper, the dress shields encountered during the surveys are most usefully placed into two distinct categories according to their construction as this is a contributory factor in their own deterioration and the damage they cause to garments.

1. Proof layer applied to a textile substrate: dress shields in this category have the proof layer applied directly to a textile layer. The rubber is either coated onto one layer of fabric or is applied so that it is sandwiched between two layers of fabric.
2. Proof layer as a self-supporting film: in this instance, the proof layer is cast as a film without a textile substrate, which is then attached to a textile layer by stitching around the edge of the crescent shape. Essentially the textile acts as a comfort cover for the proof layer. The proof film can have a layer of textile on one side only or is sandwiched between textiles layers, one on either side.

Material analysis

Material analysis of the proof layer present in the shields was conducted to confirm or refute the initial assumption that its major constituent was rubber. The component materials of 21 samples were identified using Fourier transform infrared–attenuated total reflectance (FTIR–ATR) spectroscopy.[3] Azide tests for the presence of sulphur were carried out to indicate the presence of vulcanised rubber (TCC 2002). The results of these tests indicated that the dress shields tested were, in fact, made from rubber and to a greater or lesser extent, vulcanised rubber. The results supported the information found in advertisements from contemporary haberdashery catalogues (Richard Evans and Company c.1900–10; Kleinerts Rubber Company 1936) and a patent from the period (Canfield 1881), which indicated the use of natural rubber. Analysis also identified the presence of starch in the films of the self-supporting proof layers, possibly as a filler to extend the rubber, as an additive to manipulate the properties of the rubber or to make the shields more comfortable to wear.

The inherent instability of rubber has been extensively researched and published (Loadman 1993; Blank 1988; Fisher 1957), and the susceptibility of rubber to oxidation, leaving it inflexible and brittle, is particularly well documented. The problems involved in the preservation of rubber will not be further discussed per se, but will be considered within the context of the deterioration of dress shields and the effect of the deteriorating vulcanised natural rubber on the surrounding textiles of the host garment.

Deterioration observed

Nine different types of deterioration, causing both physical and chemical changes to shields and their host garments, were identified in the course of the collection surveys. These types are categorised here by the likely cause of damage. For clarity, the causes of deterioration that are related to use of the garment are described first, followed by deterioration to dress shields that are the direct result of its component materials.

components. Further testing would be useful to establish the practical applications and viability of this solution.

The prevention of contamination of materials local to the shields

This may provide the most practical and pragmatic solution. It may be useful to isolate the garments with dress shields, storing them separately to prevent passive deterioration of other garments in their vicinity, as though in quarantine. The use of materials to absorb the pollutants and create a barrier layer between the shield and the garment (something as simple as Melinex polyester sheet or acid-free tissue), if changed regularly, could prove a most effective solution. Trials of such a solution could prove very valuable for institutions with limited resources.

Documentation

During the collection surveys it was highlighted that the presence of the dress shields in the garment was not documented at either institution. The importance of making a record of their existence is exemplified by the results of the survey, which implies that shields may eventually be lost or, at the very least, lose some of their significant components.

Removal

As discussed at the start of this paper, a reported instance of the removal of dress shields from a garment contributed to the reason for the research of this subject. The results of the research could create a climate of temptation! This clearly is a significant ethical issue and cannot be sufficiently discussed in the space available here. It is necessary, however, to draw attention to the lack of information available about dress shields, other than the objects themselves. As noted previously, removing them may compromise their place in textile history.

Dress shields are fascinating objects in their own right; they are physical manifestations of human behaviour, social mores and a personal connection to the garment's wearer. As such they perform a vital role as objects of material culture and thus contribute greatly to their host garment as a 'tool for historical and contemporary socio-cultural investigation' (Taylor 1998: 338). Removing them permanently from a garment would be a far-reaching decision that could compromise the integrity of an artefact or collection. Despite the solution being matchless in its success at reducing potential damage to garments by shields, the author does not favour it.

Conclusion

It has been shown that dress shields show a variety of types of deterioration, not all of which are active. Use-related damage may continue to deteriorate over time, but this is unlikely to threaten the integrity of the garment. It is, however, affecting the continuing survival of the dress shields themselves as objects and as examples of indispensable dress paraphernalia of the period. The research has shown that the deterioration of the rubber component of the dress shields does pose a threat to the host garment. The accelerated thermal ageing experiments demonstrated that over time, the deterioration products of the oxidation of natural rubber catalyse acid hydrolysis of cotton, causing yellowing and the breakdown of polymer chains leading to loss of tensile strength.

The results of the research demonstrate that dress shields must be identified and recorded in textile collections and their construction and any deterioration documented. Moreover, the research suggests preventive conservation measures to be an effective way to mitigate the inevitable damage they will cause to dress artefacts.

Acknowledgements

This paper is based on research conducted as part of an MA in Textile Conservation completed 2004 at the Textile Conservation Centre. The author would like to thank Sarah Howard and Alison Carter both from HCCMAS, as well as Anthea Jarvis and Miles Lambert from the Gallery of Costume, Platt Hall, Manchester City Galleries for allowing her to conduct the collection surveys and for support, help and advice. Thanks also to Dr Paul Garside, AHRC Research Centre for Textile Conservation and Textile Studies, University of Southampton and the Departments of Chemistry and of Engineering (from the same institution) for their expertise, time and the use of equipment. A special thanks to Dr Paul Wyeth, Textile Conservation Centre and Sarah Howard again for help, advice and guidance with the research project.

Notes

1. Information acquired through preliminary research of dress shields by oral communication with curators, conservators and collections care professionals.
2. *Ibid.*
3. Carried out by Dr Paul Garside, AHRC Research Centre for Textile Conservation and Textile Studies, University of Southampton.
4. From verbal communication with Yvonne Shashoua, National Museums of Denmark.
5. *Ibid.*

References

Bamberger, J.A., Howe, E.G. and Wheeler, G. (1999) 'A variant Oddy test procedure for evaluating materials used in storage and display cases', *Studies in Conservation* 44: 86–90.

Bhat, N.V, Dharmadhikari, S.N, Kulkarni, S.D. and Wani, S.N. (1990) 'Effects of perspiration on the fine structure of cotton fabrics', *Textile Research Journal* 60(4): 240–44.

Blank, S. (1988) 'Rubber in museums: a conservation problem', *AICCM Bulletin* 14: 53–93.

Cain, M.E. (1977) 'Effect of perspiration simulants on the degradation of uncovered natural rubber threads', *Textile Institute and Industry* January: 28–31.

Canfield, I.A. (1881) *Arm-pit Dress Shields*. British Patent No. 2307.

Chong, C.L., Chan, K. and Chow, F.S. (1994) 'Overcoming yellowing problems with cotton fabrics', *American Dyestuff Reporter* 83(May): 53.

Daniels, V. (1995) 'Starch adhesives', in *Starch and Other Carbohydrate Adhesives for Use in Textile Conservation*, P. Cruickshank and Z. Tinker (eds), 11–13. London: United Kingdom Institute for Conservation of Historic and Artistic Works.

Fisher, H.L. (1957) *Chemistry of Natural and Synthetic Rubbers*. New York: Chapman & Hall Ltd.

Grattan, D.W. (ed.) (1993) *Saving the Twentieth Century: The Conservation of Modern Materials*. Ottawa: Canadian Conservation Institute.

Kleinerts Rubber Company (1936) *Kleinerts Wholesale Price List* (unpublished price list in the collection of the Gallery of Costume, Platt Hall, Manchester).

Levitt, S. (1986) 'Manchester Mackintoshes: a history of the rubberised garment trade in Manchester', *Textile History* 17(1): 51–70.

Loadman, M.J.R. (1993) 'Rubber: its history, composition and prospects for conservation', in Grattan 1993, 59–80.

Lovett, D. (2003) *The Deterioration of Polyurethane (PUR) Foam with Reference to Foam Laminated 1960s Dresses*. Unpublished MA dissertation, Textile Conservation Centre, University of Southampton.

Oddy, W.A. (1975) 'The corrosion of metals on display', in *Conservation in Archaeology and the Applied Arts. Preprints of Contributions to the IIC Congress, Stockholm, 2–6 June 1975*, 235–7. London: International Institute for Conservation of Historic and Artistic Works.

Richard Evans and Company (c.1900–10) *Richard Evans and Company, Trimming Manufacturers and Importers Ltd* (unpublished haberdashery catalogue in the collection at the Gallery of Costume, Platt Hall, Manchester).

Rogerson, C. (2003) *Soil and Stain Identification and Removal* (unpublished report issued as a part of MA Textile Conservation course at the Textile Conservation Centre, University of Southampton).

Sanford, M. and Ordonez, M. (2002) 'The identification and removal of deodorants, antiperspirants, and perspiration stains from white cotton fabric', in *Strengthening the Bond: Science and Textiles. Preprints of the North American Textile Conservation Conference April 2002*, V.J. Whelan (ed.), 119–31. Philadelphia, PA: NATCC.

Shashoua, Y. and Thomsen, S. (1991) 'A field trial for the use of Ageless in the preservation of rubber in museums collections', in Grattan 1993, 363–72.

Taylor, L. (1998) 'Doing the laundry? A reassessment of object-based dress history', *Fashion Theory: The Journal of Dress, Body and Culture* 2(4): 337–58.

TCC (2000) *Testing Materials for Use in Storage and Display* (unpublished manual issued at a lecture at the Textile Conservation Centre, University of Southampton, 18 April 2000).

TCC (2002) *6.2 The Azide Test. Laboratory Manual* (unpublished manual issued at the Textile Conservation Centre, University of Southampton).

Tétrealt, J. (1994) 'Display materials: the good, the bad and the ugly', in *Exhibitions and Conservation. Preprints of the Conference, Edinburgh, 21–22 April 1994*, 79–87. Dundee: Dundee Art Gallery.

Tímár-Balázsy, Á. and Eastop, D. (1998) *Chemical Principles of Textile Conservation*. Oxford: Butterworth Heinemann.

Suppliers

- BioRad FTS 135 FTIR spectrometer, fitted with a Specac Golden Gate ATR accessory, over the range 4,000–700 cm^{-1}, with 32 scans at a resolution of 4 cm^{-1}. The spectra were subsequently manipulated using Thermo Galactic GRAMS 32/5.21 software.
- Heraeus Instruments oven: Severn Sales, 1 Lodge Road, Kingswood, Bristol BS15 1LD, UK.
- Poplin cotton shirting (152 cm width): Whaleys (Bradford) Ltd, Harris Court, Great Horton, Bradford, West Yorkshire BD7 4EQ, UK.
- 150 mm Pyrex brand hybridisation tubes with Teflon seal: Sigma Aldrich, The Old Brickyard, New Road, Gillingham, Dorset SP8 4 XT, UK.

The author

Anna Hodson has undertaken a Historic Scotland Internship in Textile Conservation at two venues, the National Museums of Scotland, Edinburgh and the Marischal Museum, Aberdeen, as well as a North West Arts Board funded residency at the Gallery of Costume, Platt Hall, Manchester. In September 2005 she began a Mellon Fellowship in Textile Conservation at the National Museum of the American Indian, Smithsonian Institution, Washington DC, USA.

Address

- Anna Hodson, National Museum of the American Indian/Smithsonian Institution, Cultural Resources Center, MRC 541, 4220 Silver Hill Road, Suitland MD 20746, USA (hodsona@si.edu).

Conservation applications: object studies

A global challenge: the search for conservation solutions for Eero Aarnio's Globe/Ball chair

Joelle Wickens

ABSTRACT The foam and adhesive degradation found in the 1968 Eero Aarnio Globe chair in the care of the Victoria and Albert Museum (London) is used as a case study for the examination of issues related to the conservation of 20th- and 21st-century foam-upholstered furniture. The history, construction and degradation of the chair are briefly summarised. The removal and replacement of degrading foam is presented as the most common previous approach to the conservation of other 20th-century foam-upholstered furniture. Three reasons why this approach may not be ideal in today's upholstery conservation environment are discussed and two other conservation possibilities are outlined. The issues that need to be examined with respect to these other possibilities are explained but results or a final conclusion for the chair has not yet been considered.

Keywords: adhesive, degradation, foam, Globe or Ball chair, upholstery, 20th-century design

Introduction

In 1963, inspired by his new empty flat in Helsinki, Eero Aarnio, a Finnish furniture designer, created the first prototype for what has come to be known as both the Globe chair and the Ball chair (Fig. 1).[1] In 1966, this piece of functional art was introduced to the wider world by the German division of the Asko Furniture Company at the Cologne Furniture Fair; in retrospect it is this event which is known to be the occasion that launched Eero Aarnio and his designs onto the international furniture market. Immediately following the launch and to the present day, the chair has been used as a prop in international marketing schemes, television series, movies and celebrity publicity photos. It has become an icon of the Pop era (Aarnion *et al.* 2003). It has been – and is still being – collected by museums as an object that represents furniture design of the late 1960s and early 1970s and for these museums it has become a conservation challenge without currently clear solutions.

One of these museums is the Victoria and Albert Museum (V&A), London, which, in 1968, received delivery of a Globe chair direct from the manufacturer Asko. It was purchased for Modern Chairs 1918–1970, an international exhibition presented by Whitechapel Art Gallery in association with *The Observer* and arranged by the Circulation Department at the V&A, and to this day it has only been stored and displayed in a museum environment. Due to the apparent nature of the materials used in its construction, however, it is rapidly degrading. The results of this degradation present modern-day upholstery conservators with significant challenges and ethical dilemmas. These challenges and dilemmas are one element of the investigation being carried out by the author in search of a potential conservation solution for the V&A's Globe chair and it is these issues that will be further discussed here.

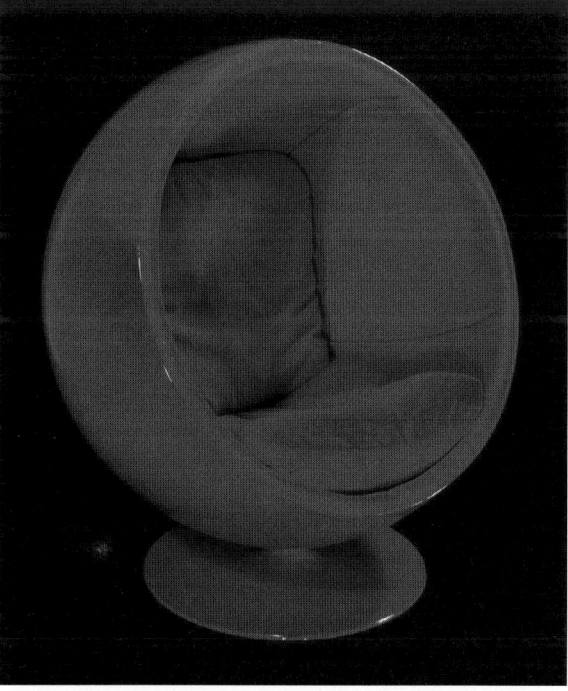

Figure 1 Eero Aarnio's Globe chair much as it would have appeared when it was first purchased by the V&A Circulation Department in 1968 (V&A Images/Victoria and Albert Museum, Museum number Circ. 12-1969) (Plate 56 in the colour plate section).

Construction and degradation of the chair

In its simplest form, the Ball chair is a fibreglass shell with a foam-upholstered interior which is mounted on a metal pedestal. The shell itself is constructed with an outer layer of gel

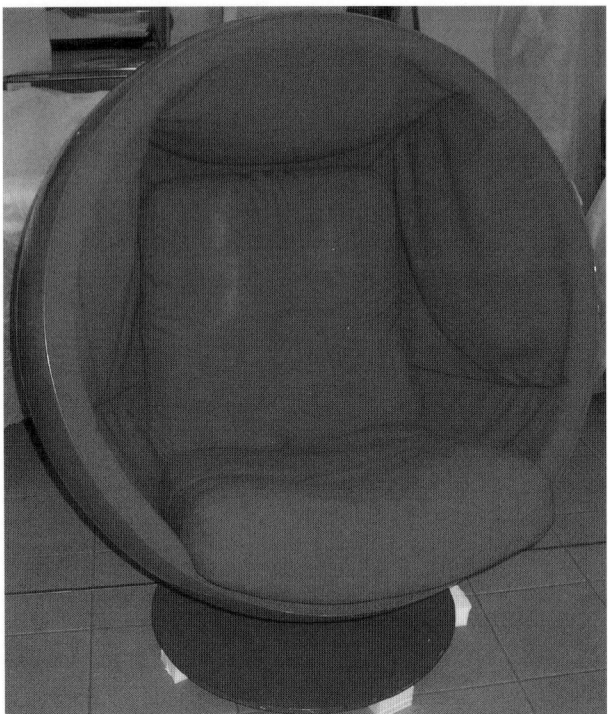

Figure 2 The condition of the V&A's Eero Aarnio Globe chair on 16 December 2004 (reproduced by kind permission of the Victoria and Albert Museum, Museum number Circ. 12-1969) (Plate 57 in the colour plate section).

Figure 2 shows the primary and most immediately obvious degradation issue for the V&A's Globe chair. The bond between the top cover fabric and the foam has failed in all areas not covered by the loose cushions. This bond failure has caused the fabric to become detached from the chair in many locations. Access to the surfaces of the foam and fabric in the areas where the bond has failed is severely limited due to the construction of the chair. Research suggests, however, that their condition is very similar to the condition of the surfaces shown in Figure 3. This photo shows the obverse of a foam wedge and the reverse of the top cover fabric removed from a Ball chair of a date similar to that of the V&A's in which the bond adhering the fabric to the foam has also failed.[3] The darker areas on both the foam and the fabric are what remain of the now hard and brittle adhesive originally used to secure the fabric to the foam. Additionally, there are small areas where the fabric-to-foam bond did not fail and small bits of foam have broken away from the padding and remain adhered to the underside of the fabric.

The other significant degradation issue – which is certainly less immediately obvious but becomes clear upon closer investigation – is the fact that in certain areas the foam in both the wedges and the loose seat cushion has become permanently compressed and/or deformed. In the case of the wedges, the bottom two foam triangles have been compressed by the pressure of the loose seat cushion and the foam in the loose seat cushion has taken on the contours of the chair shell. The foam itself remains rather resilient and bounces back when temporarily compressed with a finger, but when the pressure on the deformed areas described above was released for three days during examination of the chair no significant recovery was noted.[4]

There are also signs of degradation in a chipped gel coat and fibreglass on the chair shell and cracked paint on the pedestal, but it is the first two issues that are of primary concern when approaching the chair as an upholstered object. Therefore, how the conservation of similar elements in other artefacts has been carried out and how else the conservation of these elements in the Globe chair might be approached have been and are still being considered.

coat and an inner layer of non-woven fibreglass. It has been stabilised with four metal rods. The shell is upholstered with five triangular foam wedges. The top cover fabric is adhered to the entire obverse surface and the two longer edges of the reverse of each foam wedge. The reverse of each wedge is adhered to the inside of the fibreglass shell and the top cover fabric is further secured with stitching along the front edge of each wedge. The chair is finished off with two loose cushions of which the innermost layer is foam. The foam is surrounded by a layer of wadding and is then encased in a tightly woven cotton cover. The cushions are finally covered with a show cover made of the same fabric that is used to cover the triangular foam wedges.[2]

Figure 3 The obverse of a foam wedge and the reverse of the top cover fabric removed from a Ball chair of a date similar to that of the V&A's Globe chair. The dark brown areas on the foam and fabric are the still present but degraded adhesive (Plate 58 in the colour plate section).

Removal

Historically, the most common conservation approach for a chair with degrading foam upholstery has been to remove the degrading foam and replace it with new foam of either upholstery or conservation grade.[5] Research suggests, however, that this is not the ideal conservation approach by today's standards. Prior to about 1970 it was fairly standard museum practice to send upholstered museum objects to upholsterers' workshops where they would be stripped to their frames and then re-upholstered with all-new materials (Anderson 1990; Fairbanks and Nylander 1987; Gill 2004). This approach is still current practice outside the field of conservation but since the 1970s, upholstery conservators have been working to develop techniques that will make this a thing of the past. Conservators and historians have also been working to piece together the historical knowledge that was lost due to the removal of many original materials. Therefore, to remove original and/or historic foam upholstery materials and replace them with new materials would only be taking a step backwards along the road the world of upholstery conservation has built for itself.

Secondly, research indicates that the V&A's Globe chair is the earliest and perhaps only example of such a chair that still retains all of its original materials and construction elements, having never been altered by human hand. Admittedly, in the modern world, foam, fabric and fibreglass are extremely commonplace and therefore it might be argued that the previous point is really of little significance. In the field of upholstered furniture, however, where objects are constantly used, worn and renewed, an object that has passed through almost four decades with no human alterations is significant and particularly so if it is likely to be the only one of its kind that has done so. The fact that in this case the object is made of materials with which we come into contact on a daily basis does not negate that significance. Accepting this, removing the original foam upholstery materials and releasing original construction elements would irreversibly alter an as-yet untouched artefact and in so doing destroy part of what makes the object significant. While almost any conservation treatment is going to introduce a level of irreversible change, perhaps the replacement of original materials, particularly when the object is as yet untouched, should be recognised as one of the most extreme and probably least ideal options.

Thirdly, there is much evidence to suggest that a conservation approach for a 20th- or 21st-century upholstered piece of furniture that does not recognise that materials, design and production process are intimately linked is not considering the object as a whole. Harry Bertoia did not sculpt a chair out of wire because wire was the only material available to him or the material that all chairs were made of at the time. He selected wire as the material with which to create one of his designs because it made it possible for him to produce a see-through chair that complemented the open plan interiors that were popular in the 1950s. At the time, could a see-through chair like Bertoia's have been created without wire? In adapting production methods that allowed for the three-dimensional bending of wood and the welding of wood to rubber, glass or metal, Charles and Ray Eames found ways to bring to life many of their now easily recognisable designs (Rivers and Umney 2003). Without such production methods would their designs have been feasible? Gaetano Pesce took advantage of the fact that polyurethane foam could be moulded and compressed and developed into furniture which could be produced without high tooling costs, and packaged so that it could be carried home under the arm by the customer (Fiell and Fiell 1997; von Vegesack *et al.* 1996). Without polyurethane foam, could Pesce have made his designs so readily available to the general public? There are many, many more examples like these that demonstrate the links between materials, design and production process in furniture design in the 20th and 21st centuries, but perhaps no example is more convincing than the words of some current-day designers. Jo Saunders (Education Manager of the Crafts Council, London) interviewed the five designers or design teams shortlisted for the 2004 Jerwood Prize. Each designer was asked to respond to, among other questions, 'How would you describe your design and production process, from concept, through materials and technologies to the end product?' Jay Osgerby responded, 'At the model making stage we really start to focus in on the type of materials we'd like to use for a product. Obviously we've had something in mind, but generally as the process evolves we sort of restrict that choice and start thinking about material in a more specific way.' Edward Barber answered, 'On other projects you can start with a specific material and design using the constraints of that material.' Twentieth- and 21st-century furniture represents a complex set of relationships between design, materials and production processes and the conservation of such objects needs to take these relationships into account. If the material originally selected or specified by the designer, or required because of design or production reasons, is removed from an object, are we erasing part of what is significant about that object?

Thus, due to the current ethics of upholstery conservation, the fact that historic upholstered objects which retain all of their original materials are quite rare and the fact that design, materials and production process are intimately linked in 20th- and 21st- century furniture, the removal of original materials from 20th- and 21st-century foam-upholstered objects during the conservation process does not appear to be the ideal solution. What other options exist for the V&A's Globe chair and are such options more preferable than replacement?

Other possible solutions

One option is to try to re-adhere the original top cover fabric to the original foam wedges. This would obviously involve the introduction of new adhesive at the very least and might require the release of original stitching or original adhesive bonds that are still intact. It might require the cleaning of the foam and fabric surfaces identified for re-adhesion, which would be likely to remove original but degraded adhesive. It might require the addition of a consolidant as well as an adhesive. In order to determine just what would be required to re-adhere the original top cover fabric to the original foam wedges, a series of experiments will be carried out. These experiments will seek to answer the following questions:

- Is it possible to establish a good bond between 37-year-old foam and fabric and, if so, how long is the bond likely to last in an ideal museum environment?
- Can a method be developed that allows for the re-adhesion process to be accomplished without requiring the removal of elements of original construction?
- If original elements of construction must be released in order to accomplish the re-adhesion process, is it preferable to release the original stitching along the front edge of each foam wedge or the still-intact adhesive bonds located behind the seat cushions?
- Would cleaning the fabric and reducing the amount of adhesive residue that is still present on both the fabric and the foam improve the new bond?
- Is the foam strong enough to support the newly re-bonded fabric or does the foam need to be consolidated before re-adhesion takes place?
- If consolidation is necessary, it has been suggested that applying consolidants to foam is most successful

when a spraying technique is used (Rava *et al.* 2004). Would it be possible to use such a technique in the confines of an upholstered object and if not what are the alternatives?

But even if positive answers to all of these questions can be found, it is most likely that the foam wedges will continue to degrade and during the process the foam is likely to lose its current shape. Such a degradation result would negate any previously successful restoration of the profile of the chair and suggests that additional conservation options need to be investigated.

Another possibility that might provide a longer lasting profile restoration is to insert a rigid layer between the original foam and fabric. This layer could be shaped to the original profile of the chair and the original fabric secured to it. Such a solution would introduce a new material to the chair and how this introduction would alter the chair as an historic object needs to be considered. Assuming such an alteration would be acceptable there are certainly potential advantages to such a treatment. One would be that as the foam beneath the rigid shell continued to degrade, its change in shape would not affect the profile of the chair. Additionally, the profile of the bottom two wedges, beneath the loose seat cushion, could be restored and the cushions could be placed back on the chair without fear of continued compression of the profile beneath. Three likely drawbacks are that first, encasing the foam in a rigid shell might introduce a microenvironment that would accelerate the degradation of the foam (van Oosten 2002). Secondly, the rigid shell and the top cover fabric might respond differently to fluctuations in humidity levels and this unequal response could result in the development of wrinkles in the fabric (Graves 1990). Thirdly, a material with which to construct such a shell has not yet been found. The material would need to be thin enough to add no depth to the profile, initially flexible and with the ability to be moulded to the original profile of the chair as well as ultimately rigid and strong enough to support the weight of the fabric, adhesive and the seat cushions. Clearly there is much experimentation that needs to be carried out before it can be determined that the addition of a rigid layer is an acceptable and effective conservation solution and again, such experimentation is underway.

Conclusion

Research suggests that while removal and replacement of degrading foam has been an historically acceptable solution with regard to the conservation of 20th- and 21st- century foam-upholstered furniture, today such a solution may no longer be ideal. The current goals of upholstery conservation, the significance of original upholstery materials and construction elements, and the relationships between design, materials and production methods in the 20th and 21st centuries indicate that other solutions need to be found. Two possible alternative conservation approaches for the Eero Aarnio Globe chair in the collection at the V&A are being investigated in order to consider what these other solutions might be. Whether these other conservation solutions will be successful in the short and long term has not yet been determined. Whether the replacement of original degraded materials with upholstery or conservation grade substitutes does more to honour the relationships between design, materials and production processes in 20th- and 21st-century furniture than the alternative conservation possibilities being investigated is still being considered. It is not expected that clear solutions for all 20th- and 21st-century foam-upholstered furniture will be defined but it is hoped that at the conclusion of the investigation, upholstery conservators will be one step closer to developing conservation methods that will make it possible to retain some of the 20th-century synthetic materials used to upholster furniture in the recent past.

Acknowledgements

The author would like to thank the Textile Conservation Centre and the Victoria and Albert Museum for permission to publish. Many thanks to the Stockman Family Foundation, the Textile Conservation Centre Foundation, the Pasold Research Fund, the Zibby Garnett Travelling Fellowship, the National Association of Decorative and Fine Arts Societies, the Furniture History Society and the University of Southampton for funding her research to date and thanks to Kathryn Gill, Dr Maria Hayward and Dr Paul Wyeth for supervising her research.

Notes

1. The name most commonly used for this chair is the Ball chair. The Asko invoice for the V&A's chair, however, lists it as the Globe chair. Therefore, when referring specifically to the V&A's chair, the term Globe chair will be used but when referring to the chairs in general the term Ball chair will be employed.
2. A more detailed description of the construction of Ball chairs can be found in J. Wickens, 'Documenting the Globe: recording and conserving modern upholstery techniques and materials before they disappear', which is expected to be published in 2006/7 in the postprints of the conference, *The Forgotten History: Upholstery Conservation*, Birgitta Forum, the Faculty of Arts and Sciences, Linköping University, Sweden, 12–13 May 2005.
3. The foam pad pictured in Figure 3 was supplied with four others by Michael Marks of 20th Century Marks, a 20th-century furniture dealer located in Little Burstead, Essex, England.
4. The author carried out a detailed examination of the V&A's Globe chair in the Textile Conservation Studio of the V&A on 13–16 December 2004.
5. Supporting information for this comment was gathered by the author during a series of interviews with curators and conservators of upholstered furniture in Europe and North America primarily in the spring and summer of 2004. All notes taken during the interviews are held by the author.

References

Aarnion, E., Kalha, H., Korvenmaa, P. *et al.* (2003) *Oleta Pyöreä Tuoli: Eero Aarnion 60-Luku* [*Assume a Round Chair: Eero Aarnio and the 60s*]. Helsinki: Helsingin Taidehalli/Taideteollinen Korkeakoulu [Kunsthalle Helsinki/University of Art and Design Helsinki].

Anderson, M. (1990) 'The history and current direction of minimally intrusive upholstery treatments', in Williams 1990, 13–28.

Fairbanks, J. and Nylander, J. (1987) 'Introduction', in *Upholstery in America and Europe from the Seventeenth Century to World War*

I, E. Cooke (ed.), 11–12. New York: Norton & Co.

Fiell, C. and Fiell, P. (1997) *1000 Chairs*. Cologne: Taschen.

Gill, K. (2004) 'The development of upholstery conservation as a practice of investigation, interpretation and preservation', *Reviews in Conservation* 5: 3–22.

Graves, L. (1990) 'Conservation of an 1811 sofa with nearly-intact original upholstery', in Williams 1990, 356–76.

Rava, A., Verteramo, R. and Chiantore, O. (2004) 'The restoration of a group of works of art by Piero Gilardi', in *Modern Art, New Museums: Contributions to the IIC Bilbao Congress*, A. Roy and P. Smith (eds), 160–64. London: International Institute for Conservation of Historic and Artistic Works.

Rivers, S. and Umney, N. (2003) *Conservation of Furniture*. Oxford: Butterworth-Heinemann.

van Oosten, T. (2002) 'Crystals and crazes: degradation in plastics due to microclimates', in *Plastics in Art: History, Technology, Preservation*, T. van Oosten, Y. Shashoua and F. Waentig (eds), 80–89. Munich: Anton Siegl.

von Vegesack, A., Dunas, P. and Schwartz-Clauss, M. (eds) (1996) *100 Masterpieces from the Vitra Design Museum Collection*. Weil am Rhein: Vitra Design Museum.

Williams, M. (ed.) (1990) *Upholstery Conservation*. East Kingston, NH: American Conservation Consortium, Ltd.

The author

Joelle Wickens is currently working at the Textile Conservation Centre on a PhD thesis entitled *Investigating Possibilities for Re-adhering Fabric to Degraded Polyurethane Foam: One Step Toward Restoring the Profile of a 1960s Upholstered Chair – Eero Aarnio's Globe Chair, a Case in Point*.

Address

Joelle D.J. Wickens, Textile Conservation Centre, University of Southampton, Park Avenue, Winchester, Hants SO23 8DL (Joelle.Wickens@soton.ac.uk).

A study of sequins on a Cantonese opera stage curtain

Angela Cheung

ABSTRACT Objects associated with Chinese Cantonese opera form one of the major collections of the Hong Kong Heritage Museum. The sequins on one of the Cantonese opera stage curtains are disintegrating seriously; the curtain itself was used as a backdrop during the performance of the opera. In order to understand the possible causes for this degradation and hopefully to devise the proper strategies for remedial treatment and storage, the composition of the sequins was studied using a variety of analytical techniques including Fourier transform infrared spectroscopy (FTIR), x-ray fluorescence (XRF) spectroscopy and scanning electron microscopy with energy-dispersive x-ray spectroscopy (SEM–EDX). The analyses were carried out on broken sequins that had become detached from the curtain. The results indicated that cellulose nitrate, silver and gelatin were present. As there was little information about the fabrication and the past history of the stage curtain, the complete mechanism for its degradation could not be deduced. It is believed, however, that the deterioration of the sequins is due to the presence of unstable cellulose nitrate, accelerated by the unfavourable environment in which the curtain had been stored. Preventive measures were adopted to retard the deterioration rate of the sequins while considering other long-term preservation strategies.

Keywords: sequins, Cantonese opera, cellulose nitrate, gelatin, vibrational spectroscopy, energy-dispersive x-ray (EDX) spectroscopy

Introduction

Cantonese opera (also known as Guangdong drama) is one of the major forms of Chinese opera in southern China. The art had become popular by the end of the Ming dynasty (1368–1644) and is well received in the Guangdong and Guangxi provinces as well as in Hong Kong. Cantonese opera is a traditional Chinese art form that involves music, singing, martial arts, acrobatics and acting, to depict the country's history, traditions, culture and philosophies. To promote this art form in Hong Kong, the Hong Kong Heritage Museum has set up a Cantonese Opera Heritage Hall (exhibition gallery) to display objects associated with the opera. The range of artefacts is diverse, including opera costumes, headdresses and accessories, musical instruments, librettos and scripts, wooden trunks for costumes, wooden weapons, etc. Within this collection, a large proportion of the textile artefacts, including costumes and curtains, are sequined. When worn on stage, the sequined costumes were designed to produce a glittering effect under the stage lighting. One sequined Cantonese opera stage curtain was noted to have serious problems with the condition of its sequins when it was recently accepted into the collection. This phenomenon had not been observed on similar textiles decorated with sequins.

The Cantonese opera stage curtain

This particular stage curtain was made in the period 1940 to 1950. It measures 228 cm in height and 174 cm in width. The curtain has a hanging sleeve at the top and is comprised of two layers. The base layer is made of two pieces which can be drawn to either side from the centre. On top of the base curtain, there are two side-flaps with tassels. The curtain was intended to be hung on stage as a backdrop during the performance. It is densely decorated with disc-shaped sequins of 5 mm diameter. The non-iridescent side of each sequin is sewn onto the fabric with the iridescent side facing outwards. The sequins form a dragon pattern in the middle portion of the curtain. A total of six colours of sequins was observed on the stage curtain, namely, bright pink, purple pink, light pink, gold, purple and black. The purple and black sequins were used on the hanging sleeve and the eye of the dragon pattern respectively, and were found in smaller numbers than the other colours. The base fabric of the curtain itself is made of a rayon/silk blend, lined with rayon at the back while the two side-flaps are made of cotton. As more sequins were used on the two side-flaps, a stronger fabric (cotton) was required for the lining (instead of rayon and silk) to bear this additional weight (Fig. 1).

Condition of the stage curtain

The stage curtain is structurally sound but suffers from loss of sequins in various areas. It is torn around the middle area of

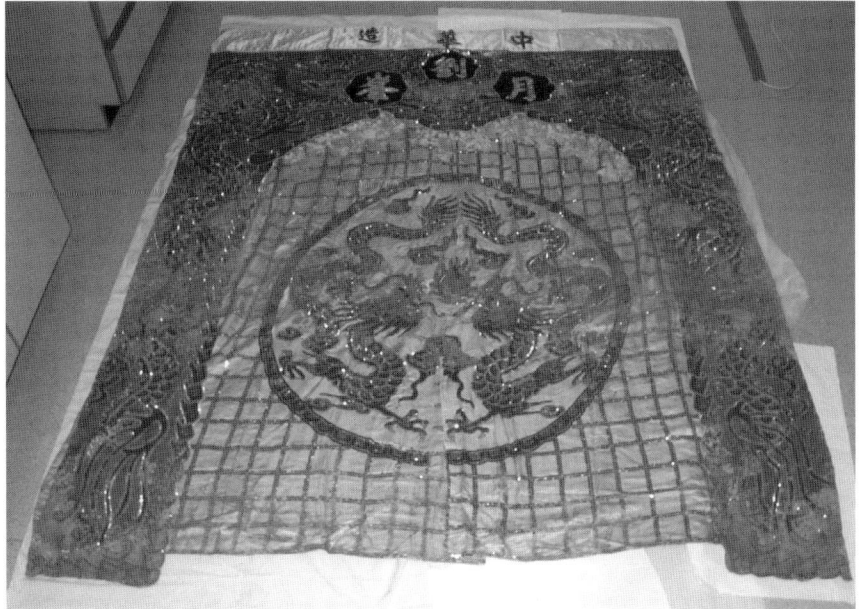

Figure 1 Cantonese opera stage curtain (Plate 59 in the colour plate section).

pink fabric and slightly split along the edges. The sequins are very brittle and many are lost each time the curtain is handled. The sequins exhibit various degrees of deterioration including surface crazing, blistering, disintegration, shattering and sticking together; some powdery substances were observed on their surface (Figs 2 and 3). The sequins in the worst condition were found in the following locations:

- The upper part near the name of the performer;
- The two side-flaps especially near the bottom;
- Around the dragon tail and head;
- The circle around the dragon pattern in the middle of the pink fabric;
- The bottom portion of pink fabric.

As can be seen in Figure 1, the pattern of loss of sequins appears to be symmetrical, which suggests that it may have been folded in the past and exposed to unfavourable storage conditions. The curtain also has pink and brown stains in numerous areas, but not limited to those areas sewn with sequins. Stains are also seen on the back of the lining where there are no sequins. Such staining is apparently caused as a result of colour transferring from the sequins when the curtain was folded for storage.

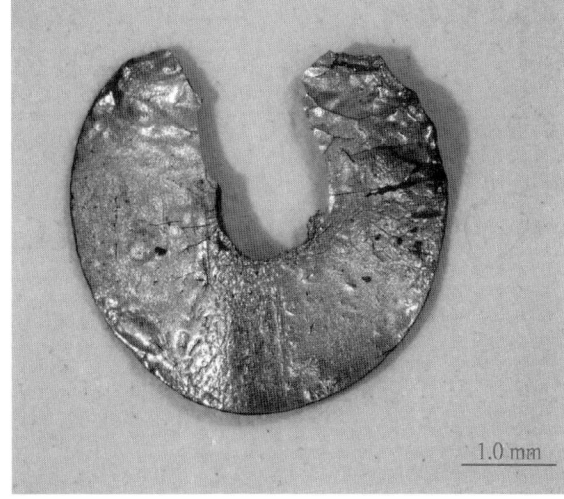

Figure 2 Condition of a degraded sequin (Plate 60 in the colour plate section).

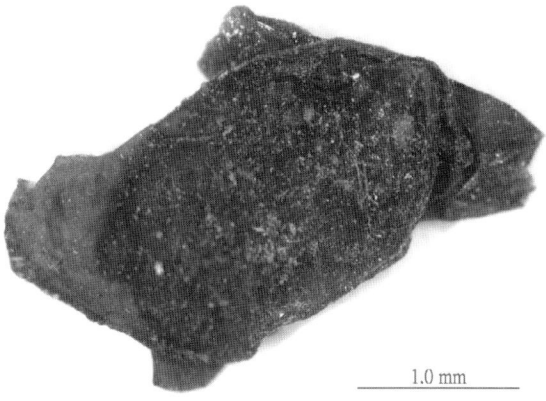

Figure 3 Condition of a degraded sequin (changed to amber in colour) (Plate 61 in the colour plate section).

Cellulose nitrate

Cellulose nitrate is formed by the reaction of cellulose $(C_6H_7O_2(OH)_3)_x$ and nitric acid (HNO_3) in the presence of sulphuric acid (H_2SO_4) (Reilly 1991). Sulphuric acid is added to control the rate and extent of the reaction. Nitrate groups replace the hydroxyl groups of the cellulose; the extent of nitration affects the physical and chemical properties of the resultant cellulose nitrate (1).

$$(C_6H_7O_2(OH)_3)_x + 3HONO_2 + H_2SO_4 \rightarrow$$
$$(C_6H_7O_2(ONO_2)_3)_x + 3H_2O + H_2SO_4 \qquad \text{Reaction (1)}$$

The greater the nitrogen content, the more unstable the cellulose nitrate (Williams 1994a; Stewart et al. 1995).

Cellulose nitrate deteriorates as a result of thermal, chemical and photochemical reactions. The O-NO$_2$ bonds cleave and produce the NO$_2$ radical, followed by chain scission or ring opening when exposed to ultraviolet (UV) radiation (Selwitz 1998). Nitrogen dioxide will be formed as the major degradation product. In the presence of moisture, nitrogen dioxide (NO$_2$) reacts to form nitrous acid (HNO$_2$) and nitric acid (HNO$_3$).

$$2NO_2 + H_2O \rightarrow HNO_3 + HNO_2 \qquad \text{Reaction (2)}$$

These acids further accelerate the degradation of the material, acting as catalysts for the cleavage of the nitrate group and also attacking the polymer in general by breaking it down into shorter chain fragments (Fig. 4).

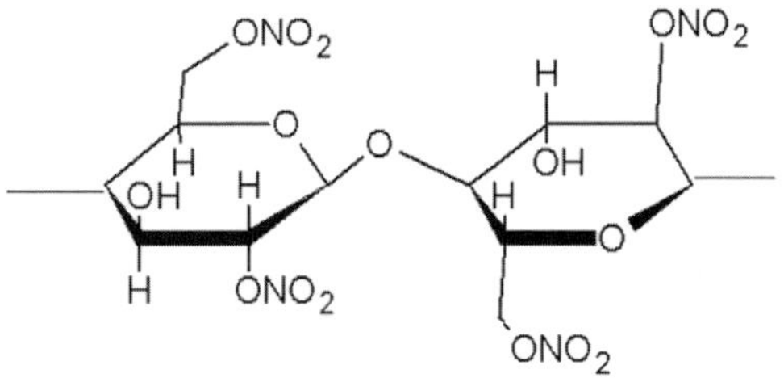

Figure 4 Cellulose nitrate.

Analytical techniques

The compositions of the sequins were studied using various analytical techniques and chemical tests in order to identify the possible causes of their degradation.

Fourier transform infrared spectroscopy (FTIR)

FTIR is a tool for identifying the types of chemical bonds in a molecule by producing an infrared (IR) absorption spectrum. When IR radiation interacts with a sample, radiation of the same frequency as that of inter-atomic bond vibrations is absorbed. Every molecule will have its own characteristic IR spectrum, akin to a fingerprint, which corresponds to the specific types of bonds found within it; hence the nature of the molecule can be deduced. Attenuated total reflectance (ATR) is one of the sampling methods available to use with the technique. The sample is placed in contact with a crystal of high refractive index, where total internal reflection will occur along the crystal–sample interface. The IR beam enters the crystal and is internally reflected along the length of sample; it interacts with the surface of the sample through a phenomenon known as the evanescent wave effect, before leaving the crystal and being recorded by a detector.

x-ray fluorescence spectroscopy (XRF)

XRF is a non-destructive analytical technique used to provide information on the elemental composition of materials.

When a sample is irradiated with x-rays, radiation can either be absorbed by the component atoms or scattered through the material. If the energy of an x-ray is sufficient, an electron from the inner shells will be ejected to give rise to an unstable state. An electron from the outer shell will then drop down to fill this vacancy, losing its excess energy as an x-ray of characteristic frequency. Each element has a unique set of energy levels and as a result the energies of x-ray fluorescence are specific to the elemental composition of the material in question.

Scanning electron microscopy with energy-dispersive x-ray spectroscopy (SEM–EDX)

SEM–EDX can also give an elemental analysis of the sample. The surface of the sample is scanned using a beam of high energy electrons. The incident electrons will give rise to different types of emitted particles or radiation depending on the type of interaction between the electron beam and the sample. Secondary electrons and backscattered electrons emitted from the sample may be captured to yield high magnification images of the sample surface. Furthermore, the electron beam on the sample will also generate other signals, such as x-rays with characteristic energies, in a similar manner to the XRF technique. With a built-in energy-dispersive x-ray (EDX) spectrometer, the elemental composition of sample can be identified from the x-ray spectrum.

Experimental method and results: analytical techniques

Light microscopy

When the sequins were examined under the stereomicroscope, it was noticed that they were composed of three layers. The top layer was the iridescent surface (with colorant); the middle layer was shiny and appeared to be metallic; the base material was a gelatin-like substance. The iridescent surfaces were readily soluble in acetone, revealing the metallic layer below. The colour of the base substrate ranged from pale yellow to light brown and amber. This layer dissolved in water to form a jelly-like substance and was readily soluble in hot water.

Fourier transform infrared spectroscopy (FTIR)

The sequins were analyzed using a Nicolet Nexus 470 FTIR spectrometer and ATR was employed as the sampling technique. The spectra were recorded over the range 4,000–400 cm^{-1}, with a resolution of 0.5 cm^{-1}, averaging the data over 64 scans. Sequins of various colours were analyzed by FTIR and the results summarised in Tables 1 and 2. The spectra indicate that cellulose nitrate is present in the sequins. The peaks at 1,636 and 1,277 cm^{-1} correspond to N-O stretching, 1,017 cm^{-1} is assigned to C-O bending and 845 cm^{-1} to N-O bending. The additional peaks observed from the spectrum at around 3,269, 1,500–1,550 and 1,440–1,460 cm^{-1} suggest the presence of an amide group (Derrick et al. 1999). From spectra

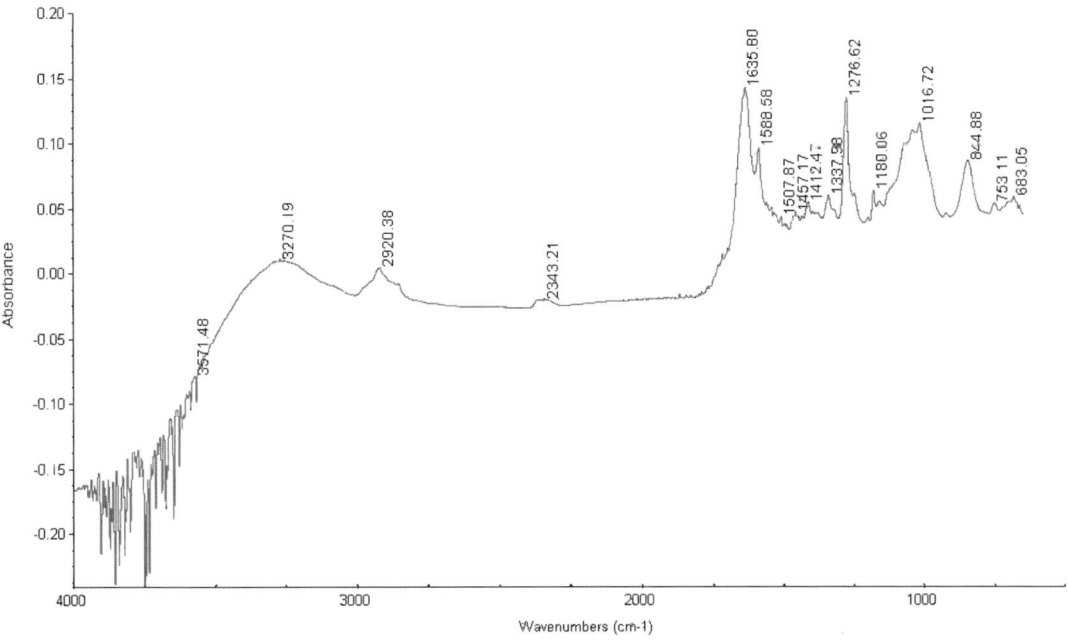

Figure 5 FTIR spectrum of a pink sequin (iridescent side) (Plate 62 in the colour plate section).

of the base material (using a sequin from which the colorant and the metallic layer had been scraped off), it appears that the peak near 1,540 cm^{-1} is due to C-N and N-H vibrations, and C-H bending near 1,450 cm^{-1} which suggests a high possibility that this layer contains gelatin.

As some of the sequins were so highly degraded and discoloured as to appear amber, only the substrate could readily be observed by light microscopy. The FTIR spectrum of the degraded sequins shows the peaks at 1,613, 1,316, 1,015 and 760 cm^{-1}, which are likely to arise from the presence of oxalate. The peak at 1,277 cm^{-1} is absent, which suggests that the N-O bond is not present in the degraded material (Fig. 5).

XRF

The analysis on the sequins was conducted with a SEA 200 XRF spectrometer, using measurement conditions of 50 kV for 600 seconds, with a 5 mm collimator. Using this equipment, which can detect elements from sodium to uranium, silver was detected in the sequins.

SEM–EDX

The sequins were analyzed by SEM–EDX using a Quanta F 200, in order to detect the possible elemental composition of sequins at different stages of deterioration, including those elements of lower atomic number which cannot be detected by XRF. The elemental analysis suggests the presence of carbon, oxygen, nitrogen, silver, sulphur, calcium and chlorine in the sequins. For the badly degraded sequins, however, nitrogen, silver and sulphur are absent (Fig. 6).

Experimental method and results: chemical tests

Diphenylamine spot test

The diphenylamine spot test was used to confirm the presence of cellulose nitrate in the sequins. A solution of 0.5% diphenylamine in 90% sulphuric acid was employed as the reagent; this reagent is highly corrosive. Sulphuric acid reacts with the

Table 1 Principal peaks from FTIR spectrum of the iridescent sides of sequins.

Sequin	Wavenumber (cm^{-1})
Pink	3,270; 2,920; 1,636; 1,589; 1,338; 1,277; 1,016; 845
Light pink	3,246; 2,921; 1,637; 1,277; 1,061; 845
Purple pink	3,253; 2,923; 1,636; 1,541; 1,457; 1,277; 1,016; 845
Yellow	1,639; 1,278; 1,033; 833; 695
Purple	3,269; 1,639; 1,278; 1,017; 846
The surface where the colorant and middle layer were scraped off (base material)	1,628; 1,541; 1,448; 1,332; 1,236; 1,079

Table 2 Principal peaks from FTIR spectrum of the non-iridescent side of sequins.

Sequin	Wavenumber (cm^{-1})
Pink	3,270; 2,920; 1,637; 1,589; 1,412; 1,276; 1,016; 844
Light pink	3,225; 2,921; 1,636; 1,276; 1,016; 842
Purple pink	3,269; 2,923; 1,636; 1,541; 1,457; 1,277; 1,015; 846
Yellow	1,639; 1,278; 1,005; 839
Purple	1,636; 1,276; 1,015; 843
The surface where the exterior layer of the base of sequin was scraped off (base material)	3,291; 1,653; 1,636; 1,559; 1,540; 1,457; 1,279

125

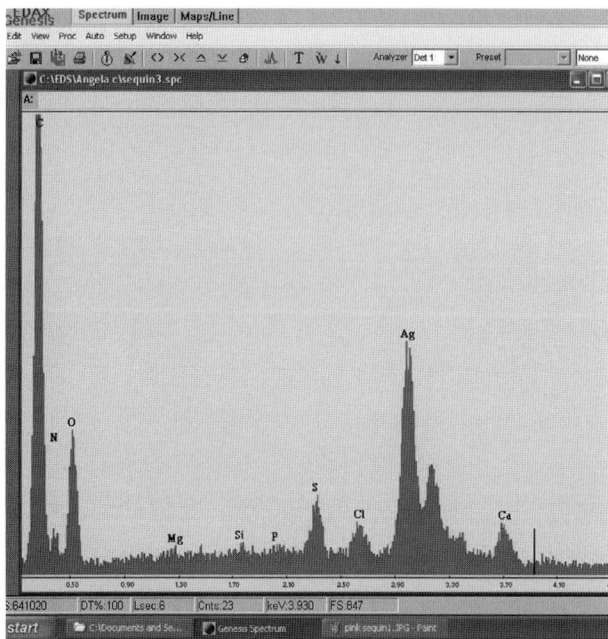

Figure 6 SEM–EDX spectrum of a light pink sequin (Plate 63 in the colour plate section).

cellulose nitrate to liberate nitrogen oxide. The diphenylamine is then oxidised by the nitrogen oxide to form a quinoid-type blue dye. As a result, if cellulose nitrate is present, a blue-violet colour will develop (Williams 1994b).

The sequin was placed on a microscopic slide and a drop of reagent was added. A blue colour developed on the sequin within seconds, confirming the presence of cellulose nitrate.

Ehrlich's reagent test

This test is commonly used to identify the gelatin in photographic emulsion, and requires the following solutions:

- 4-dimethylaminobenzaldehyde (5 g) dissolved in 100 cm^3 propan-1-ol
- 0.01 M copper sulphate solution
- (20 vol) hydroxide peroxide solution which must be freshly prepared
- 2 M sulphuric acid
- 4 M sodium hydroxide solution

If gelatin is present, a pink coloration will develop. The sequin showed a positive result for this test.

Discussion

The analyses and chemical tests carried out on the sequins indicate that cellulose nitrate, gelatin and silver are present. Trace amounts of elements such as calcium, sulphur and chlorine are also detected. The signs of degradation of the sequins, such as blistering, disintegration, browning, etc, are similar to those recorded in the literature for cellulose nitrate degradation (Blank 1990; Derrick *et al.* 1993; Matsumura *et al.*

2002). It is well known that cellulose nitrate is unstable and will degrade over time, therefore the overall deterioration of the sequins may be a direct result of cellulose nitrate deterioration. The rate of degradation of cellulose nitrate-based materials depends on the composition of the material (additives, stabilisers, etc.) as well as its past storage history.

Degradation of sequins

Cellulose nitrate is prone to degrade under unfavourable environmental conditions, such as high temperature, high humidity and exposure to sunlight. Gelatin is also very susceptible to damage in the presence of moisture. As this Cantonese opera stage curtain was used frequently in performances, it is not unreasonable to deduce that it had been stored inappropriately in the past and subjected to unfavourable environments of high and possibly fluctuating humidity, high temperature and poor ventilation, resulting in the rapid deterioration of the sequins. The contaminants introduced during the manufacturing process and the additives used in the fabrication of the cellulose nitrate are also likely to have contributed to this degradation. As noted in the SEM–EDX analysis, calcium, sulphur and chlorine are present. Calcium may have come from the additives or stabilisers of cellulose nitrate, whereas sulphur and chlorine may be trace elements left after either the fabrication process of the cellulose nitrate or from the colorants.

The FTIR spectra of the badly degraded sequins are different from the less heavily degraded ones. The nitrate bond of the cellulose nitrate breaks down during deterioration and degradation products in the form of oxalate may result. The badly degraded sequins contain no cellulose nitrate at all. The presence of silver (Ag), as detected by both XRF and SEM–EDX, could originate from the reflective surface of the sequins. Nitrogen dioxide, released by the degradation of cellulose nitrate, is known to react with the very fine silver particles, such as those found on the surface of the sequins.

$$2NO_2 + Ag \rightarrow NO + AgNO_3 \quad \text{Reaction (3)}$$

$$2HNO_3 + 2 Ag \rightarrow AgNO_3 + AgNO_2 + H_2O \quad \text{Reaction (4)}$$

Furthermore, the corrosive acidic gas formed in reaction (2) will attack the fabric itself.

$$2NO_2 + H_2O \rightarrow HNO_3 + HNO_2 \quad \text{Reaction (2)}$$

Evidence of such damage is particularly noticeable in the middle portion of the pink curtain where there is a peach-like pattern. This area is seriously torn and stained brown in colour.

Preservation strategies

The early detection of cellulose nitrate and the adoption of appropriate strategies can help to slow down the overall deterioration of the object and minimise further damage (Sutcliffe and Jenkins 2003). Following the detection of the unstable cellulose nitrate, pH indicator paper such as Cresol red (o-cresolsulphonephthalein) can be employed as a monitoring tool for the emission of the acidic nitrogen dioxide (Fenn 1995). A thorough evaluation of the application method and concentration of pH indicator dyes can be found in the literature (Matsumura *et al.* 2002).

Further research is required to more fully understand the constituents and compositions of historical sequins, and the preventive approaches required to slow down the extent of their deterioration. In the interim, the stage curtain is currently stored in an acid-free archival box under a controlled and stable museum environment, maintained at 23 °C and 60% RH.

Conclusion

The composition of the sequins under investigation is different from those of modern sequins, which are usually made of polyvinyl chloride as a base material. The core substrate of the sequins from the curtain was found to be gelatin with the additional presence of silver and cellulose nitrate. The presence of cellulose nitrate might be the root cause of the severe deterioration of the curtain. As no effective stabilising method is yet available to halt the degradation of cellulose nitrate materials, preventive measures are currently being adopted in order to slow down the overall deterioration of the sequins. A long-term storage and treatment strategy for the stage curtain will be drawn up after further research when, hopefully, the physical and chemical properties of the sequins are better understood.

Acknowledgements

The author wishes to thank S.W. Chan, Chief Curator at Central Conservation Section (CCS) and Alice Tsang, Curator at CCS for their professional advice and support. The advice offered by Evita Yeung, Curator at CCS is gratefully acknowledged. Thanks are also due to W.F. Lai for conducting the analytical tests such as SEM and XRF, and to Carol Tang for suggesting the chemical test for gelatin from the point of view of a paper conservator.

References

Blank, S. (1990) 'An introduction to plastics and rubbers in collections', *Studies in Conservation* 35: 53–63.

Derrick, M., Stulik, D. and Landry, J.M. (1999) *Infrared Spectroscopy in Conservation Science*. Los Angeles, CA: Getty Conservation Institute.

Derrick, M., Stulik, D. and Ordonez, E. (1993) 'Deterioration of cellulose nitrate sculptures made by Gabo and Pevsner', *Saving the Twentieth Century: The Conservation of Modern Materials*, D.W. Grattan (ed.), 169–82. Ottawa: Canadian Conservation Institute.

Fenn, J. (1995) 'The cellulose nitrate time bomb: using sulphonephthalein indicators to evaluate storage strategies', in Heuman 1995, 87–92.

Heuman, J. (ed.) (1995) *From Marble to Chocolate: The Conservation of Modern Sculpture*. London: Archetype Publications.

Matsumura, M., Eastop, D. and Gill, K. (2002) 'Monitoring emissions from cellulose nitrate and cellulose acetate costume accessories: an evaluation of pH indicators dyes on paper, cotton tape and cotton threads', *The Conservator* 26: 57–69.

Reilly, J.A. (1991) 'Celluloid objects: their chemistry and preservation', *Journal of the American Institute for Conservation* 30: 145–62.

Selwitz, C. (1988) *Cellulose Nitrate in Conservation*. Marina del Rey, CA: Getty Conservation Institute.

Stewart, R., Littlejohn, D., Pethrick, R. *et al.* (1995) 'Degradation studies of cellulose nitrate plastics', in Heuman 1995, 93–7.

Sutcliffe, H. and Jenkins, A. (2003) 'Compensation for loss: ethics and practice in the conservation of two Schiaparelli evening coats and the replication of missing ornamental elements', *The Conservator* 27: 51–63.

Williams, R.S. (1994a) *Display and Storage of Museum Objects Containing Cellulose Nitrate*, CCI Notes 15/3. Ottawa: Canadian Conservation Institute.

Williams, R.S. (1994b) *The Diphenylamine Spot Test for Cellulose Nitrate in Museum Objects*, CCI Notes 17/2. Ottawa: Canadian Conservation Institute.

The author

Angela Cheung currently works as an Assistant Curator (Conservation) in textile conservation in the Central Conservation Section, Leisure and Cultural Services Department in Hong Kong.

Address

Angela Cheung Yuen Kuen, Central Conservation Section, 4/F, Hong Kong Heritage Museum, 1 Man Lam Road, Shatin, Hong Kong (aykcheung@lcsd.gov.hk).

Wet look in 1960s furniture design: degradation of polyurethane-coated textile carrier substrates

Tim Bechthold

ABSTRACT In the second half of the 1960s, plastics were used increasingly in furniture design, particularly in Italy. Polyurethane, which can be produced in a variety of different formulations, began its successful career in industrial applications. Textile carrier substrates coated with polyurethane are typical of this era. Towards the end of the 1960s, new processing technologies and products enabled the manufacture of micrometer-thin polyester polyurethane coatings for elastic fibres, knitwear and non-woven fabrics made of artificial fibres. Their characteristics led these textiles to be called 'wet look' or 'crinkled patent'. As a result of the extreme thinness of the coatings and the sharp decrease in hydrolytic stability with age, within just a few years of manufacture, these products became less resistant to cracks and their adhesive qualities were greatly reduced. Such degradation leads to discoloration, brittleness, cracking and detachment of the coating. The deteriorated material is highly susceptible to mechanical stress. Three main aspects of degradation in polyester polyurethane coatings are assessed: the influences of light, oxygen and hydrolytic reactions. A method of consolidating this kind of coating is also presented and the manner in which such objects can be stored and displayed in museums is discussed.

Keywords: wet look, crinkled patent, polyurethane, textile coating, transfer process, design

Introduction

The following text is based on a diploma thesis (Bechthold 2002). The fundamental ageing processes and methods of conservation of textiles with thin polyurethane coatings, also known as 'wet look' or 'crinkled patent', are discussed, using Joe Cesare Colombo's Tube chair (designed in 1969), as an example.

History and technical development

In the second half of the 1960s, furniture designers, especially in Italy, increasingly utilised plastics that until then had not been in widespread use in furniture manufacture. In particular, polyurethane (a fully synthetic material derived from the polyaddition of a polyisocyanate and a polyol, which can be produced in various formulations) became more and more important. Its range extends from foam systems such as flexible, semi-flexible, high-resistance and integral skin foam, to elastomeric fibres and well-known materials for outer surfaces, such as DD-lacquers. Textile carrier substrates coated with polyurethane are typical of this era.

Towards the end of the 1960s, the development of novel techniques for using reversal and transfer processes for polyurethane coatings provided new methods of coating textiles. Until that point, the coating had been applied directly to the textile using a coating or spraying process; now it was possible to avoid undesirable shrinkage of the carrier textile using specially treated transfer papers, thus enabling the manufacture of micrometer-thin polyurethane coatings. This process made possible thinner and cheaper coatings on knitted goods, elastic tissues and non-woven fabrics made of artificial fibres (Fig. 1).

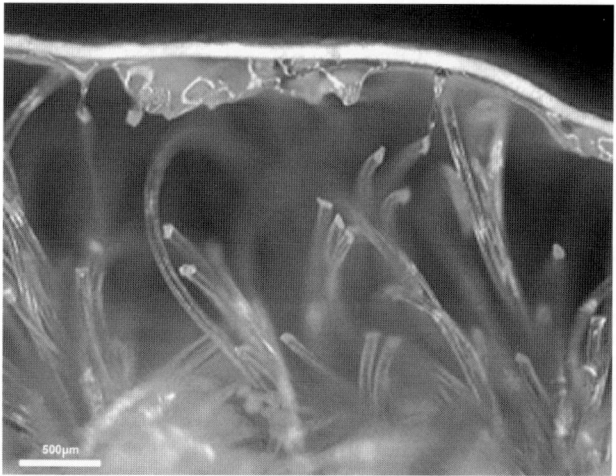

Figure 1 Cross-section of a polyurethane-coated web textile (Plate 64 in the colour plate section).

In addition to its low weight, the new product had the following advantages over previous coatings: improved air and water vapour permeability, a high level of flexibility at low temperatures without the use of plasticisers (flexible in temperatures as low as −40 °C) and a high degree of softness and elasticity. Further positive aspects were the high level of abrasion resistance and tenacity, good bonding with chemical fibres and a leather-like feel.

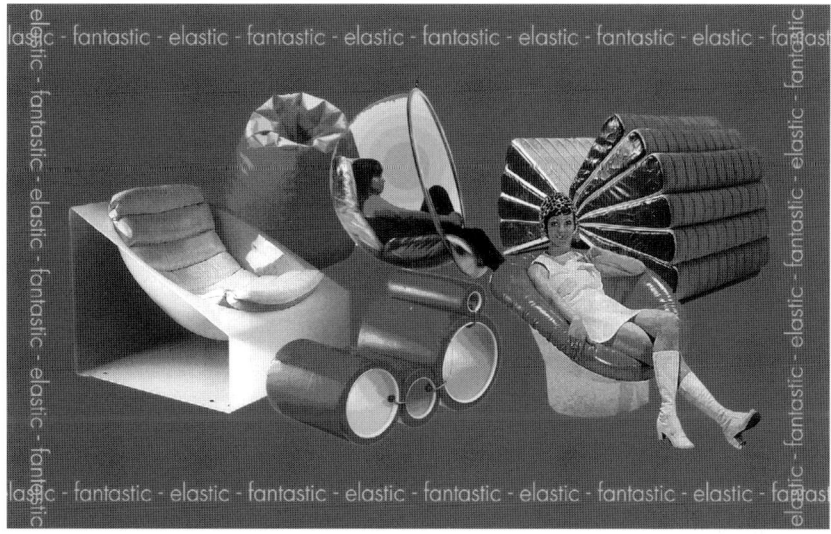

Figure 2 Tube chair, Joe Cesare Colombo, 1969 (Die Neue Sammlung, München) (Plate 65 in the colour plate section).

Towards the end of the 1960s, the number of products available for coating textiles increased and polyurethane coatings became more widespread in outer garments, protective work clothing, upholstery and the manufacture of bags. The acceptance and use of textiles coated with polyurethane in the world of fashion opened up whole new areas of implementation. There were great expectations coupled with tremendous enthusiasm and a strong sense of optimism for the future of the product. The prior focus in textile coating, on breathable fabrics and overall feel, now gave way to an interest in colour, a glowing look to outer surfaces combined with considerable elasticity and softness. As a reflection of their characteristics, these materials were referred to as 'wet look' or 'crinkled patent'. These highly polished, futuristic wet-look surfaces embodied perfectly the spirit of the 'space age' (Fig. 2). The Tube chair, a multi-part chair designed in 1969 by the famous Italian designer Joe Cesare Colombo, is one of the best-known products covered in this wet-look material.

The setback

The furniture industry, which, with the production of cold-foam polyether upholstery, had been able to produce highly supple furniture, became increasingly interested in this sort of upholstery, a process that reached its zenith around 1970. 'At the last furniture trade fair in 1970, armchairs covered in wet-look, leather-like polyurethane material were on show at over 20 stands. They cost more than normal leather chairs with split leather' (Kunststoff-Berater 1970). For the most part, the manufacturers' information made only limited or indirect mention of the less attractive qualities of the material that led to a serious setback in the market, especially in the field of upholstery, a few years later.

As a result of the extreme thinness of the coatings and the sharp decrease in hydrolytic stability with age, the surfaces became less resistant to cracking and were therefore more prone to damage. Whereas advertising had initially been euphoric in its praise of the high level of wear and tear the furniture could withstand, the increasing number of complaints were even mentioned in an article in *The Times* in 1974: 'Research experts are unable to predict beyond eighteen months how the fashionable polyurethane-coated fabrics will wear, but this has not stopped manufacturers and retailers selling such suites at £300 or more with no warning to the public' (Katz 1978). Both in their original and later states, the polyurethane-coated fabrics of the late 1960s are fundamentally impaired by exogenous and endogenous ageing factors that can lead to such phenomena as discoloration, embrittlement, cracking and detachment. As far as they are still in existence, these fabrics are seriously endangered.

Main aspects of degradation

The characteristically thin, extremely versatile polyurethane coating consists of soft segments, generally polyester polyols, and hard segments such as urethane and carbamide species. Characteristics of the materials, such as hardness, tenacity and resistance to cracking, depend for the most part on the functionality of the components used, on the proportions of ingredients and on the nature of the soft and hard segments in the polyurethane polymer. The length of the chain in the soft segment is of crucial importance, particularly for the soft and versatile coatings. An increase in the molecular weight of the soft segments increases the elasticity of the polymer. Since the hard segments are able to develop inter-chain hydrogen bonds, any reduction in their number leads to a decrease in the extent of cross-linking in the polymer, with corresponding negative effects on the stability of the material itself. Because of the required softness and the leather-like feel, crinkled patent textiles as a rule possess a high proportion of soft segments, leading to amorphous regions of polyester polyols. Their sensitivity with regard to hydrolytic swelling and diffusion is correspondingly high. In addition to exogenous influences such as light, oxygen, moisture, temperature, atmospheric pollutants and microbial infestation, characteristics stemming from production processes and induced by the materials themselves may also be responsible for deterioration of the polymer. Generally, degradation is a result of a combination of all these factors.

The three principal factors implicated in the fundamental ageing processes of polyurethane coatings – light, oxygen and hydrolysis – are discussed below.

Light

The effect of light on the polyurethane polymer leads to the activation of carbonyl groups that eventually disassociate. The macromolecule generally breaks at the urethane (C-N) bond, leading to the formation of new amino and carboxyl groups in the polymer, with the release of carbon monoxide and carbon dioxide. With oxygen present, peroxyl radicals are formed, which in turn can react with the organic substance of the polymer and become hydroperoxides. Under the influence of heat, light or a variety of other factors, which may act as catalysts, these disintegrate into new radical species. This process has an autocatalytic effect, as illustrated by the acute danger of shortwave light. The influence of light can also lead to crosslinking reactions within the polymer, however, which may predominate. This in turn leads to the hardening of the material and shrinkage in volume. The material loses its toughness and elasticity; micro-cracks spread and its overall resistance to breaking decreases. The outer surface becomes rough and hydrophilicity increases. The material becomes more prone to decomposition processes, which finally results in peeling.

Oxygen

As a rule, extremely soft polyurethane coatings of this type exhibit a high level of oxygen diffusion, and thus significant rates of oxidation. This is made possible by the large proportion of amorphous regions in the polymer as well as the extremely thin nature of the film itself. The flexible segments are converted into less flexible ones by the oxidation of hydrocarbons into polar, hydrophilically functional groups. The coatings initially become more sensitive to moisture and, as oxidation increases, so the crystallinity increases and the physical characteristics change. Breakages occur in the amorphous areas and the length of the polymer decreases. The yellowing of the material is caused first and foremost by the influence of oxidation. In the polymer, aromatic isocyanates such as diisocyanatodiphenylmethane (MDI) are principally responsible.

In the case of MDI, the methylene bridge is particularly sensitive to oxidation. Under the influence of light, auto-oxidation initiates the decomposition of the molecule at this point, with quinoid structures forming. The yellowing effect comes about as a result of the formation of carbonyl chromophores and the loss of mechanical qualities results from the scission of the chain. The absorption of light thus shifts to longer wave frequencies. As a result of structural processes in the material itself, the influence of photo-oxidative processes leads to a decrease in the high level of ductility so characteristic of new polyester polyurethane. Its level of sensitivity to hydrolytic and mechanical influences increases. The increasing cross-bonding causes tension within the material, which manifests itself in the emergence of cracks and the subsequent drawing apart of the edges of these cracks. At an advanced stage, the network of cracks becomes finer, as do the parts of the coating that flake off.

The catalytic effect of light on oxidative processes can be seen quite clearly on the Italian Libro easy chair designed by Gruppo Dam, 1970 (Fig. 3). Those zones protected from light have scarcely yellowed at all and are noticeably less damaged. The sharp, yellowed edges of the uncovered sides and the corresponding network of cracks are quite clearly recognisable. The discoloured surface is extremely sensitive to mechanical strain. This effect is quite clearly visible in damage from a pressure point in the area of the yellowed edge. The network of cracks that has arisen here is spreading out increasingly into the dark, yellowed part, whereas the area that has been protected from light is almost fully elastic and only slightly deformed. Areas subject to a high level of mechanical stress exhibit an increased tendency to detach.

Hydrolytic processes

In addition to photo-oxidative processes, the lack of resistance to hydrolytic reactions must be considered an important factor in the ageing process of polyurethane coatings based on polyester polyols. The hydrolysis takes place predominantly in the amorphous areas, since water can diffuse more quickly into these regions. Ester and amide bonds in the polymer are particularly susceptible to hydrolytic reactions.

The polyester molecules in polyurethane are in equilibrium with the products from which they are formed (dicarbon acids and glycol). The effect of moisture, acids or alkalis can cause some of the ester bonds to break down to yield these original components. This tendency increases with the hydrophilicity of the specific polyester from which the material is constructed. Diethylene glycolapidates, which at the end of the 1960s were used frequently in the synthesis of polyesters, as well as the fashionable trend towards ever softer polyurethane coatings, produced with a high proportion of polyester polyols, lead to exceedingly hydrophilic and correspondingly sensitive end products. In addition, it must be assumed that when combined with foam upholsteries, the hydrolytic decomposition of the coatings was further accelerated as a result of catalyst residues in the foam upholstery, such as organometallic compounds like tin dioctate and dibytil tin dilaurate. As a result of their acidity, the carboxyl groups that are formed during hydrolysis also act as catalysts for further hydrolytic reactions. Even the hydrolysis of just a few polyester groups can lead to the marked loss of mechanical function in polyurethane coatings. For this reason, low molecular weight breakdown products such as hydrogen chloride (HCl), and phthalic acid can also accelerate the speed of hydrolytic phenomena. This means that the products released by decomposition of nearby materials such as PVC can possibly also act as catalysts, increasing the speed of hydrolytic reactions.

In the case of the Libro easy chair, the high level of sensitivity of the polyurethane coating, as opposed to other materials, is demonstrated by the whitish imprint of a hand on the edge of the coating (Fig. 3). It can be assumed that as a result of sweat impregnations, a breakdown of the polyester to the influence of light and hydrolytically effective materials, the experiments conducted also addressed swelling properties, the interaction of hydrolytically aged with non-hydrolytically aged materials, and microbial contamination. Although the distinctive swelling property of a polyurethane coating is for the most part reversible, due to the marked increase in the number of local flow zones, increased danger from mechanical stress must be considered.

The fact that under normal climatic conditions (relative humidity 50%, temperature 17 °C) microbial contamination was detected was both surprising and alarming. It leads to discoloration and embedded structures of hyphae in fine networks of cracks. It can be supposed that the resulting higher surface roughness results in an increasing dust accumulation. Finally, this encourages microbial contamination again (Fig. 4).

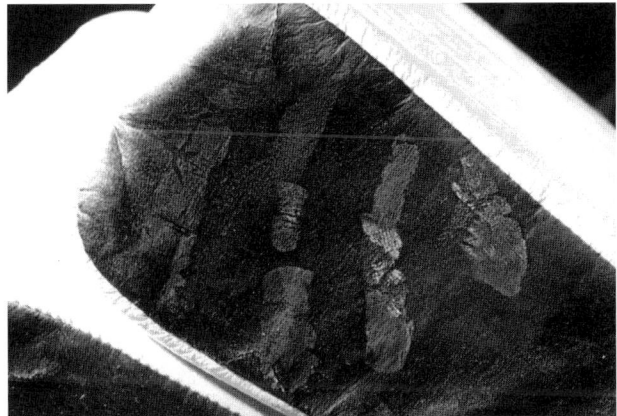

Figure 3 Libro chair: whitish fingerprints, polyurethane coating.

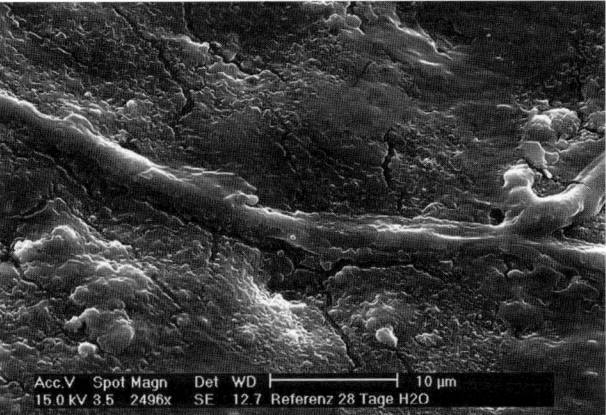

Figure 4 Tube chair: SEM-micrograph, net of hyphae, polyurethane coating.

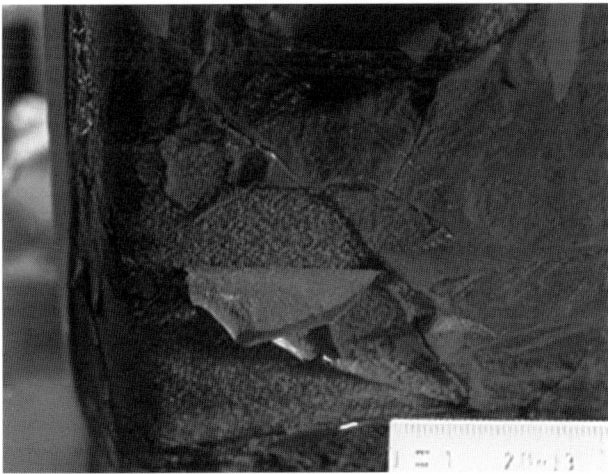

Figure 5 Tube chair: detachment, polyurethane coating (Plate 66 in the colour plate section).

components occurred, and that the whitish remains are adipic acid anhydride.

The hydrolytic influence of substances was researched in two parallel studies for HCl and ammonia (NH_3). It became evident that the effect of both acidic and alkaline materials can lead to considerable structural changes in the polymer. As demonstrated by the use of x-ray diffraction (XRD), in both cases scission of the polyester components occurs. In addition

Conserving wet-look artefacts

The final aspect of this paper considers aspects of the conservation of crinkled patent objects, particularly those arising from the Tube chair. The coating, which has become loose in many places and is flaking, is in grave danger of deteriorating further (Fig. 5). Sufficient interaction between the adhesive and adherent must be produced by means of suitable consolidation treatments. With regard to their physical and chemical nature, those areas consolidated should correspond as far as possible to the structure of the original material, i.e.:

- The adhesive must not penetrate the grain of the fabric too deeply.
- The physical qualities of the adhesive such as Young's modulus (E) and the thermal coefficient of expansion should be similar to the aged polymer film.
- The adhesive must be able to be applied as thinly as the original adhesive (10–15 μm).
- It must not increase the rate of the ageing process.
- It must not impair the original function of the coating.
- It should be reversible and allow the measures to be corrected later if necessary.

Considering the requirements listed, it would appear that, in theory, aqueous polyurethane dispersions are a suitable adhesive. As a result of their linear construction and their cross-linking by means of hydrogen bonds, they display physical qualities that are very similar to the original coating. It can be safely assumed that there is correspondingly good bonding in the original. Solvent-free processing prevents possible interactions that are difficult to control, such as the formation of tension cracks or other such problematic reactions. Just like the original coating, the aqueous polyurethane dispersions are thermoplastic, and as a result of their high coating qualities can be processed very well. The high degree of flexibility and elasticity of the film once it is dry ensures that the textile qualities of the original coating are preserved. Attention must be paid to the hydrolytically sensitive nature of the polyol components. The use of an aqueous polyether-polyol-based dispersion is better suited for this purpose, but is crack-proof only to a limited

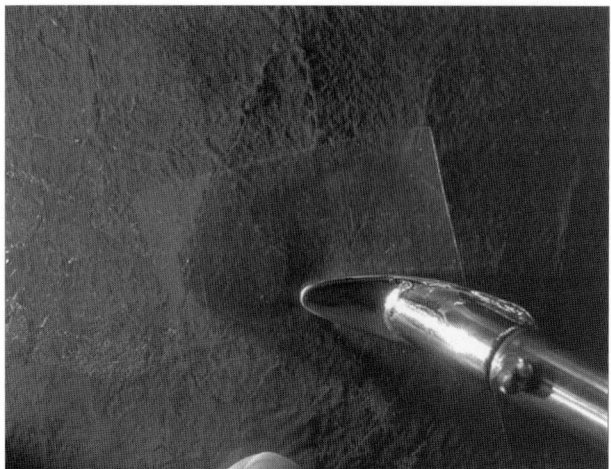
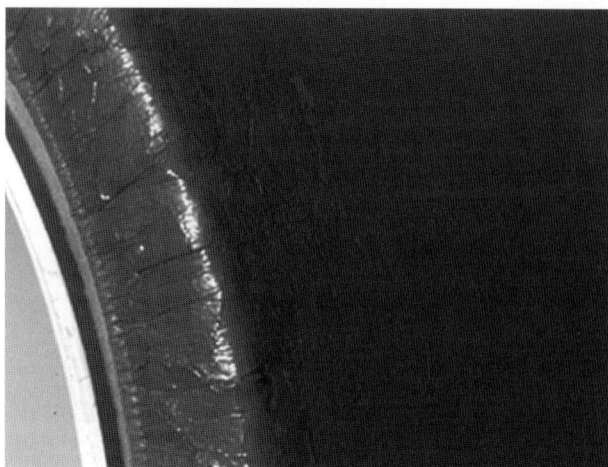

Figure 6 Tube chair: consolidation treatment (Plate 67 in the colour plate section).

Figure 7 After consolidation treatment (Plate 68 in the colour plate section).

degree and has limited protection against oxidation (Figs 6 and 7).

For the treatment an aqueous, finely dispersed anionic polyurethane dispersion, Bayer Isovin v, was used. The reverse side of the flaking coating was wetted with a brush. By means of a heated spatula it was fixed at approximately 40 °C. A piece of Melinex served as a separation layer to avoid undesirable adhesion to the spatula. Any excess dispersion was removed with cotton swabs slightly wetted with distilled water. The use of polyurethane systems with a lower glass transition (T_g) point than the coating allows adjustments to be made subsequently.

The consolidation treatment was carried out in defined areas of the chair in 2002. Although the chair has now been on display for almost two years, the consolidated coating is stable and without any undesirable aesthetic or mechanical changes. Nevertheless it must be noted that due to a lack of experience in conservation with these materials, there are no data available with regard to long-term behaviour for this kind of adhesive.

Storage

Originally the Tube chair was stored dismantled in a simple cotton bag, organised like a Russian doll with the tubes nested inside each other in order of decreasing size. To minimise deleterious influences, the chair should ideally be stored in a dark, low-oxygen level environment with a temperature of around 7 °C. In order to reduce the risk of both hydrolytic and microbial decomposition processes, a humidity level of 45% RH is desirable. A fair proportion of the damage identified on the textile coating of the Tube chair stems from mechanical influences, which may in the short term occur through improper handling or in the long term as a result of continuous strain. This sort of damage can be avoided simply through adequate transportation methods or suitable cordoned-off areas in storage rooms and at exhibitions, as well as by orderly storage procedures.

Conclusion

Since the end of the 1960s, polyester polyurethane has been used widely as a coating material for elastic textiles, typically for covering soft foam furniture. It was shown that these systems are quite unstable and degrade relatively rapidly over time. Most importantly, external influences can often lead to chemical reactions that result in irreversible structural changes. In this context, preventive conservation plays a crucial role. This includes the following four aspects.

1. Examination and documentation of materials and technology.
2. Scientific analysis of materials and environments.
3. Stabilisation and adjustment of storage conditions to object requirements.
4. Regular monitoring.

The typically degraded wet-look surface is often flaking and loose. At that point a consolidation treatment is inevitable. In the case of wet-look furniture such as the Tube chair and the Libro easy chair, consolidation treatments as described above are time-consuming but appropriate. These consolidation treatments are nearly impossible, however, when dealing with objects that lack a well-defined shape – for example, another classic Italian design, Il Sacco, a sack-like seat filled with pellets.

Acknowledgements

The author would like to thank the following institutes and people without whom this work would not have been possible: Christof Krekel (Staatliche Akademie der Schönen Künste, Stuttgart); Andreas Burmester (Director, Doerner-Institut, München); Thea van Oosten (Instituut Collectie Nederland, Amsterdam); Johann Koller (Doerner-Institut, München); Andrea Kaser (Doerner-Institut, München); Ursula Baumer (Doerner-Institut, München); Dr Steppner (Bayerisches Landeskriminalamt, München); Reiner Letsch (Prüfamt für Kunststoffe, Technische Universität, München); Susanne Huber (Bayerisches Landesamt für Denkmalpflege).

References

Bechthold, T. (2002) *Polyurethanes in Furniture Design of the 1960s. Technology, Degradation, Conservation – with a Special Focus on Polyurethane-coated Textile Carrier Substrates* Diploma thesis, Technical University, Munich.

Katz, S. (1978) *Plastics, Design and Materials*, 126. London: Eastview Editions.

Kunststoff-Berater (1970) Fachzeitschrift über die Fortschritte in der Anwendung und Verarbeitung synthetischer Stoffe und Fasern, 7/1970, 678.

Supplier

Polyurethane dispersion Isovin v: Kremer Pigmente, Aichstetten, Germany.

The author

Tim Bechthold is currently Head of the Conservation Department at Die Neue Sammlung, Staten Museum of Applied Arts, Design (since 2002) in the Pinakothek der Moderne, Munich. He trained as a carpenter before studying furniture conservation at the Göring Academy for Furniture and Sculpture Conservation, Munich. He then worked as a freelance furniture conservator before studying at the Technical University, Munich at the chair of Conservation Science, Arts Technology and Conservation. He graduated on the presented paper in 2002.

Address

Tim Bechthold, Head of Conservation Department, Die Neue Sammlung, Staten Museum of Applied Arts, Design in the Pinakothek der Moderne, Munich, Türkenstr. 15, 80333 München, Germany (bechthold@die-neue-sammlung.de).

Storage issues for contemporary textile art: a solution for one example

Rosemary Baker

ABSTRACT Textile-based art is often composed of unusual materials combined in novel ways. The resultant structures may present curators and conservators with new challenges when trying to devise suitable storage systems for them. This paper discusses one solution for a multi-medium artwork, *Monsoon Capital* by Shelly Goldsmith, which provides suitable long-term support within a reasonably compact storage box.

Keywords: modern textile art, storage, long-term stability

Introduction

Modern textile art raises novel issues for the conservator faced with its storage in three respects: first, by the presence of unusual and innovative materials; secondly, by the use of unexpected combinations of materials and thirdly, by the often unconventional shape, size and construction of the final object. A textile conservator normally has previous experience of devising storage systems for items such as flat textiles, costume and accessories which include mixed materials and awkward shapes. The breadth of experience with these problems by individuals within the profession means that it has been possible to draw up broad outlines for storage systems (Robinson and Pardoe 2000). By contrast, the nature of modern textile art is such that the artist seeks to push the boundaries of the expected in form and materials and therefore each piece presents storage issues hitherto unknown. In addition, it is probable that older objects still present in collections, after 50 years or more, are a select group of the more robust; some of the more transient and fragile artefacts having been lost through rapid degradation. Some recent studies on accelerated degradation of textiles have suggested that deterioration is not linear but rather a stepwise process. Moreover, there may be phases early in the life of an object when degradation is relatively fast followed by a point beyond which the degradation rate slows (Jordan 2004). Storage systems, as part of a preventive conservation strategy that help to preserve objects and reduce the rate of initial degradation processes in particular, are therefore imperative for newly acquired objects in collections.

The Whitworth Art Gallery (Manchester) regularly commissions and collects contemporary pieces of textile art. The project described in this paper illustrates how issues of storing modern textile art, outlined above, were encountered and solved during a placement there. With advice from the textile conservator, a storage system was devised for a piece of contemporary textile art, *Monsoon Capital*, which made use of easily available conservation grade materials.

Monsoon Capital by Shelly Goldsmith was created in 1999 and consists of two funnel-shaped pieces linked by long continuous lengths of nylon monofilament thread. One funnel is made from silk printed with an x-ray of the artist's rib cage and the second is woven from silk and nylon monofilament which emerges from the weaving to form the continuous lengths that thread through the silk funnel. With reference to the interaction of her work with its surroundings, the artist explained (in 2000) that the warp in her recent pieces had become dominant and had been forced through the tapestry weave, hanging at the edges and suggesting tears, leaks or waterfalls (Harris 2000: 47). Thus, the three-dimensional aspect of the piece had become paramount and the interaction with its backdrop an important component of the overall effect.

The solution

A storage box was required for the piece which would hold all the component parts so that there were no stresses on any area and which would also protect the object from dust and light. The storage space in the museum is necessarily finite and therefore *Monsoon Capital* needed to occupy the least room practicable. It could not be stored in the extended linear manner in which it is displayed. The interior of the storage box is constructed from Correx (twin-walled polypropylene/polyethylene copolymer sheet) bound with Tyvek tape (pressure-sensitive adhesive on a high density heat-bonded polyethylene carrier) and joined using glue applied with a heat gun (Fig. 1). Pieces of Plastazote (closed-cell cross-linked polyethylene foam) are glued to the insides of the box to hold a tray for the main components of the work. This tray is also constructed from Correx with foam cylinders used to form wells to accommodate accessories to the piece (Fig. 2). The funnels are supported upright with a slot cut in the front of the tray to allow the piece to slide in and out. The slot is needed to accommo-

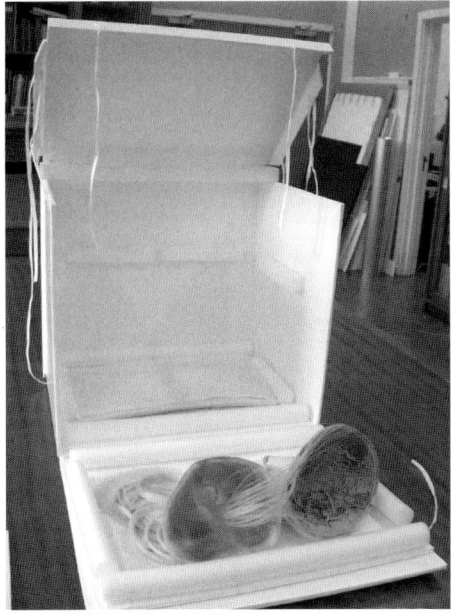

Figure 1 The interior of the box showing the fold-down front section and supports for the top tray.

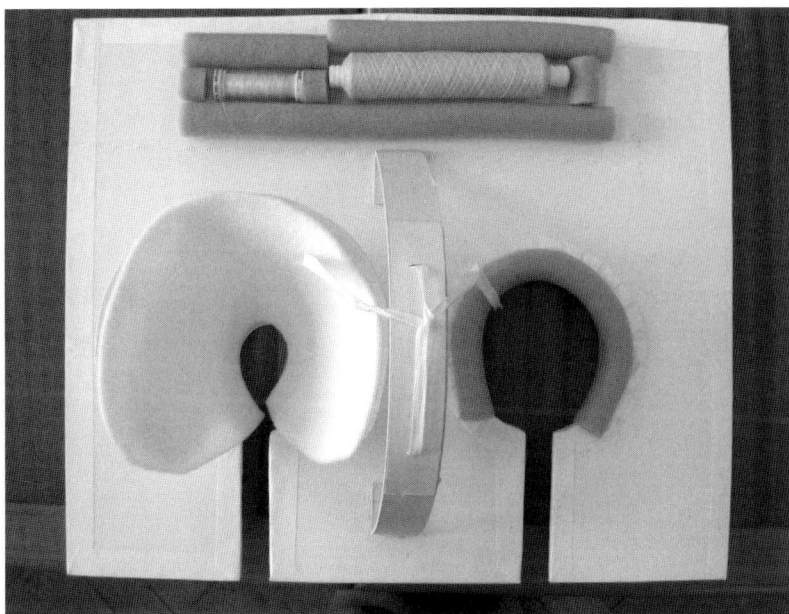

Figure 2 The top tray showing the supports for the two funnels and the warps (dimensions 440 × 290 mm).

date the nylon filaments that link the two funnels. The woven funnel is rigid enough to retain its shape without additional support but shaped pieces of acid-free card and polyester felt were inserted to provide supplementary support for the more fragile silk funnel. An arched piece of acid-free card is secured between the two funnels, lifting the nylon filaments away from the tops of the funnels to avoid potential distortion to their edges in the long term; cotton tape holds them in place. The front of the box is designed to fold down to allow easy access to the interior.

Two further trays, positioned on the floor of the finished box, were made to hold, respectively, the spare nylon monofilaments and the monofilaments that form an integral part of the piece (Fig. 3). These are constructed from tubes of Ethafoam (open-celled, rigid expanded polyethylene foam) secured to Correx sheets to form lipped edges and act as spacers. The assembled box allows the two funnels to be adequately supported while being suspended above the trays holding the other components (Figs 4 and 5).

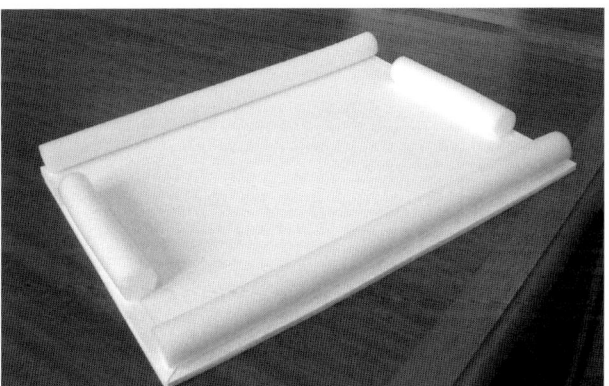

Figure 3 One of the lower trays showing foam tubing used as edging/spacers (dimensions 440 × 290 mm).

Conclusions

Although general principles can be applied to the storage of modern pieces, including the use of inert materials and provision of adequate support, there were other issues to be considered in this case. The main challenge was to devise a way of allowing the linked pieces to be inserted into the storage box easily while giving appropriate support. In her comments on her work, Shelly Goldsmith makes it clear that the three-dimensional form and feelings of movement of *Monsoon Capital* are important elements that must be preserved. Maintaining the funnel shapes and linear nylon filaments so that they were not crushed or distorted was crucial. The finished box (Fig. 6) is constructed from easily available conservation grade materials using familiar techniques, but its construction

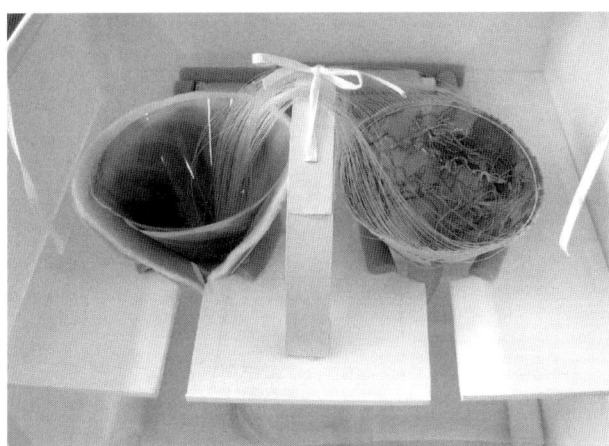

Figure 4 The top tray with *Monsoon Capital* in position.

and form took considerable development time to accomplish. The case study exemplifies that unconventional modern art in collections can place additional demands on resources for their preservation and appropriate storage. *Monsoon Capital* is a relatively new piece of textile art and its future deterioration is

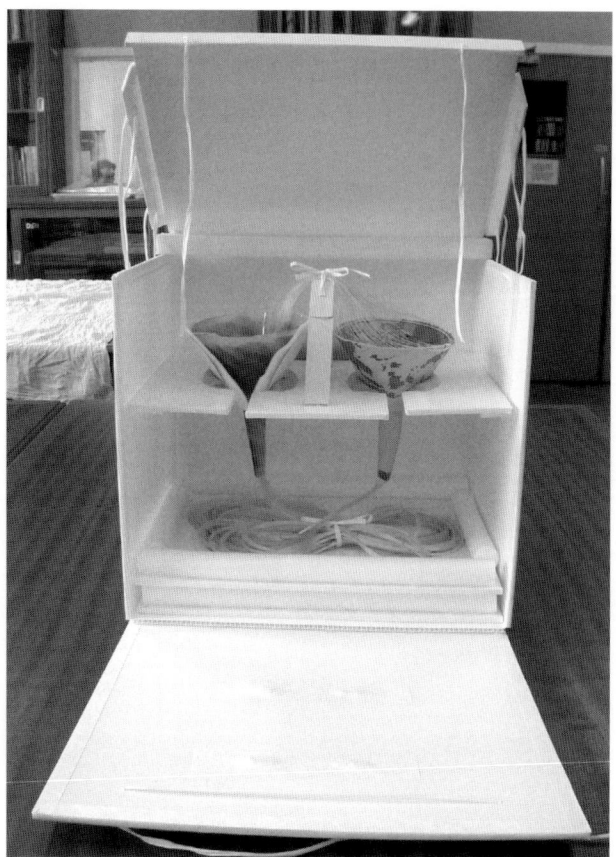

Figure 5 The interior of the box with *Monsoon Capital* in position.

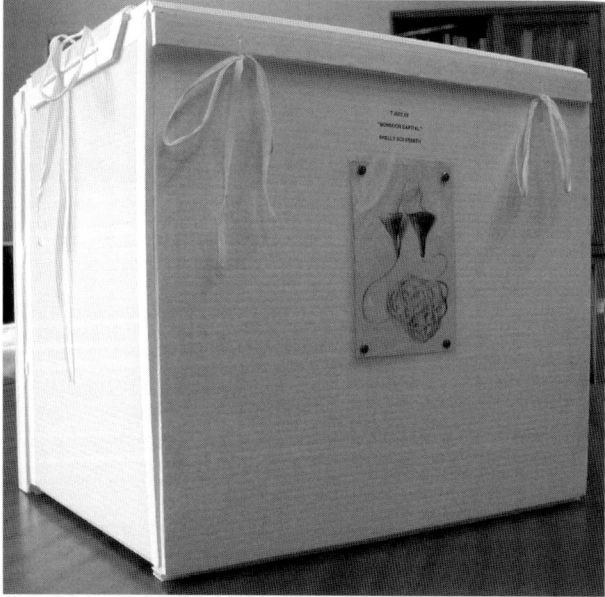

Figure 6 The assembled box ready for storage (dimensions 450 × 300 × 550 mm).

not easily predictable, but the efforts made in preventive conservation when it is new will pay dividends for its preservation in the future. The storage box will protect it from dust and light and support it without imposing stresses on the component parts of the piece, thereby reducing the agents of degradation as much as practicable.

Acknowledgements

The author would like to acknowledge the help and encouragement particularly of Ann French (Textile Conservator, Whitworth Art Gallery) and of the other Textile Department staff in this project. Many thanks are due to the Whitworth Art Gallery for permission to publish the results.

References

Harris, J. (2000) *Art Textiles of the World: Great Britain*, Vol. 2. Winchester: Telos Publishing.

Jordan, M. (2004) 'Heat and dust: George II's travelling bed traumas', in *Opening Up Open Display. Postprints of the Joint Forum of UKIC Textile and Historic Interiors Sections, 29 March 2004*, A. Cogram and M. Jordan (eds), 26–34. London: UKIC.

Robinson, J. and Pardoe, T. (2000) *An Illustrated Guide to the Care of Costume and Textile Collections*. London: Museums and Galleries Commission.

The author

Rosie Baker worked as a protein chemist for the UK National Health Service developing protein purification methods. During this time she pursued an active interest in textiles, particularly costume, and in 2003 she enrolled on the MA Textile Conservation course at the Textile Conservation Centre.

Address

Rosemary M. Baker, Textile Conservation Centre, Winchester School of Art, University of Southampton, Park Avenue, Winchester, Hants SO23 8DL, UK (rmb203@soton.ac.uk).

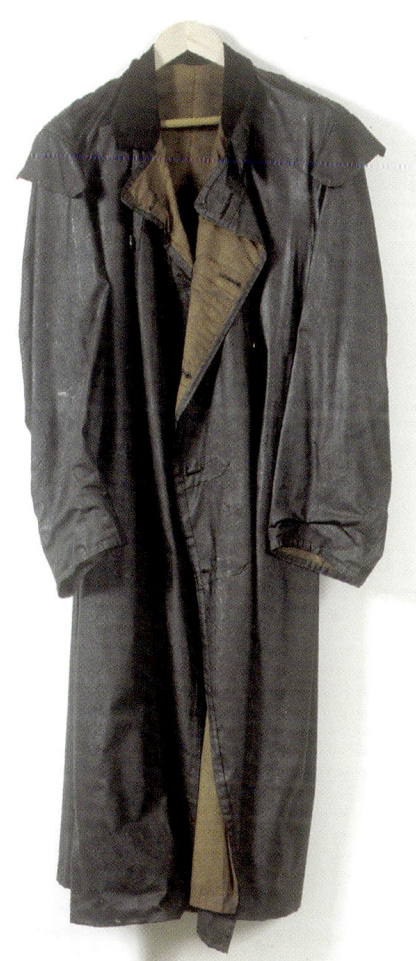

Plate 39 Sample uniform (1918): surface treated with bitumen. Stiffened, deformed and tacky (Fig. 1, p. 84).

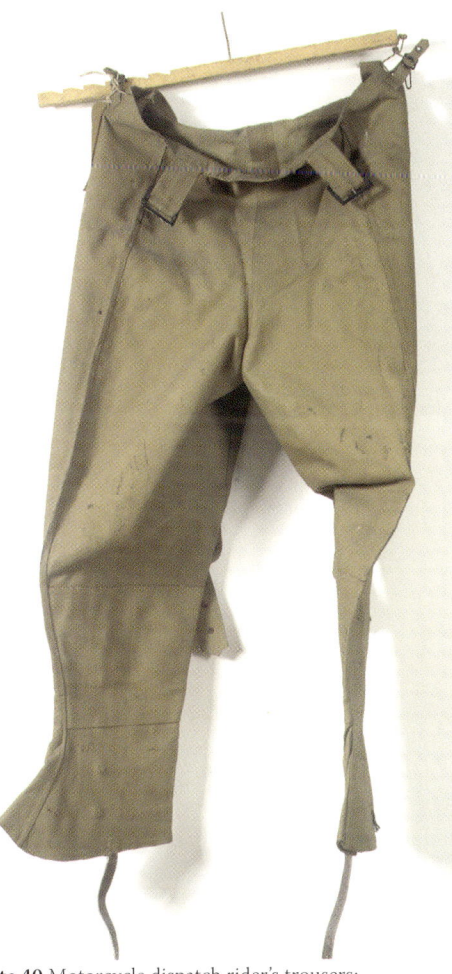

Plate 40 Motorcycle dispatch rider's trousers: waterproofed with rubber sandwich method. Stiff and deformed from poor storage (Fig. 2, p. 87).

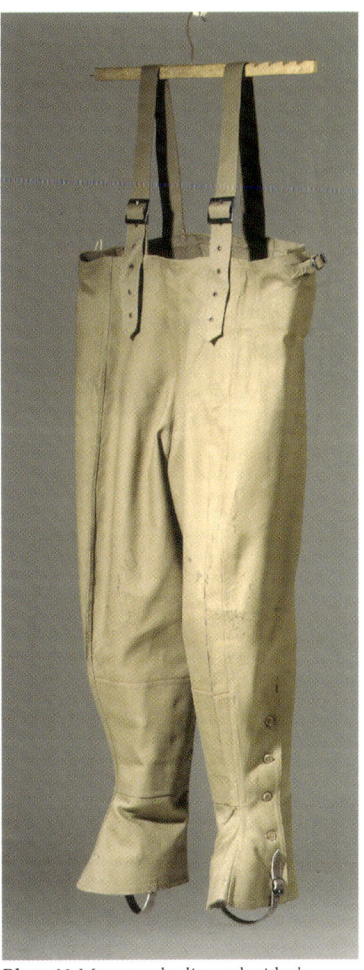

Plate 41 Motorcycle dispatch rider's trousers: original shape recreated (internally supported) (Fig. 3, p. 89).

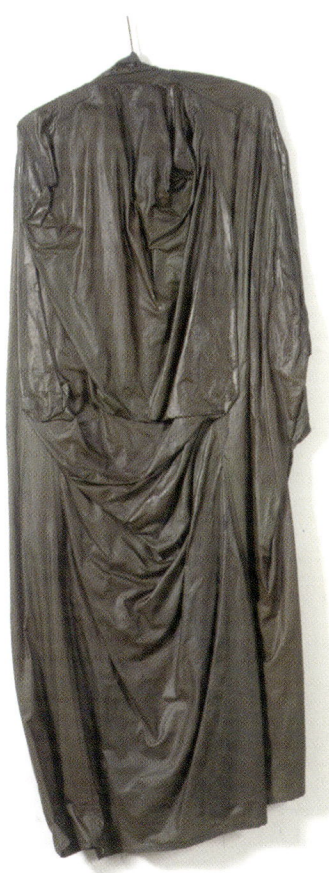

Plate 42 Oilskin uniform (1945): flat and sticky. The shape of the coat is unidentifiable (Fig. 4, p. 89).

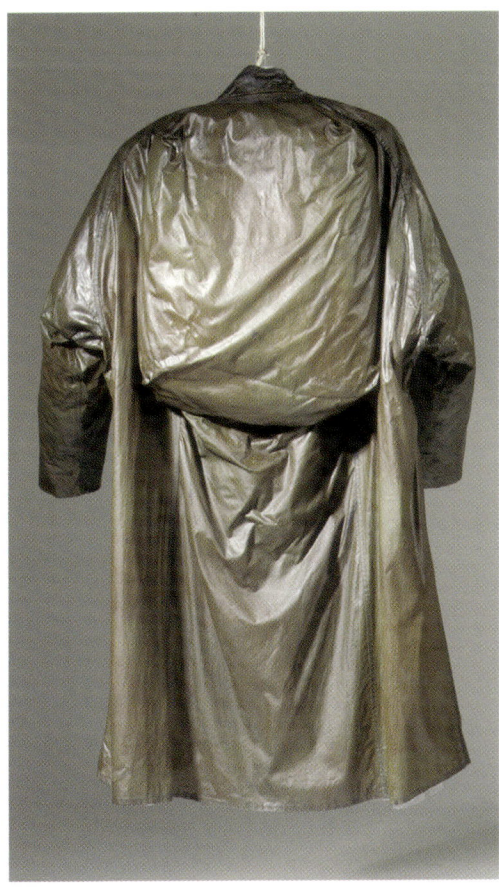

Plate 43 Oilskin uniform (1945): internally supported with bubble plastic and polyethylene (Fig. 5, p. 90).

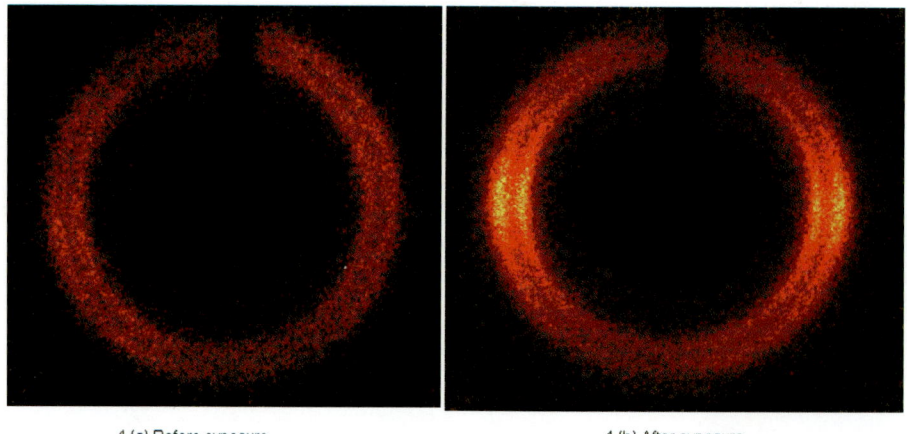

Plate 44 XRD pattern for the undyed net (a) before exposure and (b) after exposure (Fig. 3, p. 96).

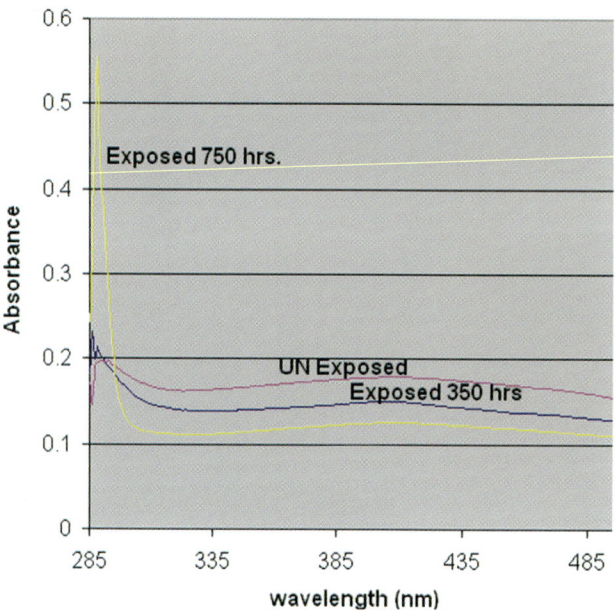

Plate 45 UV-visible spectra of undyed net (see p. 97).

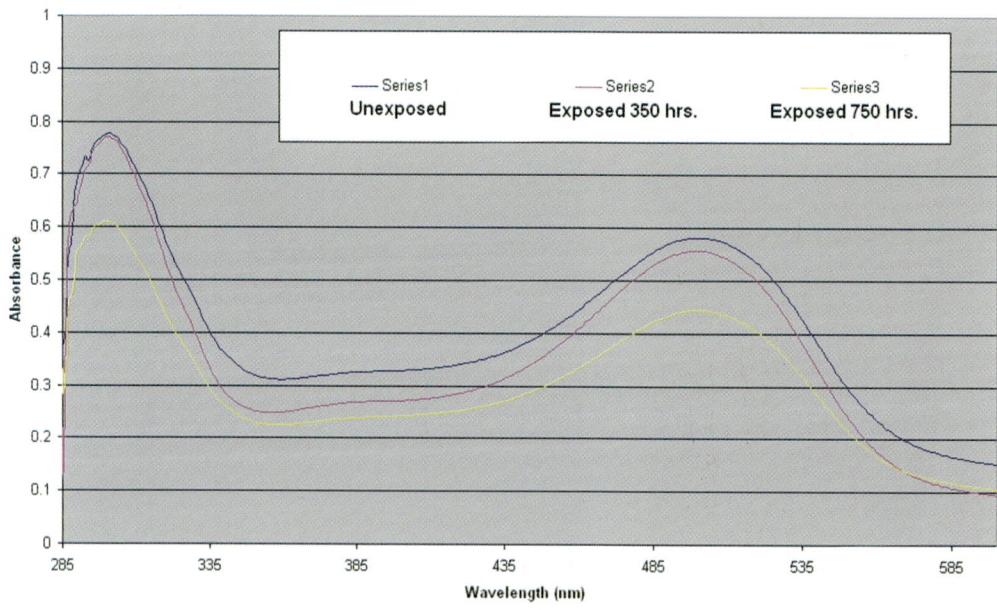

Plate 46 UV-visible spectra of net dyed with Tectilon Red 2B (see p. 98).

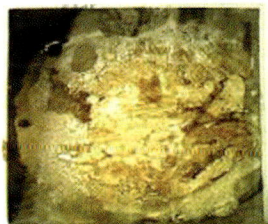 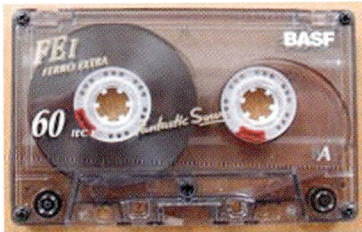 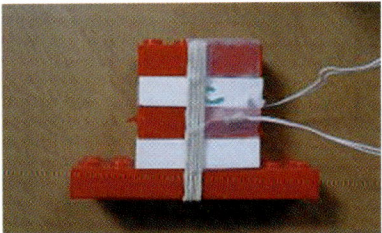

Plate 47 Sample plastics included a degraded CN negative film from the 1940s (left), polyester cassette tape from 1970s (centre) and a toy ship made from ABS LEGO bricks (right) (Fig. 1, p. 101).

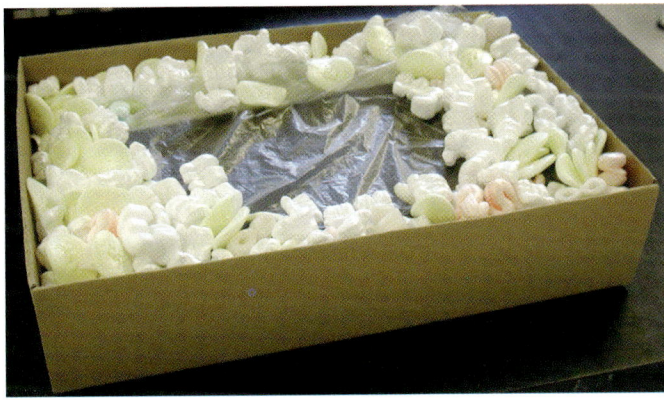

Plate 48 Sample plastics were insulated by packing them with expanded polystyrene chips prior to cooling (Fig. 2, p. 102).

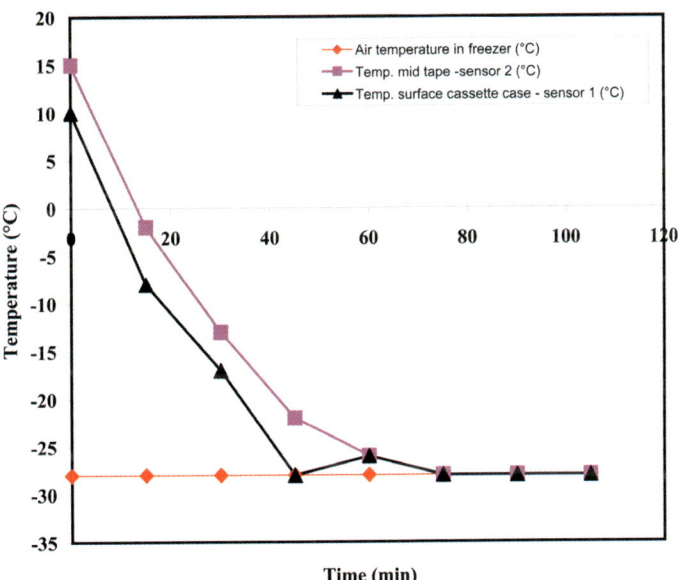

Plate 49 Rate of cooling of thin-walled cassette tape from ambient to freezer temperature (Fig. 4, p. 103).

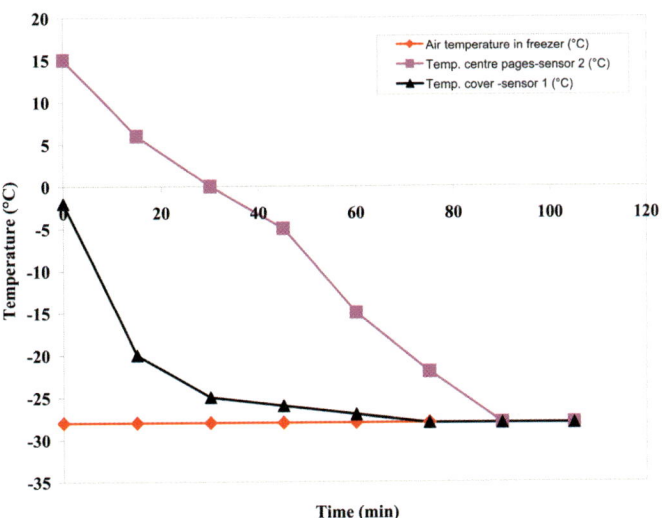

Plate 50 Rate of cooling of thick-walled photograph album from ambient to freezer temperature (Fig. 5, p. 103).

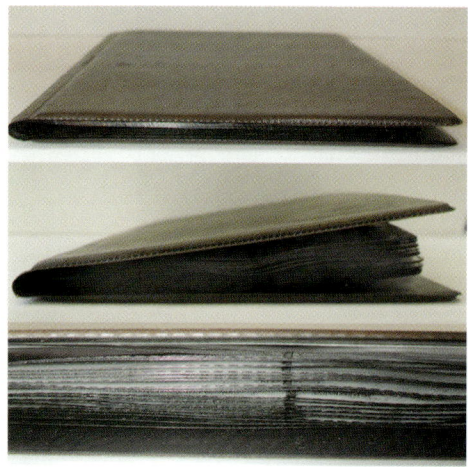

Plate 51 PVC photograph album at ambient (upper) and at freezer temperature (centre). Pages of the album became cockled and stiff on cooling (lower) (Fig. 6, p. 104).

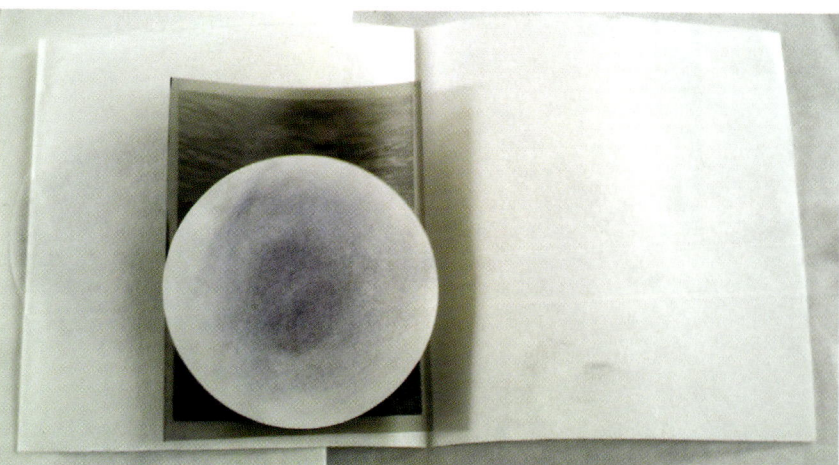

Plate 52 Methylene blue powder was applied to filter paper to detect the presence of moisture. No condensation developed on CN negative film in albums after removal from cold storage (Fig. 7, p. 105).

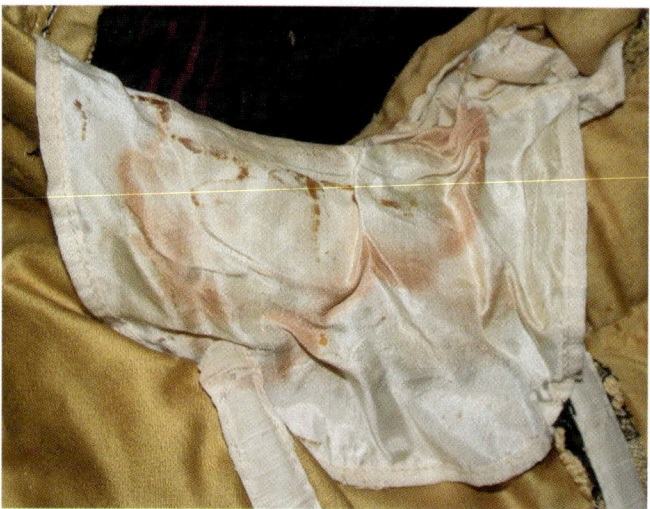

Plate 53 A dress shield showing use-related damage in the form of creasing and deformation, some waterborne staining and transferral of the proof layer (Fig. 2, p. 109).

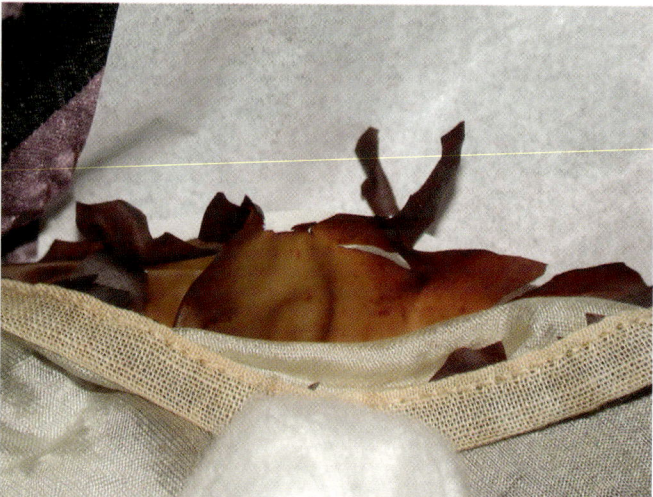

Plate 54 An example of darkening and stiffening of a self-supporting film along the line of waterborne staining that has caused physical breakdown (HCCMAS collection) (Fig. 3, p. 109).

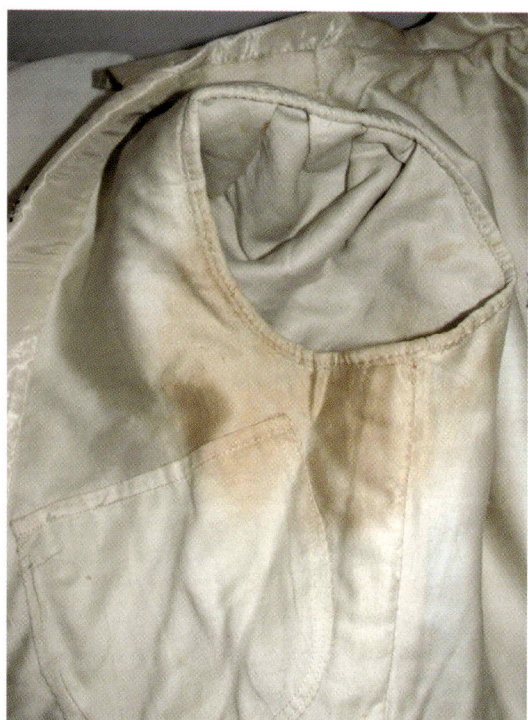

Plate 55 Uniform discoloration of the lining of a garment where dress shields have been removed (Fig. 4, p. 110).

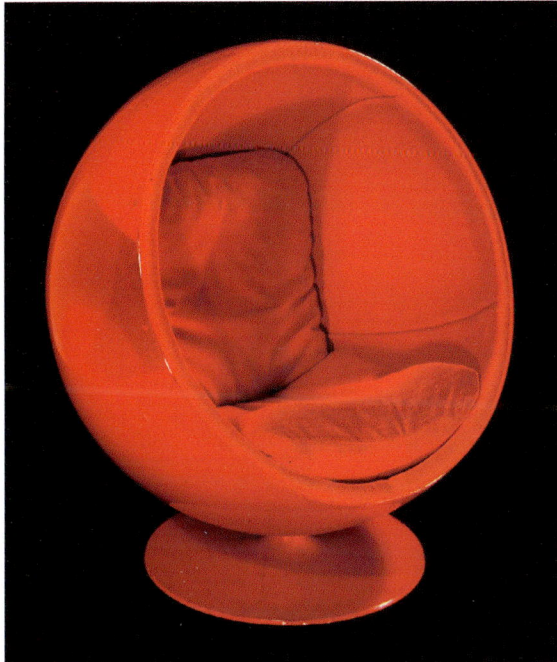

Plate 56 Eero Aarnio's Globe chair much as it would have appeared when it was first purchased by the V&A Circulation Department in 1968 (V&A Images/Victoria and Albert Museum, Museum number Circ. 12-1969) (Fig. 1, p. 117).

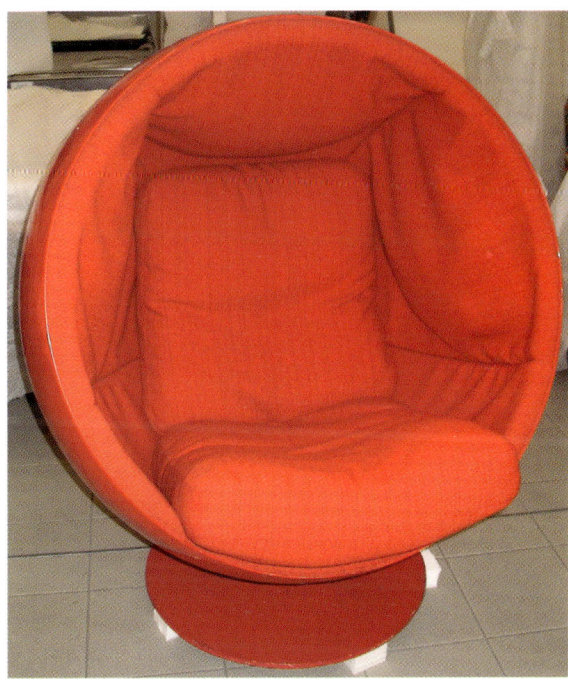

Plate 57 The condition of the V&A's Eero Aarnio Globe chair on 16 December 2004 (reproduced by kind permission of the Victoria and Albert Museum, Museum number Circ. 12-1969) (Fig. 2, p. 118).

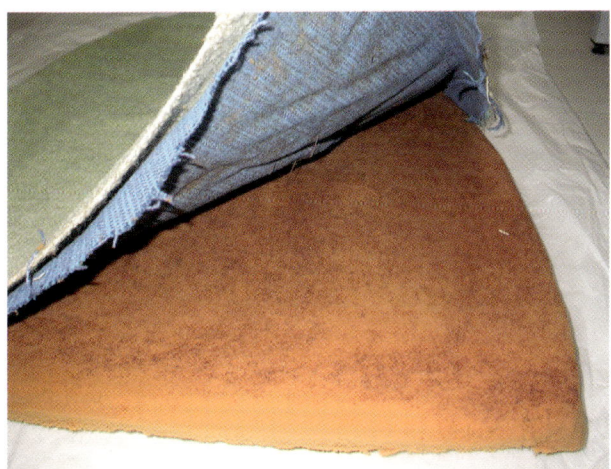

Plate 58 The obverse of a foam wedge and the reverse of the top cover fabric removed from a Ball chair of a date similar to that of the V&A's Globe chair. The dark brown areas on the foam and fabric are the still present but degraded adhesive (Fig. 3, p. 118).

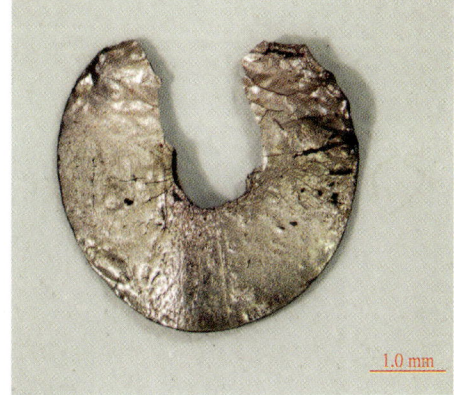

Plate 60 Condition of a degraded sequin (Fig. 2, p. 123).

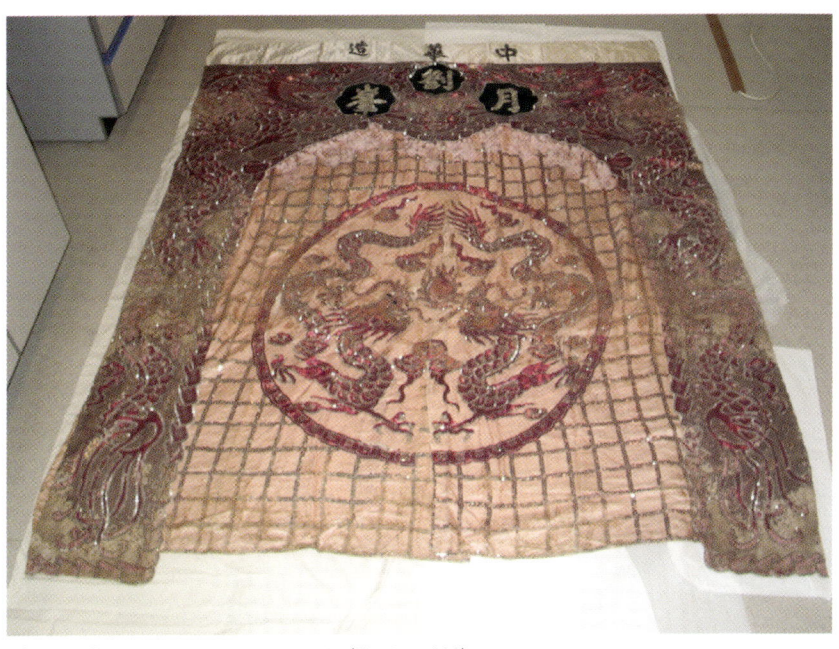

Plate 59 Cantonese opera stage curtain (Fig. 1, p. 123).

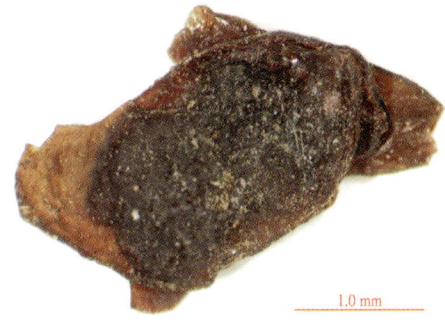

Plate 61 Condition of a degraded sequin (changed to amber in colour) (Fig. 3, p. 123).

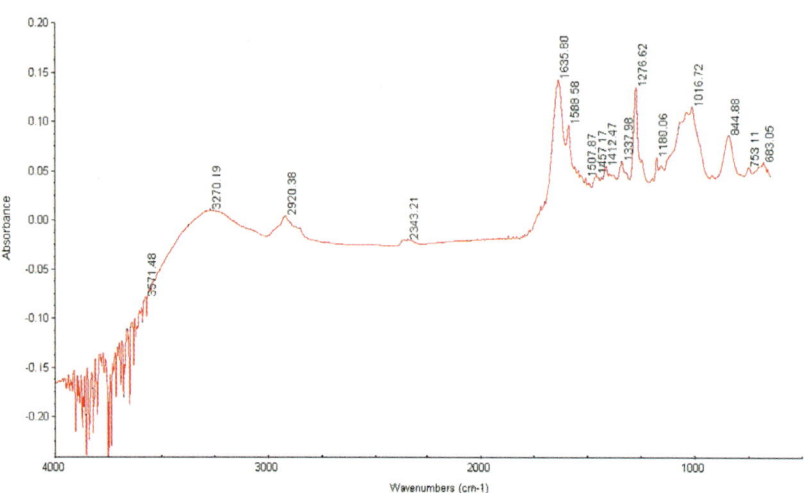

Plate 62 FTIR spectrum of a pink sequin (iridescent side) (Fig. 5, p. 125).

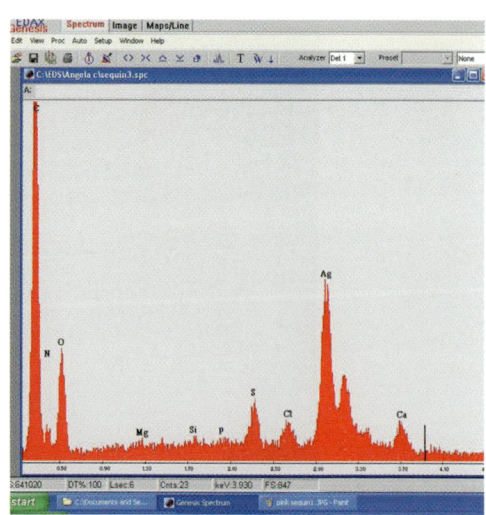

Plate 63 SEM–EDX spectrum of a light pink sequin (Fig. 6, p. 126).

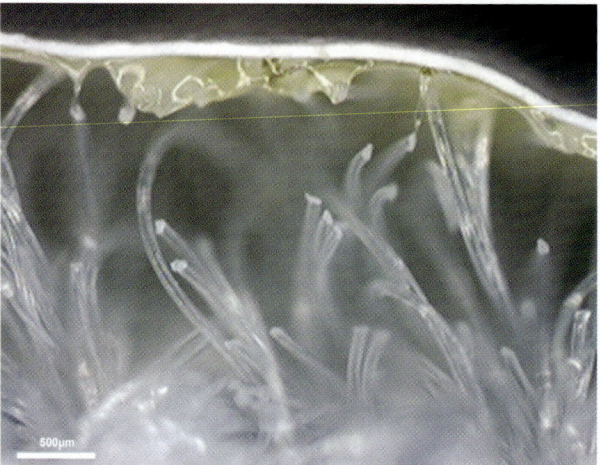

Plate 64 Cross-section of a polyurethane-coated web textile (Fig. 1, p. 128).

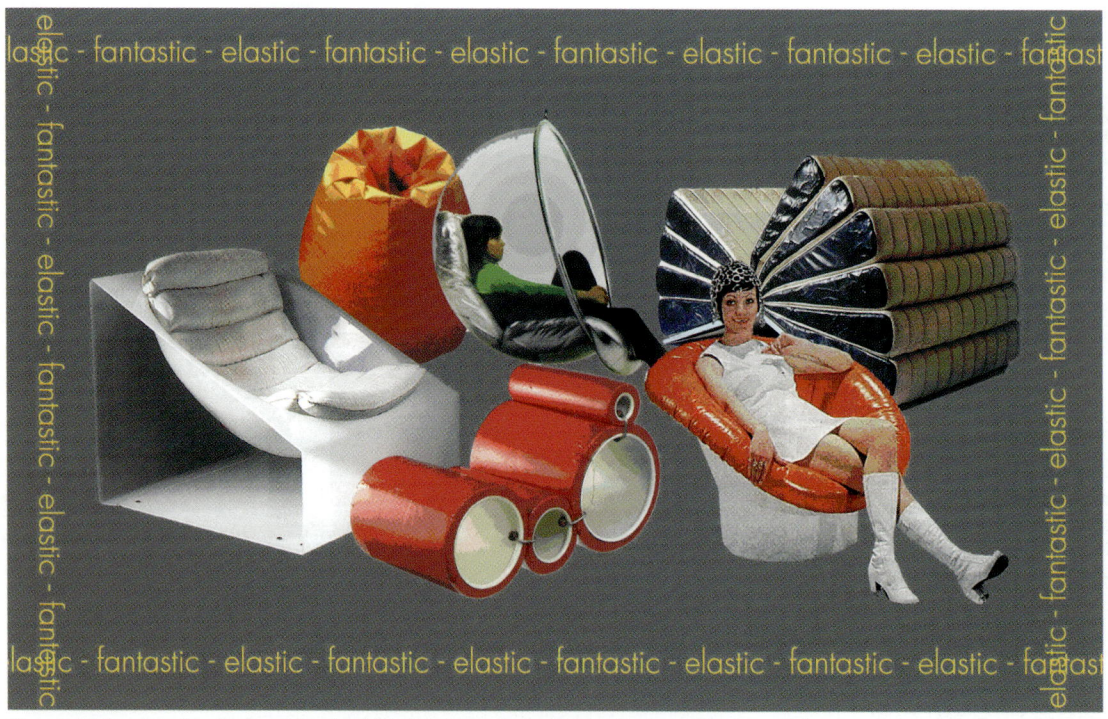

Plate 65 Tube chair, Joe Cesare Colombo, 1969 (Die Neue Sammlung, München) (Fig. 2, p. 129).

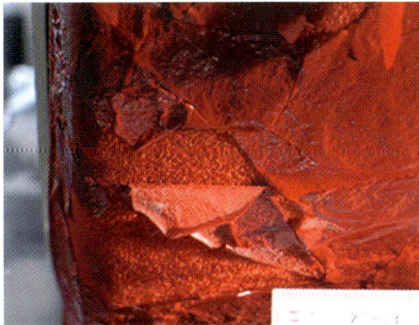
Plate 66 Tube chair: detachment, polyurethane coating (Fig. 5, p. 131).

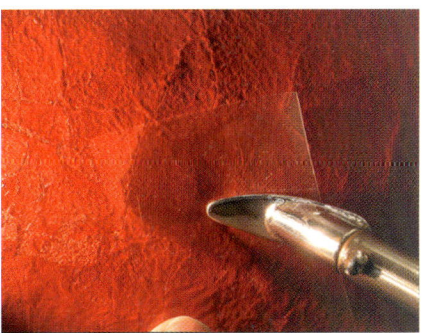
Plate 67 Tube chair: consolidation treatment (Fig. 6, p. 132).

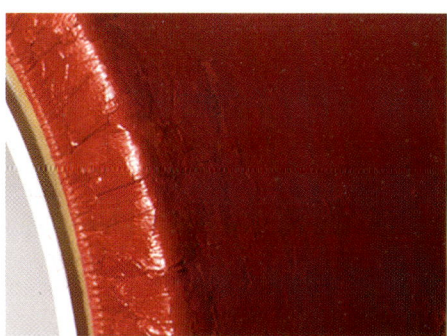
Plate 68 After consolidation treatment (Fig. 7, p. 132).

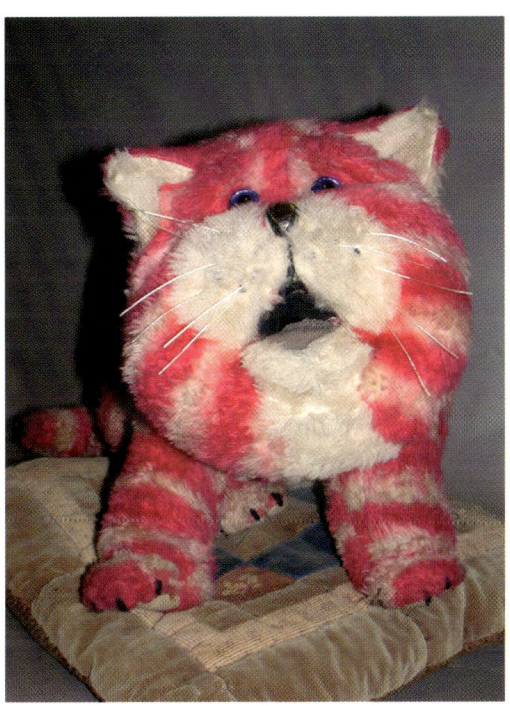
Plate 69 The Bagpuss puppet (Fig. 1, p. 137).

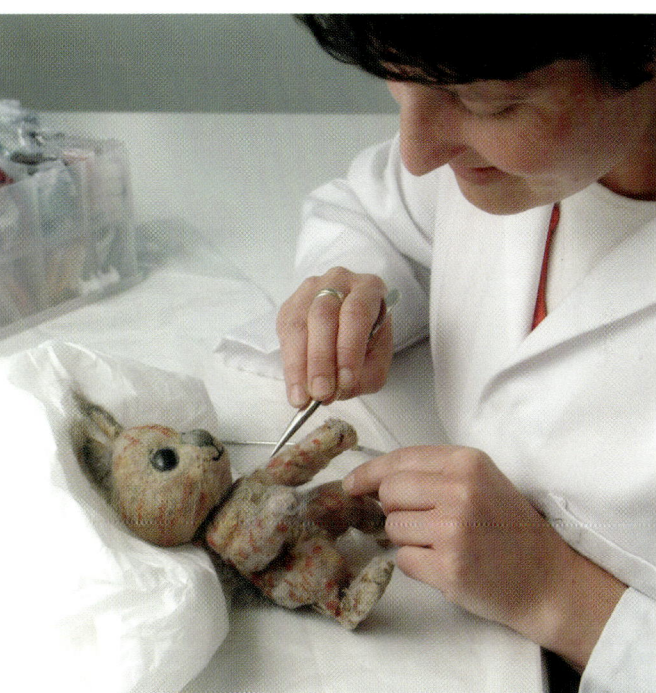
Plate 70 Tog: conservation in progress (Fig. 2, p. 138).

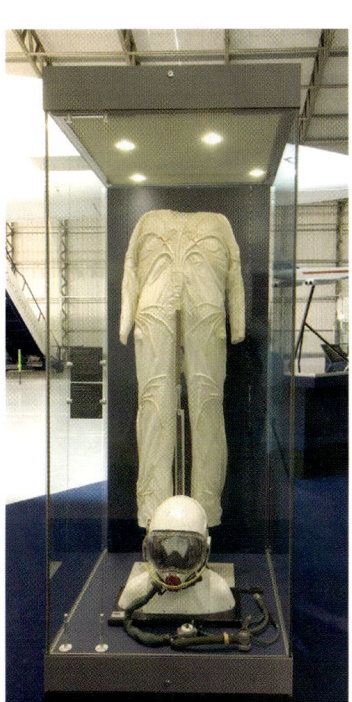

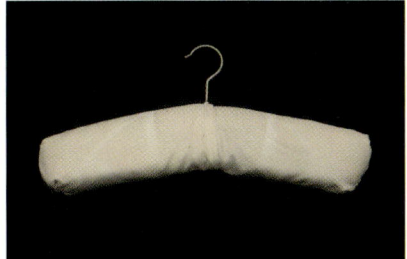
Plate 72 Hanger showing evidence of light damage (photo: © The Trustees of the National Museums of Scotland) (Fig. 2, p. 140).

Plate 71 Flight suit on display at the Museum of Flight (photo: © The Trustees of the National Museums of Scotland) (Fig. 1, p. 139).

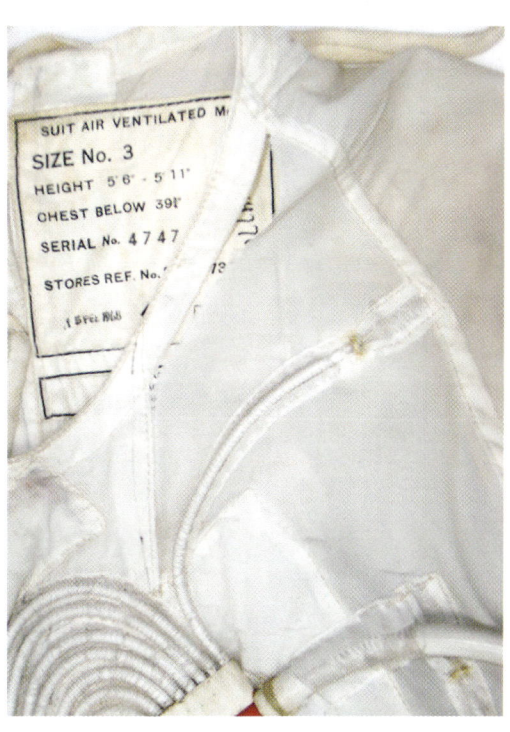

Plate 73 Splits in the nylon along the lines of machine stitch at the seams (photo: © The Trustees of the National Museums of Scotland) (Fig. 3, p. 140).

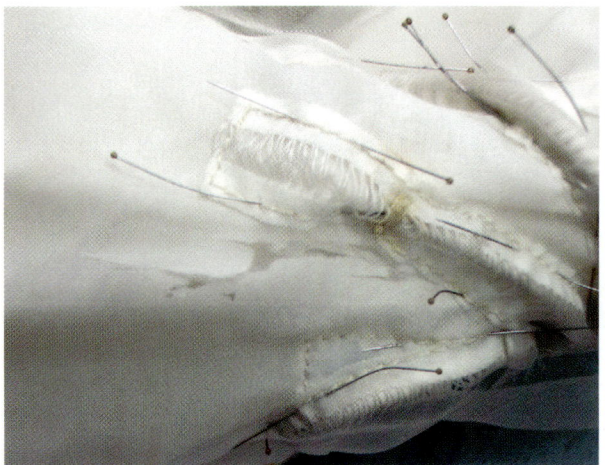
Plate 74 Splits formed across the weave like shattered silk (photo: © The Trustees of the National Museums of Scotland) (Fig. 4, p. 140).

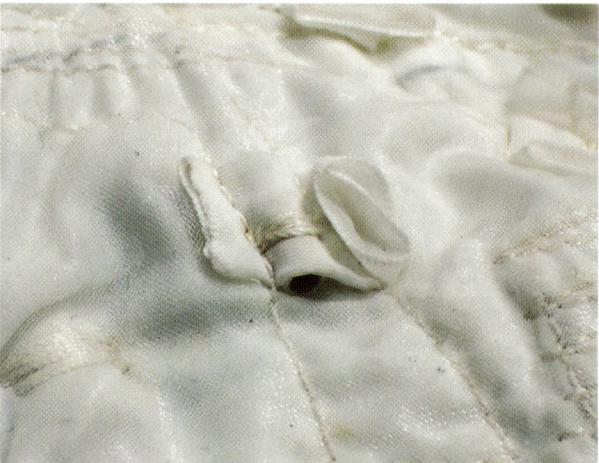
Plate 75 Pipe ends through which air was delivered to the body creating the uneven surface of the interior of the suit (photo: © The Trustees of the National Museums of Scotland) (Fig. 5, p. 140).

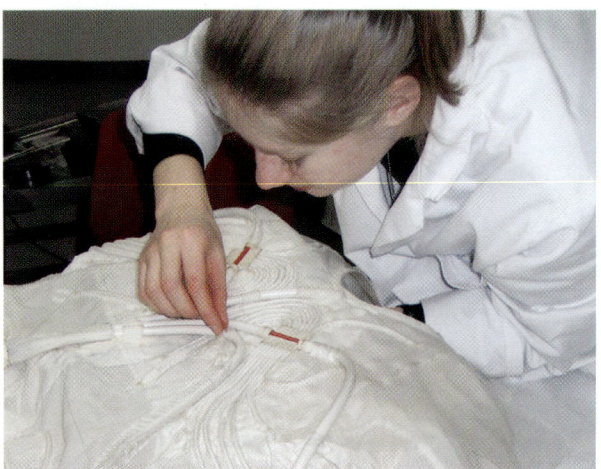
Plate 76 The sandwich treatment being carried out (photo: © The Trustees of the National Museums of Scotland) (Fig. 6, p. 140).

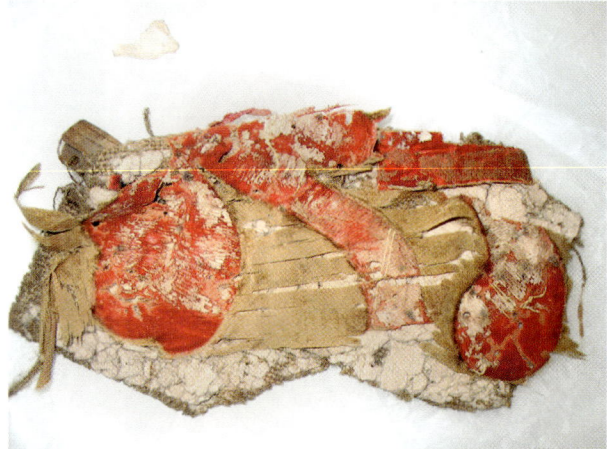
Plate 77 Fragment of carriage velvet from the Great French Carriage before restoration (Fig. 1, p. 142).

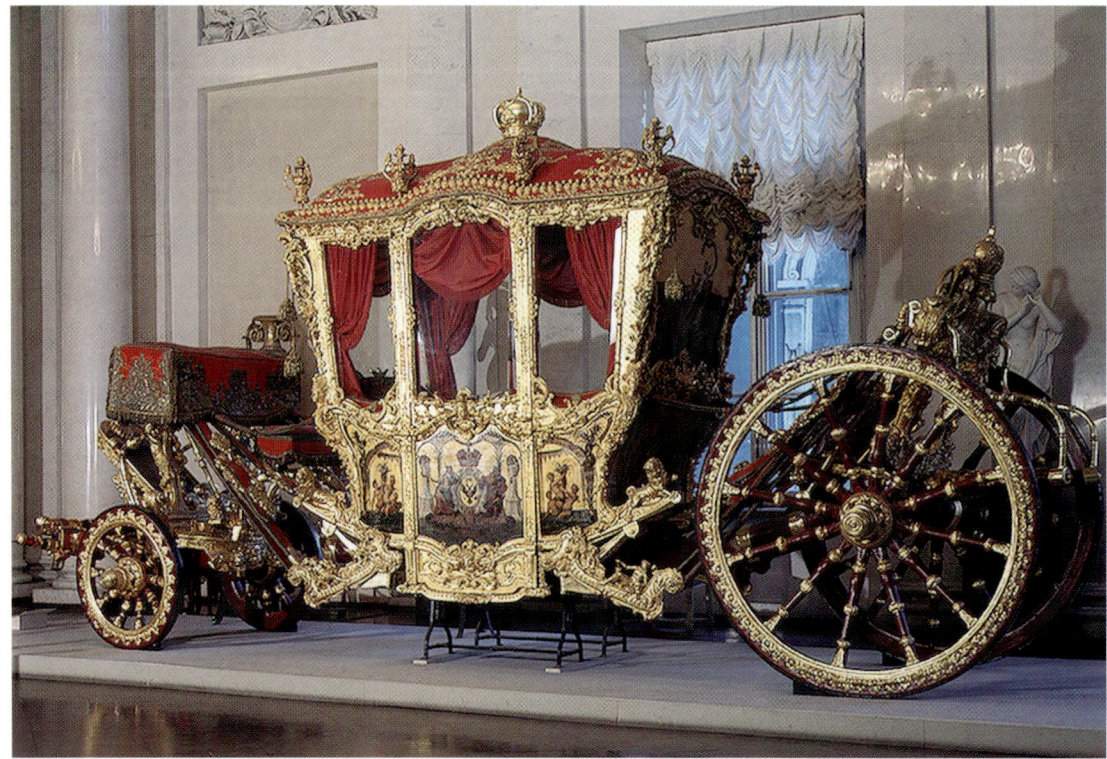
Plate 78 The carriage after restoration (Fig. 2, p. 143).

Television puppets from the 1960s and 1970s: creation, materials and conservation

Rebecca Smith

ABSTRACT This paper illustrates and briefly describes the aims, methods and results of research into the creation, materials and conservation of television puppets from the 1960s and 1970s. The main object of the research was Bagpuss, on loan to the Museum of Canterbury (UK), to determine why the puppet was in such good condition compared to other similar items of the period, and to suggest the most suitable display conditions over the course of the loan. Non-destructive Fourier transform infrared (FTIR) analysis was used to identify the materials – this revealed that the filling was polyether polyurethane and the 'fur' fabric was composed of a polyester ground with an acrylic pile; the internal structure was known to be wood and metal. It is known that polyurethane foam is particularly susceptible to oxidation, leading to catastrophic loss of form. Artificial ageing experiments demonstrated that the outer layers of the puppet offered a good degree of protection to the foam, accounting for its good condition. For long-term storage and display, a stable environment with controlled light and humidity levels was suggested.

Keywords: television puppets, Bagpuss, polyether polyurethane, Oddy tests, preventive conservation

Introduction

This paper illustrates and briefly describes the aims, methods and results of research into the creation, materials and conservation of TV puppets from the 1960s and 1970s. When Bagpuss, the puppet from the children's television series *Bagpuss*, first broadcast in 1974, was compared to other contemporary puppets that were beginning to show signs of deterioration (for example, Tog from the television series *Pogles Wood* and Larry the Lamb (Keneghan 1995)), research was undertaken to understand why Bagpuss was in such good condition and to determine if its condition would deteriorate during a ten-year loan to the Museum of Canterbury (Figs 1 and 2).

The Bagpuss puppet was studied to identify the materials used. Results of analysis revealed that polyether polyurethane (PU-PEt) had been used as the filling and that the 'fur' fabric on the exterior had a polyester ground with an acrylic pile. The internal structure (skeleton) was identified by Peter and Joan Firmin, the makers, as wood and metal.[1] In 2004, the date of the project, Bagpuss was found to be in good condition. Once the type of foam had been established, research into the properties of the foam revealed that the main cause of deterioration of polyether polyurethane is oxidation. Despite the propensity of PU-PEt foam to deteriorate, Bagpuss was not seemingly suffering in this way. Therefore, testing was carried out to establish if the 'fur' fabric used to cover the puppet acted as a buffer against the oxidation and the deterioration of the foam. Accelerated ageing (Oddy testing) was used to determine if volatile components were being released from the materials during the ageing process. Results revealed that the individual components would deteriorate over time, if the conditions in which they were stored or displayed were very humid. Further testing could be used to confirm the conjecture that deterioration of one of the components would lead to the deterioration of the other materials over a period of time.

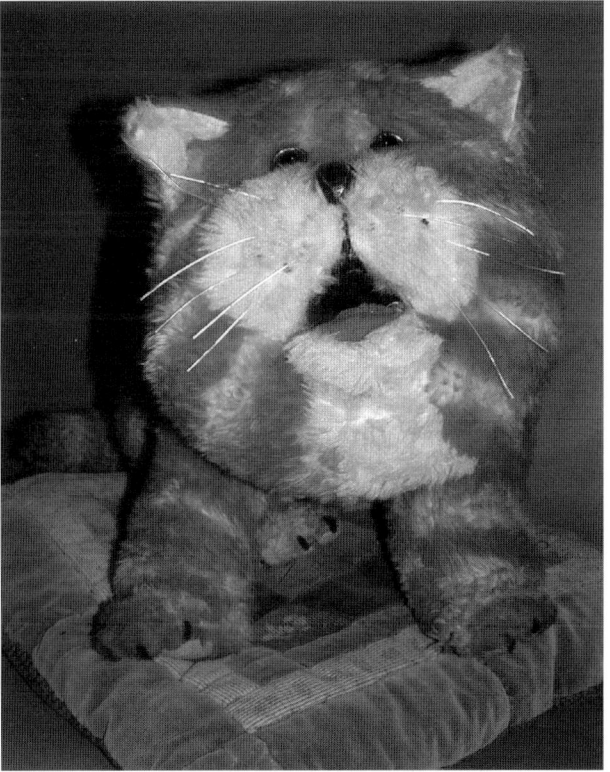

Figure 1 The Bagpuss puppet (Plate 69 in the colour plate section).

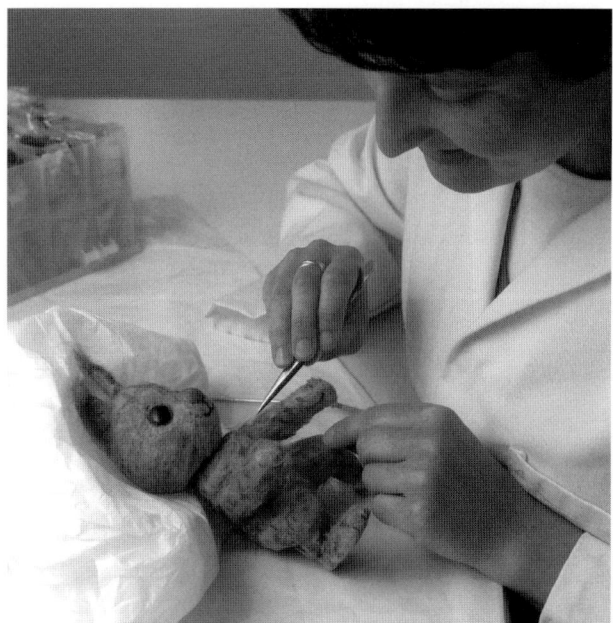

Figure 2 Tog: conservation in progress (Plate 70 in the colour plate section).

Light testing confirmed that the fur fabric, cotton jersey and black embroidery, which comprise the exterior of Bagpuss, do offer protection to the foam layer. Without this defence the foam would have been affected by light, resulting in changes to the physical and visual properties. Hence today, fortuitously, Bagpuss is still in good shape, unlike other puppets. The conservation problems that television puppets can present are very much dependent on the materials that have been used to make them. Research has led to an understanding that foam-covered puppets such as Larry the Lamb (held in the Victoria and Albert Museum collection in London) can present problems such as loss of the foam surface and deterioration of objects stored nearby, due to the release of volatile compounds. Ongoing research is being carried out to determine if the addition of the oxygen scavenger Ageless is the answer to the safe storage of puppets with a foam component (Keneghan 1995). Larry the Lamb is destined to stay in storage since little can be done to stabilise the foam at present.[2] Such problems are not limited to puppets of UK origin. PU-PEt foam obviously lends itself to the aesthetic and functional demands of puppets in children's entertainment in general. The Canadian Conservation Institute has also instigated projects to contend with the PU-PEt foam on television puppets within Canadian collections.[3]

Bagpuss was recently voted 'the nation's favourite children's programme' (BBC television poll, January 1999), an indication of the mass appeal and continuing popularity of the programme and its lead character. Although the sentimental value of well-known puppets is hard to measure, the extent of merchandising of current and past television puppets suggests that they are a notable part of contemporary Western society. With this in mind, preservation is necessary for those few puppets that are in the public domain to maintain the material evidence of our contemporary culture. Interviews with the maker of Bagpuss, Peter Firmin, revealed that he had reservations as to whether he would recreate the puppets if they were irreparably damaged. Moreover, he feels that their natural life should be preserved for as long as possible, to ensure they are enjoyed by future generations.

To maintain Bagpuss in relatively good condition during display in Canterbury Museum, preventive conservation is the key. The combination of materials and construction has served Bagpuss well thus far. The puppet's continued existence is contingent upon provision of an environment that will maintain these materials as far as practicable. The research suggests that Bagpuss should be placed in a stable environment since the materials of construction are susceptible to high light and humidity levels. Furthermore, detailed documentation of Bagpuss and other television puppets, treasured by generations of the populace, will help to note changes in their condition as well as record their existence for the future.

Notes

1. Personal interview with Peter and Joan Firmin, makers of Bagpuss, 14 July 2004.
2. Personal correspondence with Catherine Howell, Museum of Childhood at Bethnal Green, London, 14 May 2004.
3. Personal correspondence with Martha Segal, Canadian Conservation Institute, 1 March 2004.

Reference

Keneghan, B. (1995) 'Trouble in toytown: Larry the Lamb falls to pieces', in *From Marble to Chocolate: The Conservation of Modern Sculpture*, J. Heuman (ed.), 1–8. London: Archetype Publications.

The author

Rebecca Smith's final year dissertation considered the examination and analysis of television puppets from the 1960s and 1970s. Since graduating, Rebecca has gained a wide range of conservation experience working initially at the Conservation Centre, Liverpool, followed by employment in private practice, the Textile Conservation Consultancy, Stamford.

Address

Rebecca Smith, 41 Kimball Close, The Cottesmore, Ashwell, Rutland LE15 7QP, UK (rebecca_smith37@excite.com).

The treatment of the light-damaged nylon component of a flight suit used during the test flights of Concorde c.1968

Anna Hodson

ABSTRACT This paper discusses the deterioration and conservation of the woven nylon element of a flight suit held by the National Museums of Scotland (Edinburgh), to be displayed alongside Concorde at the Museum of Flight, East Fortune (Scotland). The treatment of the object was complicated by the range of materials used in its construction, including nylon 66, PVC, metal hoses and insulating tape. The nylon itself is brittle and has split along the seams, both in the direction of weave and along creases. Comparisons with similar items in the RAF Museum suggest the damage has been caused by exposure to light. Due to the extensive deterioration of the nylon component, it was determined that the most appropriate method of support was a combination of an adhesive support sandwiched with an overlay and stitch.

Keywords: flight suit, nylon 66, plasticised PVC, support, stitching, adhesive

Introduction

This brief paper characterises the deterioration of the woven nylon element of a flight suit caused by exposure to light and discusses the choice and method of treatment dictated by this complex, multimedia garment. The flight suit was recently acquired by the National Museums of Scotland to be exhibited alongside Concorde at the Museum of Flight, East Fortune (Scotland). The suit was conserved to be displayed upright on a custom-made form (Fig. 1). The garment forms the underlayer of a high-altitude flight suit used in the test flights of Concorde. A heavy metal hose is attached to the front of the suit, from which emerges a network of plasticised PVC pipes. The pipes are attached to the exterior of the full length, finely woven, nylon 66 garment, with the use of textile casings attached with machine stitch and insulating tape. The presence of the PVC pipes causes cockling and distortion of the garment and places areas of the nylon fabric under constant tension.

The nylon fabric element of the suit is brittle and split. This damage is concentrated on the chest area, although several small splits appear elsewhere. The splits have formed along the seams, both directions of the weave and along creases in the fabric. Evidence of previous display and observation of a collection of similar garments held at the RAF Museum, which do not reveal the same sort of deterioration, indicate that the damage was caused by light (Figs 2–4).

Neither adhesive nor stitching alone were found to be effective means of support; the brittle fabric was damaged by stitch holes and a good adhesive bond was impossible due to the suit's construction which created an uneven surface (Fig. 5). The nature of the object dictated that a sandwiched adhesive

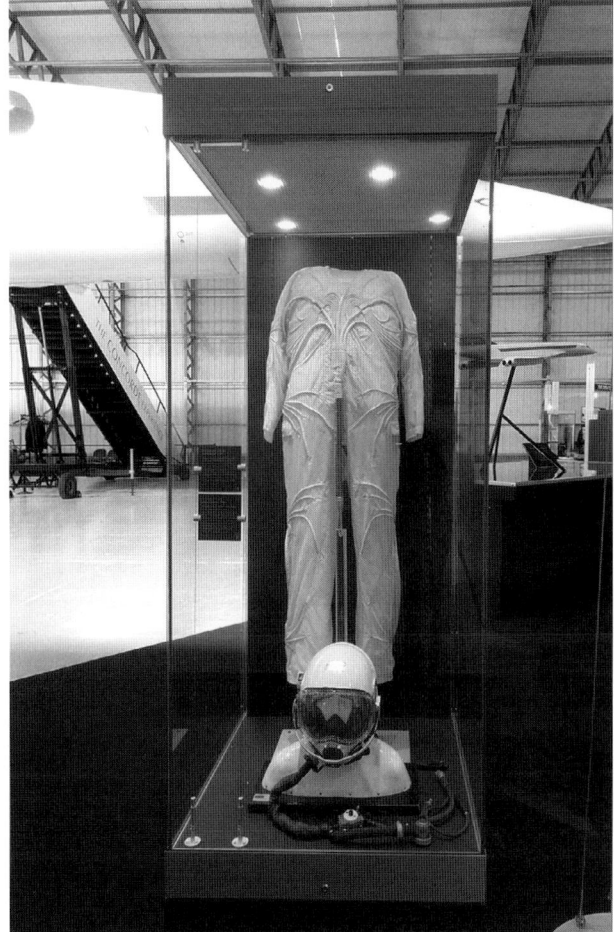

Figure 1 Flight suit on display at the Museum of Flight (photo: © The Trustees of the National Museums of Scotland) (Plate 71 in the colour plate section).

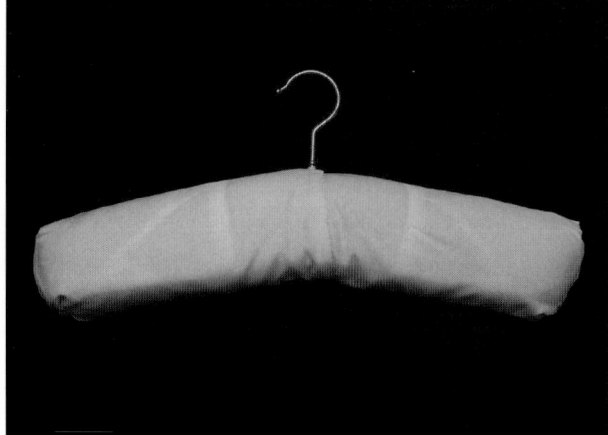

Figure 2 Hanger showing evidence of light damage (photo: © The Trustees of the National Museums of Scotland) (Plate 72 in the colour plate section).

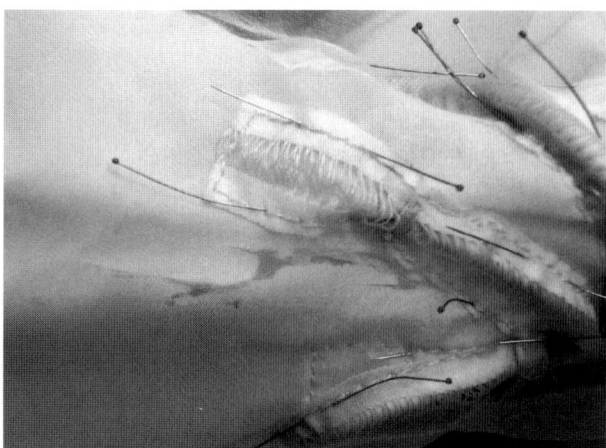

Figure 4 Splits formed across the weave like shattered silk (photo: © The Trustees of the National Museums of Scotland) (Plate 74 in the colour plate section).

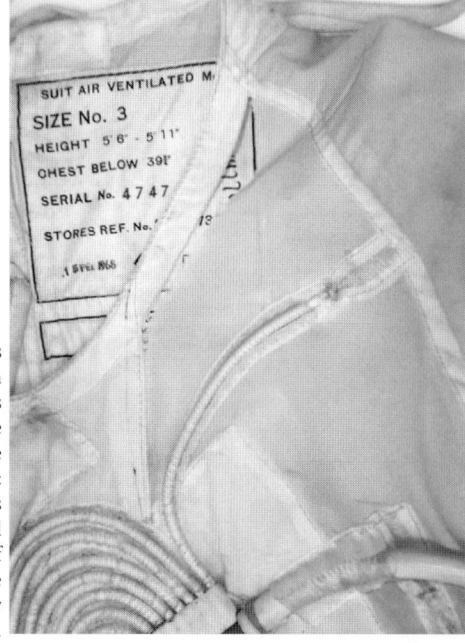

Figure 3 Splits in the nylon along the lines of machine stitch at the seams (photo: © The Trustees of the National Museums of Scotland) (Plate 73 in the colour plate section).

Figure 5 Pipe ends through which air was delivered to the body creating the uneven surface of the interior of the suit (photo: © The Trustees of the National Museums of Scotland) (Plate 75 in the colour plate section).

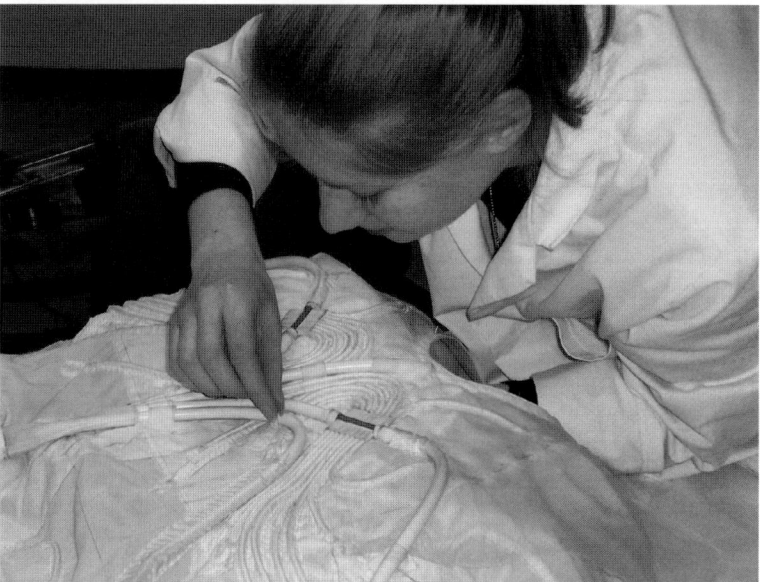

Figure 6 The sandwich treatment being carried out (photo: © The Trustees of the National Museums of Scotland) (Plate 76 in the colour plate section).

and stitched treatment be employed as an effective support treatment (Fig. 6).

The author

Anna Hodson has undertaken a Historic Scotland Internship in Textile Conservation at two venues, the National Museums of Scotland, Edinburgh and the Marischal Museum, Aberdeen, as well as a North West Arts Board funded residency at the Gallery of Costume, Platt Hall, Manchester. In September 2005 she began a Mellon Fellowship in Textile Conservation at the National Museum of the American Indian, Smithsonian Institution, Washington DC, USA.

Address

Anna Hodson, National Museum of the American Indian/Smithsonian Institution, Cultural Resources Center, MRC 541, 4220 Silver Hill Road, Suitland MD 20746, USA (hodsona@si.edu).

Modern textile materials in practice at the State Hermitage Museum

Elena Mikolaychuk

ABSTRACT The State Hermitage Museum (St Petersburg) has a rich and varied textile collection, spanning a range of regions and time periods. Modern materials are used in three main aspects of the Museum's work: restoration and conservation; furnishing of exhibition and storage facilities; decoration and embellishment of exhibition halls. The selection of modern materials is carried out on a scientific basis requiring knowledge of chemical, physical, mechanical, optical and other properties and the skills to interpret these factors. The conservation of the Great French Carriage, used for the crowning ceremonies of Russian emperors, is presented as an example of the use of these materials.

Keywords: fibre degradation, modern materials, remedial conservation, Russian heritage, synthetic fibres

Introduction

Rapid development of chemistry in the 20th century has resulted in huge numbers of artificial and modern textile products being produced. Technical progress has indeed touched the textile industry. As a consequence there is a large variety of synthetic threads, fabrics, non-woven textile materials and also dyes for textile materials. Many museums throughout the world have already started to create textile collections of 20th-century artefacts made from modern materials. The State Hermitage Museum in St Petersburg, however, having a relatively conservative outlook and collections policy, has only just only started to form a collection that includes synthetic materials. Despite this, modern textile materials have already become dominant in all spheres of life at the Museum.

First, artefacts containing textiles with large areas of damage and loss may need remedial restoration works. In such cases the historical textiles are completely replaced by new, modern components (e.g. the textile elements of the Great French Carriage – the carriage used for the crowning ceremonies of Russian emperors) (Figs 1 and 2). Modern textile materials used during conservation processes include fabrics for supporting areas of loss, for linings, threads for sewing and embroidery, imitation gold and silver threads and so on.

The second main area is the furnishing of exhibitions and storage facilities. Designers at the State Hermitage Museum frequently use fabrics for decoration of showcases, both temporary and permanent exhibitions, and also for lining and covering boxes for exhibits in storage. The third main area is the decorating and embellishing of exhibition halls with window curtains and floor coverings.

Unfortunately, not all modern textile materials are suitable for museum purposes. The choice of modern textile materials for specified purposes in a museum needs to be carried out on a scientific basis. Knowledge of chemical, physical, mechanical,

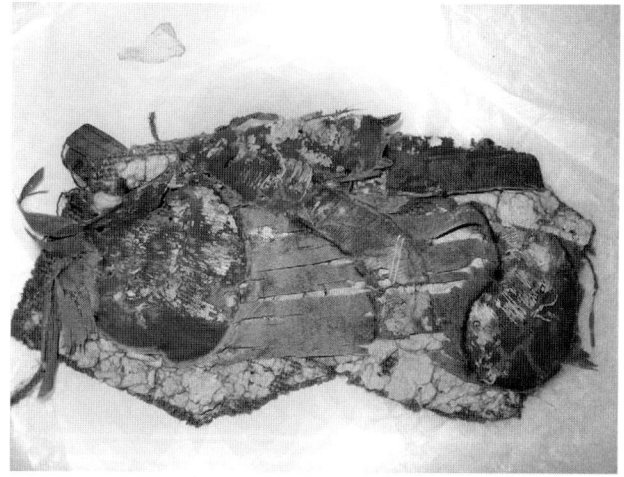

Figure 1 Fragment of carriage velvet from the Great French Carriage before restoration (Plate 77 in the colour plate section).

optical and other properties of modern textile materials – as well as the skills to interpret these properties – is required. In museum conditions, the problem becomes more complicated because modern textile materials must not harm the exhibits with which they are directly in contact. On the other hand, the modern textile material, having been chosen for use in the museum in one of three specified directions, itself becomes an exhibit, or a part of an exhibit. Thus it is necessary to take into account any probable influence it may have on other exhibits or their constituent parts. It is possible to cite many cases of incompatibility – fabrics containing sulphur within their structure or dyes, for instance, must not be combined with silver or copper otherwise undesirable consequences will result.

The mode of storage should be individually selected for each type of modern textile material. This is especially important for synthetics, as many of them are exposed to irreversible proc-

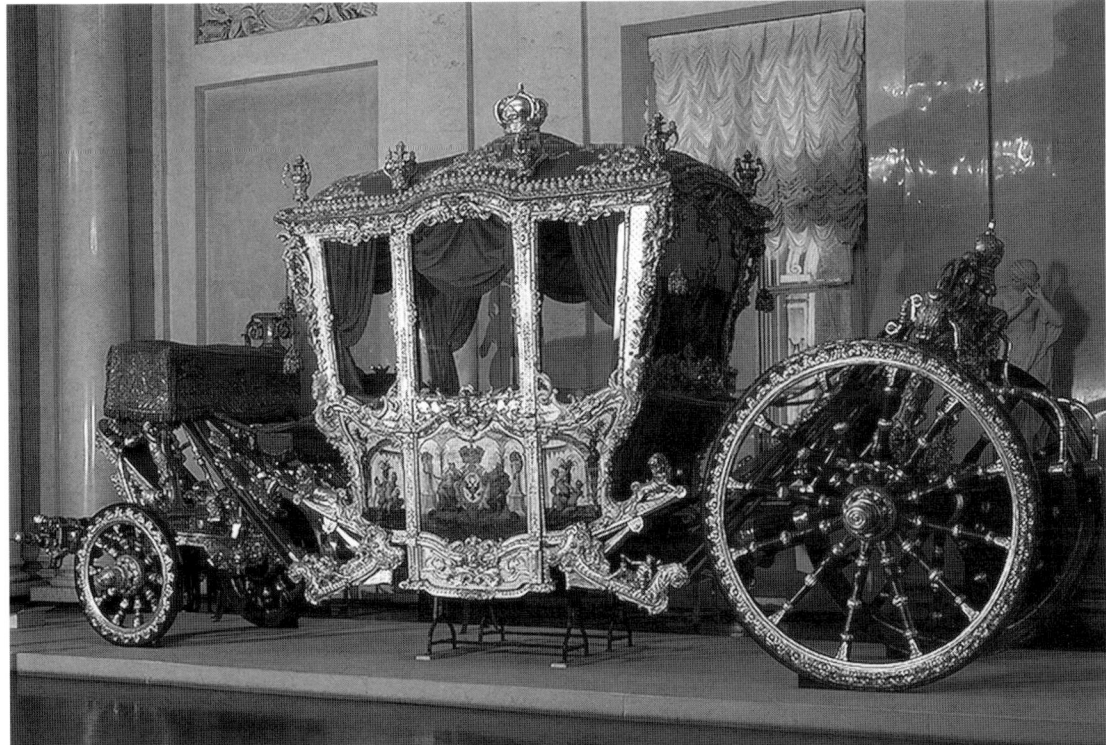

Figure 2 The carriage after restoration (Plate 78 in the colour plate section).

esses during ageing which significantly change initial properties of the polymers. For example, the colour of the polymer may change, its elasticity may alter, it may become opaque and cloudy and it can become rigid and so on. Polyamide fibres are particularly susceptible to light and this should be taken into account when illuminating a showcase. In a museum environment, the problem is exacerbated by the fact that the modern textile materials are in contact with other materials used for exhibition purposes. In each case an approach that considers how all the materials involve behave will ensure better long-term storage and exhibition of artefacts.

The author

Dr Elena Mikolaychuk is Senior Scientist, Department for Scientific-Technological Examination of Works of Art at the State Hermitage Museum, St Petersburg, Russia, where she has worked since 1975. In her current role she researches the Hermitage collections as well as acting as a consultant for museums in Russia and other countries.

Address

Elena Mikolaychuk, The State Hermitage Museum, 34 Dvortcovaya Emb., St Petersburg, 190000, Russia (mikolaychuk@hermitage.ru).

Heart Disease

© Aladdin Books 1989

Designed and produced by
Aladdin Books Ltd
70 Old Compton Street
London W1V 5PA

First published in
Great Britain in 1989 by
Franklin Watts Ltd
96 Leonard Street
London EC2A 4RH

Design: David West Children's Book Design
Editor: Zuza Vrbova
Picture Research: Cecilia Weston-Baker
Illustrator: Stuart Brendon

ISBN 0-7496-0045-4

Printed in Belgium

CONTENTS

INSIDE THE HEART	4
DISORDERS OF THE HEART	8
TESTS AND TREATMENTS	16
LIVING WITH HEART DISORDERS	24
TAKING CARE OF YOUR HEART	28
GLOSSARY	31
INDEX	32

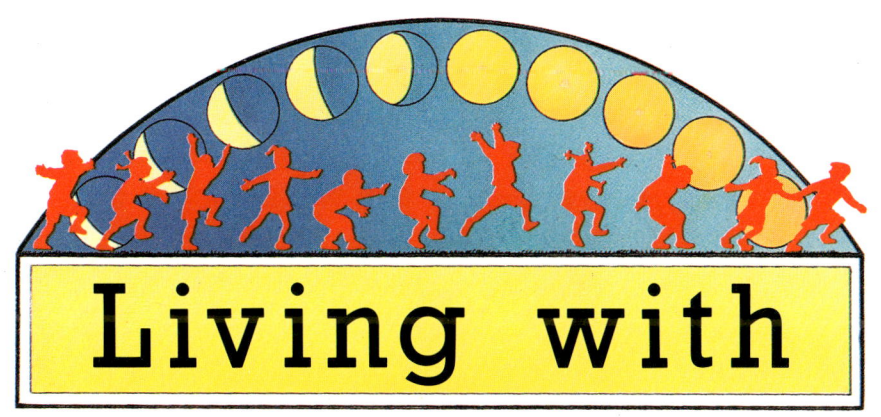

Living with
Heart Disease

Steve Parker

FRANKLIN WATTS
London : New York : Toronto : Sydney

INSIDE THE HEART

In times past, the heart was thought to be the origin of love, loyalty, bravery and emotions. We now know that this fist-sized organ, slightly to the left of the centre of the chest, is a muscular pump that pushes blood round and round the body. It is an efficient, adaptable and incredibly reliable pump. During exercise, it can increase the amount of blood it pumps to more than 4 times what it pumps at rest. In an average lifetime it pumps, or "beats", more than 3,000 million times.

The heart consists of two small thin-walled upper chambers (called atria) and two large thick-walled lower ones (ventricles). Arteries are strong, elastic-walled vessels that carry blood away from the heart. Slack, thinner-walled veins bring blood back to the heart. One-way valves between the chambers and in the vessels make sure blood flows in the correct direction. It is the closure of the different heart valves that produces the familiar heartbeat sound.

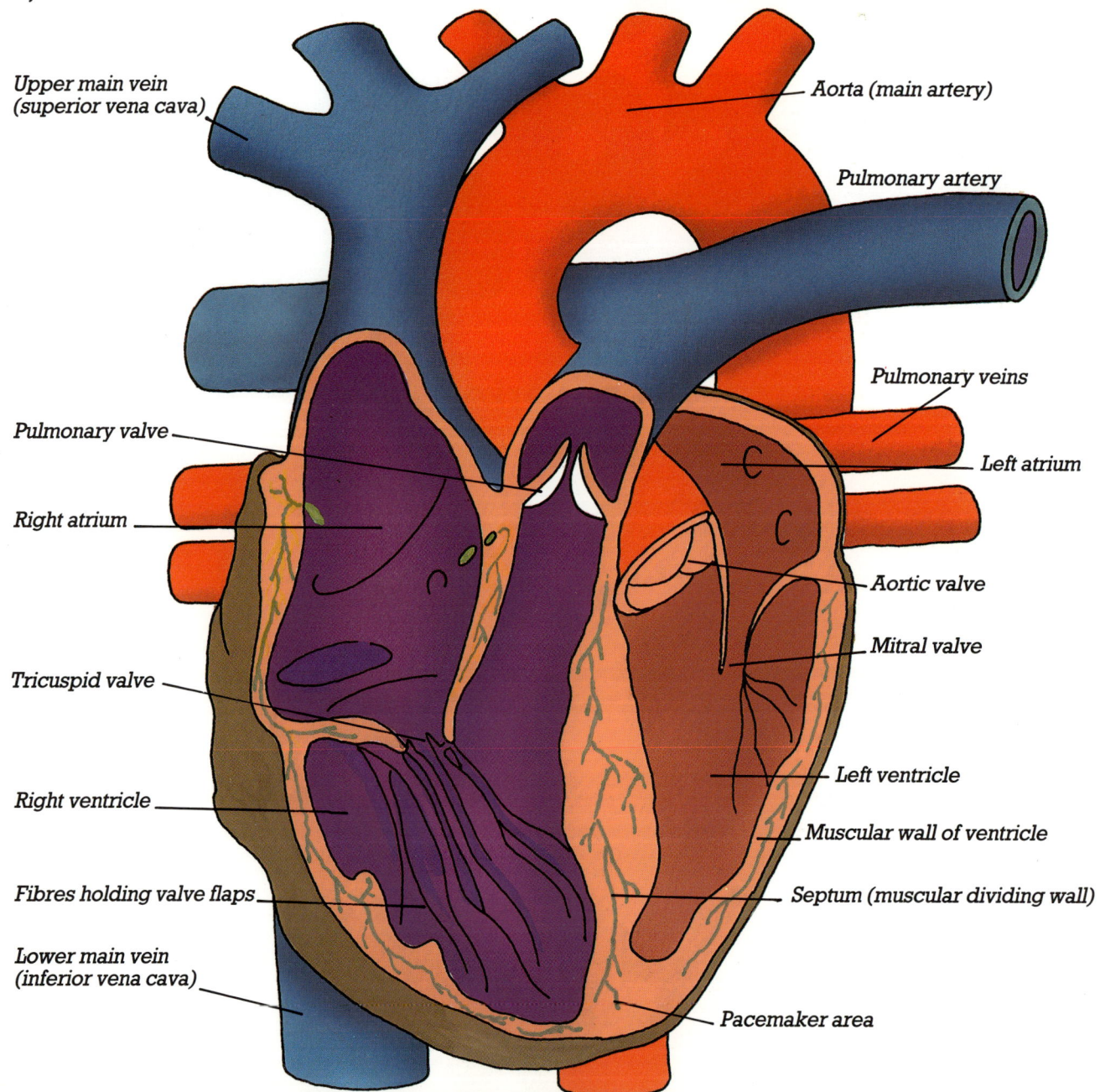

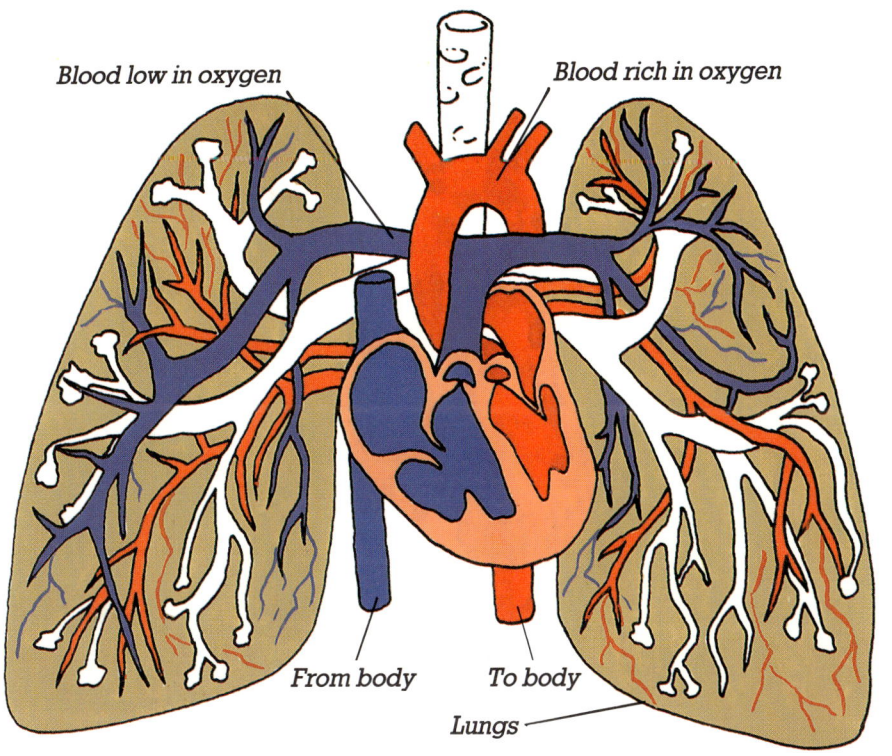

Blood low in oxygen
Blood rich in oxygen
From body
To body
Lungs

Heart and lungs

The heart is not a single pump, but two pumps side by side. Blood flows out from the left pump along the main artery, the aorta, and around the body (the systemic circulation). As it passes through the body tissues, it supplies them with the oxygen necessary for life. Low in oxygen, the blood then returns to the right pump. From there it is sent along the pulmonary arteries to the lungs (the pulmonary circulation). There it absorbs oxygen, and then flows back to the left pump to begin the circuit around the body again.

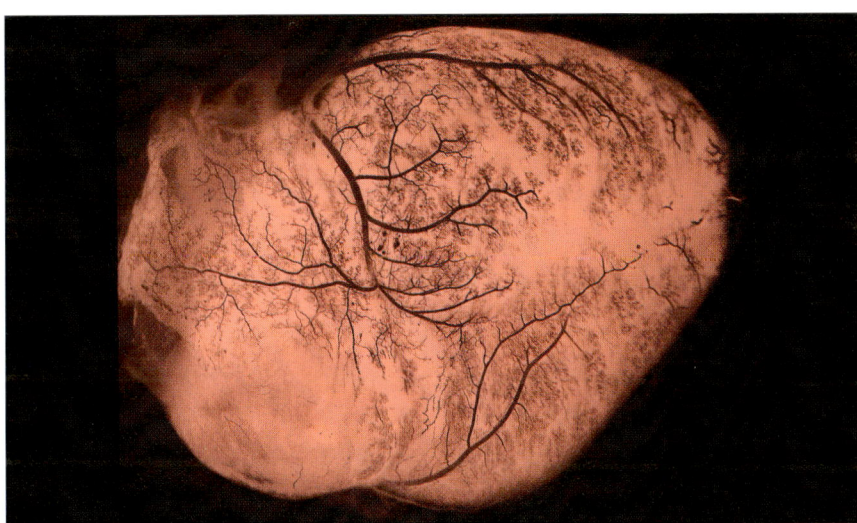

The human heart showing coronary arteries

The heart and circulation

The heart has to pump blood into the arteries, which branch and spread into every part of the body. For example, the carotid artery conveys blood to the head, the renal artery goes to the kidney, and the hepatic artery supplies the liver. Each artery divides many times and the branches become smaller, until they are microscopic tubes known as capillaries. These have extremely thin walls, and oxygen and nutrients can easily pass through them, from the blood into the tissues.

The heart cannot use the blood flowing through its chambers to nourish and oxygenate its own tissues, for several reasons. This blood is under tremendous pressure, is travelling too fast, and in the right side of the heart, it is very low in oxygen. So the heart's muscular walls have their own blood supply from the coronary arteries, which form a network over the heart's surface.

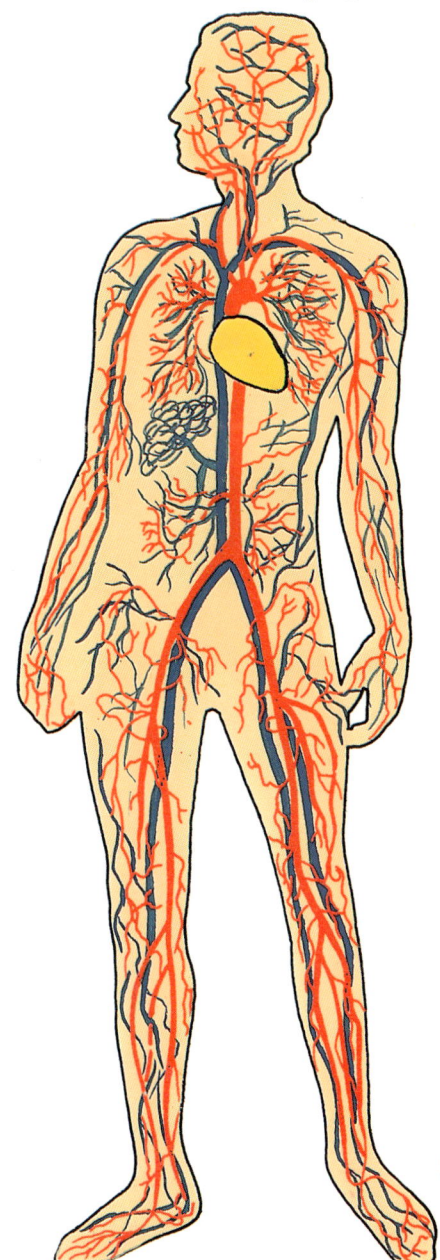

Heart muscle

The heart's walls are made of muscle and a supporting "framework" of stiffish fibres. The fibres give the heart some rigidity, to stop it collapsing completely. But the walls are also flexible so that when the muscle contracts (shortens), they can squeeze together to force blood out of the chambers.

Heart muscle is arranged in a spiral pattern inside the walls and is of a special type. It is called cardiac muscle or myocardium. Under the microscope, it does not look like other types of muscle in the body. It is composed of branching fibres which form a complicated network, unlike the neatly arranged parallel fibres of ordinary muscles. Also, unlike muscles in the arms or legs, cardiac muscle never becomes tired.

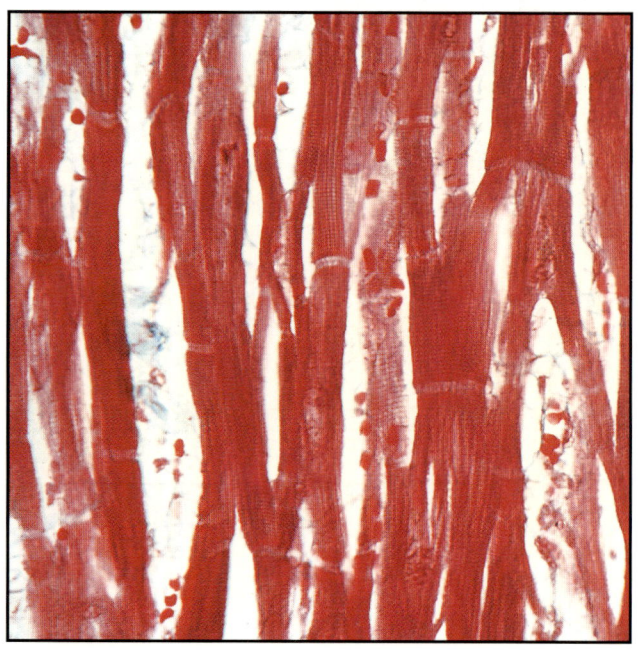

Micrograph of human heart (cardiac) muscle

One-way valves

There are one-way valves in the heart, main arteries and main veins. They make sure that blood circulates continually around the body in one direction. Each valve is made of two or more flexible flaps of fibrous tissue, like "pockets". As a heartbeat pushes blood through them the correct way, the flaps fold flat against the wall and the blood can flow through easily. When the heart refills again for the next beat, blood tries to travel from the arteries back into the heart. This attempt to flow backwards opens the "pockets" and makes them bulge together to create a good seal and prevent backflow.

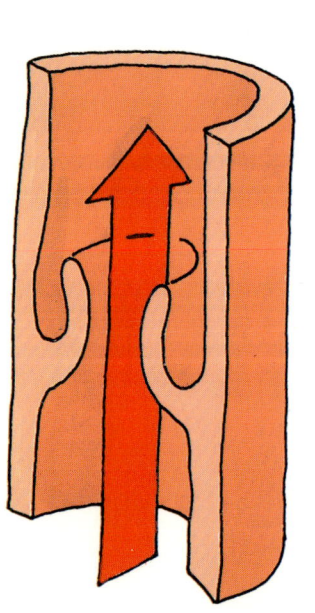 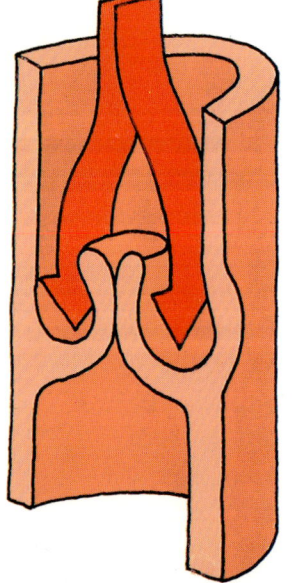

The squeeze-pump

The heart muscle contracts from the base of the ventricles upwards. This makes the space inside the chamber smaller and forces blood up and out into the aorta and pulmonary artery. You can show this by cupping your hands together to make a tight "bowl", scooping up some water, and then squeezing from the bottom upwards. The water squirts out – just as blood flows out of the heart. In an adult, the heart beats 60-72 times a minute at rest, accelerating to as many as 160-180 beats per minute during exercise, anxiety or fear.

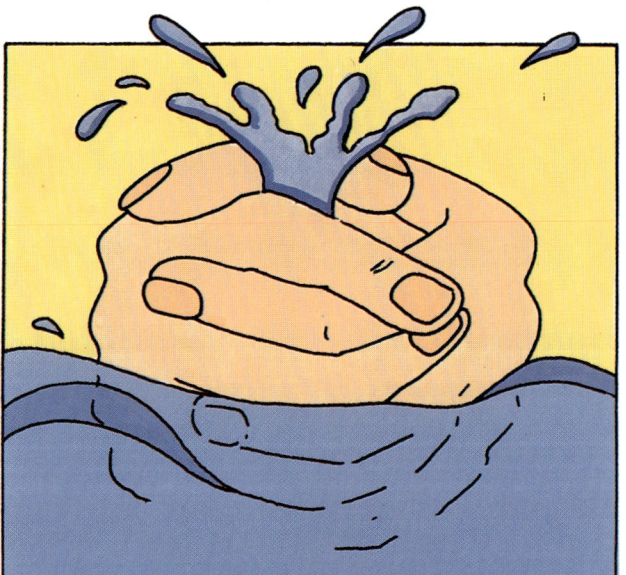

The four phases of the heartbeat

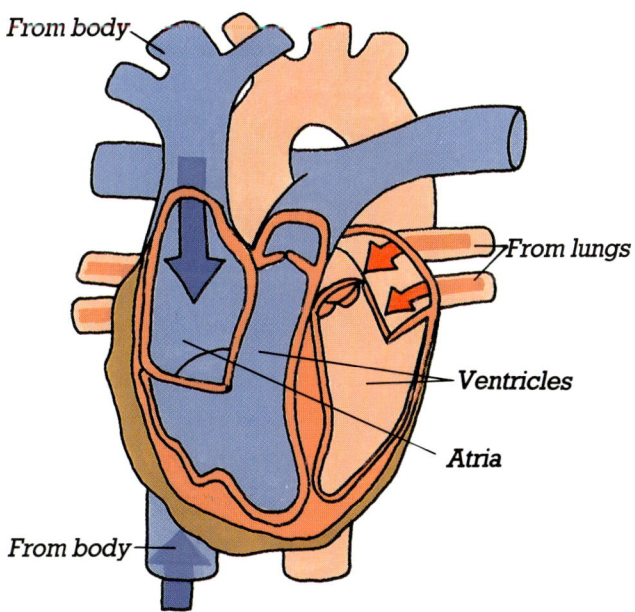

Phase 1: Blood flows along the veins into the two atria. The valves in the main arteries are shut, stopping blood from moving backwards.

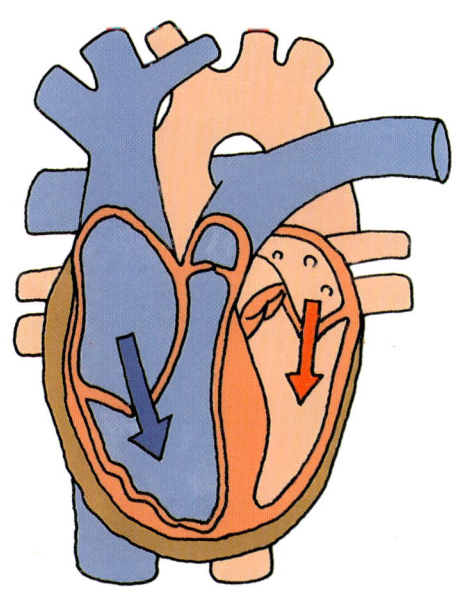

Phase 2: The ventricles, having just pumped out blood, expand and suck in blood from the atria through the tricuspid and mitral valves.

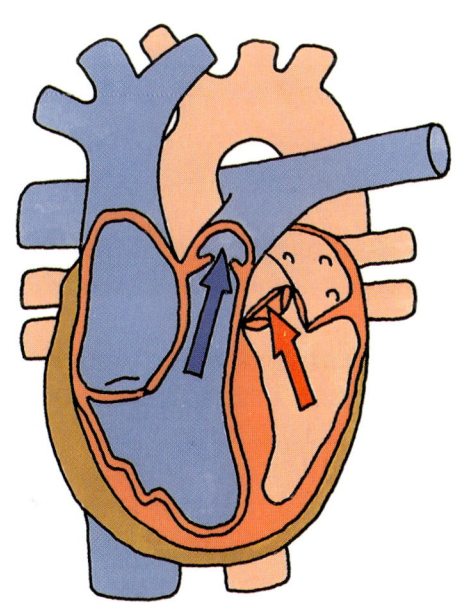

Phase 3: The ventricles begin their powerful contraction. The mitral and tricuspid valves slam shut and blood shoots into the arteries.

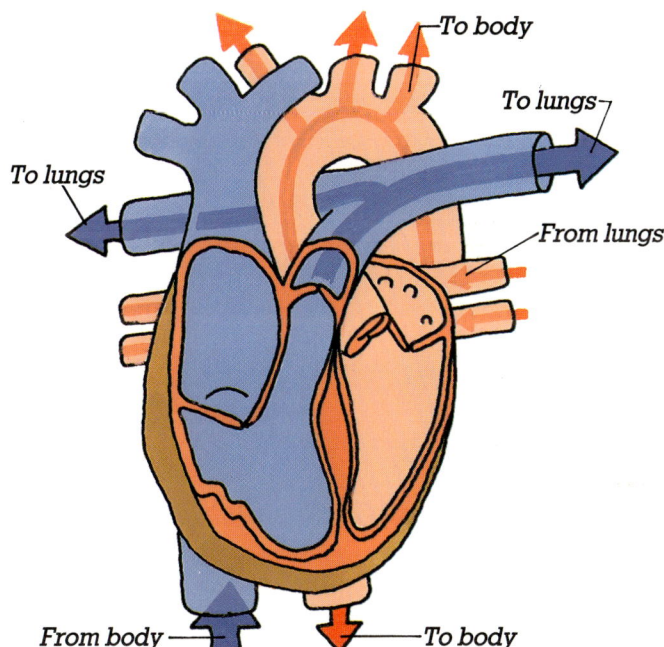

Phase 4: Blood rushes into the main arteries, pushing the blood from the previous beat ahead of it. The cycle then starts again.

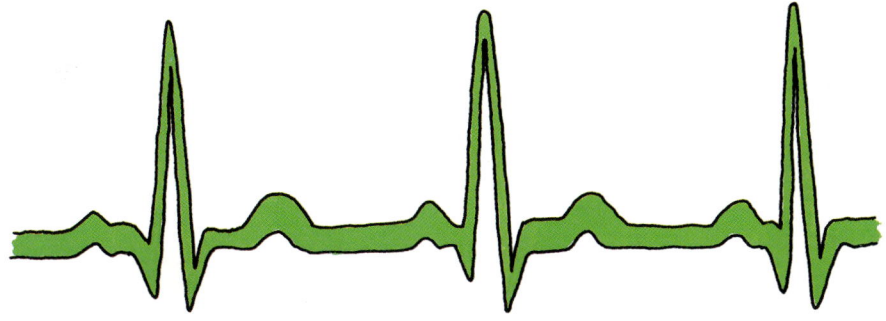

Tiny electrical nerve signals co-ordinate the heart's beat. These can be detected by sensors placed on the skin and shown as a trace in an electrocardiogram (ECG).

DISORDERS OF THE HEART

All parts of the body need a continuing supply of oxygen and nutrients, brought by the blood, in order to stay alive. So a strong regularly-beating heart and clear unblocked blood vessels are vital for good health. If the blood flow slows down, the affected organs "starve", and cannot work efficiently. If the supply ceases completely, in a few minutes the tissues start to die. This is why heart disease can have such serious effects all over the body. Heart disease is the main cause of illness and death in many Western countries. In the United Kingdom, disorders of the heart and circulation (together called cardiovascular disorders) contribute to the deaths of more than 300,000 people every year. The main single problem is coronary artery disease, which affects the arteries carrying blood to the heart muscle. Most of the time the heart functions very well. But extra demands, like a stressful lifestyle or overeating, may create problems.

Before birth

The developing baby, in the womb, has no air to breathe. Its supply of oxygen comes from the mother, as part of her systemic circulation. The pulmonary arteries to the lungs are "collapsed" and hardly any blood flows through them. Instead, the lung circulation is largely bypassed by two "short-circuits". In one, blood from the lower main vein flows into the right atrium and through a hole in the heart's central partition called the oval window (foramen ovale) to the left atrium. Blood from the upper main vein tends to go into the left ventricle and up the pulmonary artery, but then through the second short-circuit, a connecting tube called the arterial duct (ductus arteriosus) and into the main aorta (see above right).

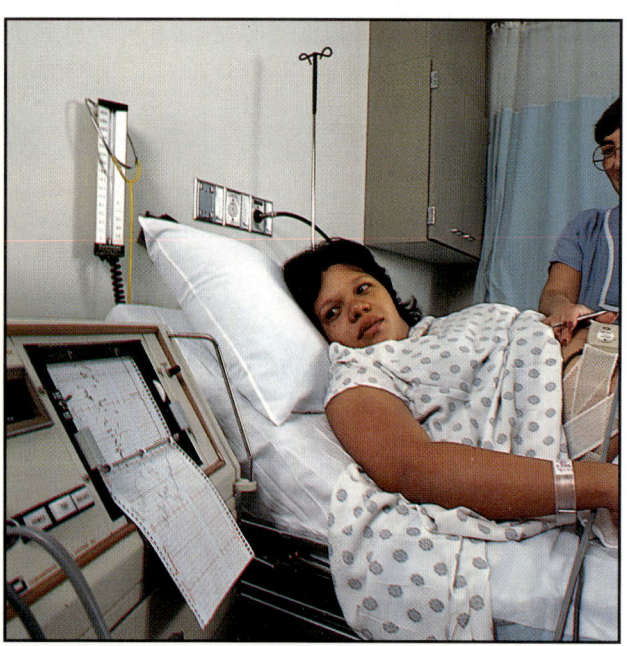

Monitoring a baby's heartbeat during birth

Changes at birth

As the newborn baby takes its first gasps of air, extraordinary changes take place in its heart and circulation. The oval window and arterial duct begin to close. This diverts blood into the pulmonary arteries to the lungs, which are rapidly "blowing up" as the baby cries and takes its first deep breaths. The pulmonary circulation is soon working well and the baby's lungs provide it with oxygen. A few weeks after birth the arterial duct has shrivelled to a thin cord, and several months later the oval window has closed.

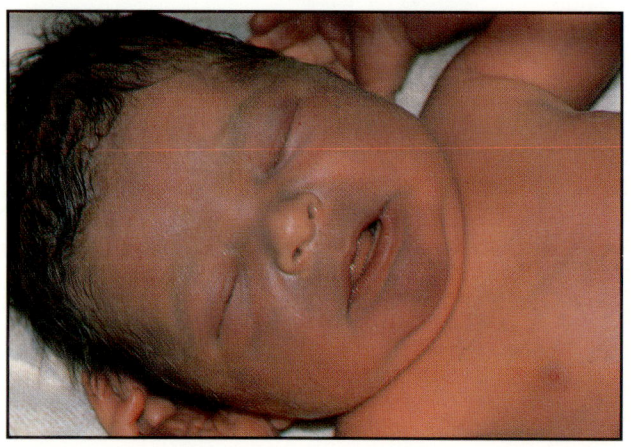

A baby born with a heart disorder.

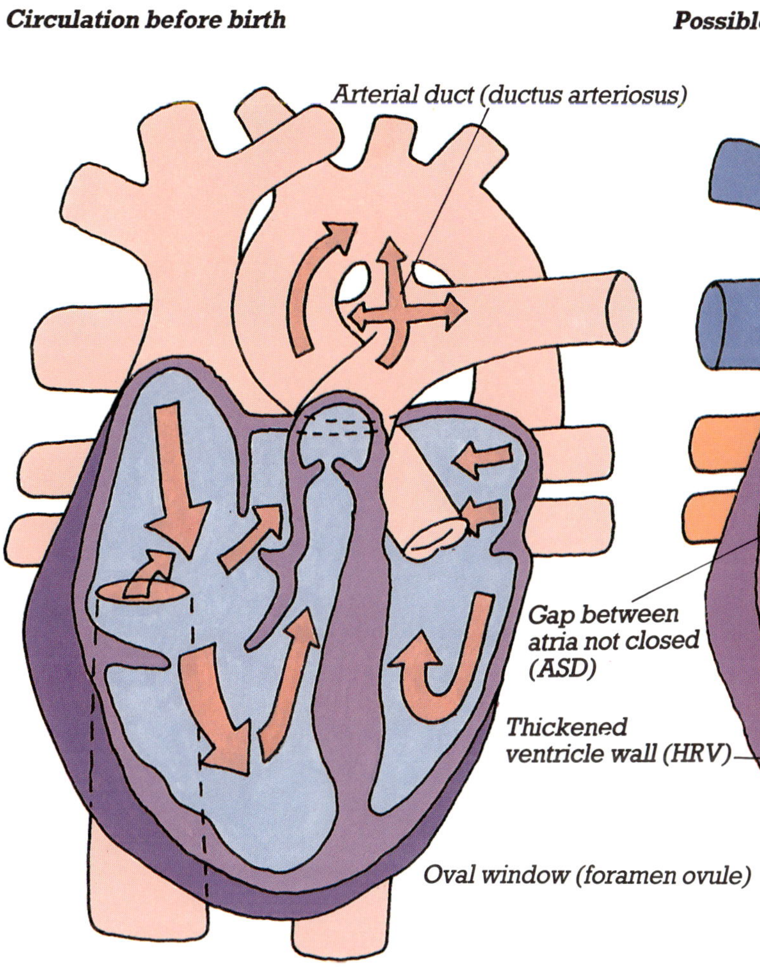

Circulation before birth

Possible problems at birth

- Arterial duct (ductus arteriosus)
- Arterial duct not closed (PDA)
- Gap between atria not closed (ASD)
- Narrowed valve (CPS)
- Thickened ventricle wall (HRV)
- Gap left between ventricles (VSD)
- Oval window (foramen ovule)

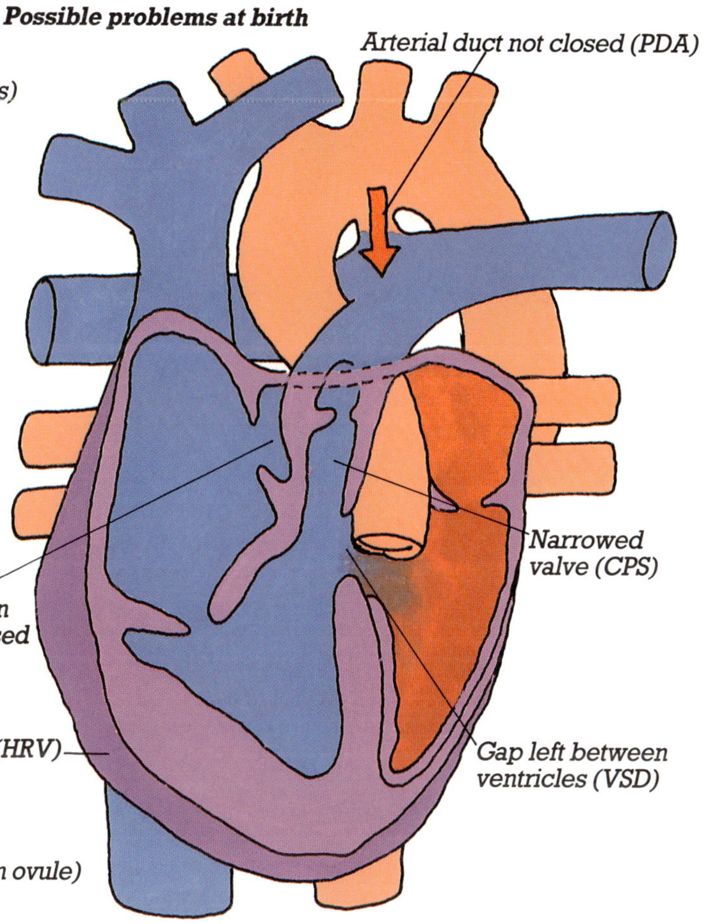

Problems at birth

During early life in the womb, the heart is simply two straight blood vessels lying together. It forms as the baby grows. In about 1 baby in 125, the heart's development is faulty, resulting in a heart disorder that is present from birth (it is congenital). Sometimes a hole remains between the two atria or the two ventricles – a "hole in the heart". A hole between the two atria is called an atrial septal defect (ASD) and a hole between the two ventricles is called a ventricular septal defect (VSD). There are other similar problems that may occur at birth. For example, after birth, the arterial duct may not close properly, leaving a patent ductus arteriosus (PDA); one of the heart valves may be thickened and narrowed, which is called congenital pulmonary stenosis (CPS), or the wall of one or both ventricles may be too thick and unable to pump properly, as in the condition known as hypertrophied right ventricle (HRV).

Depending on the problem, there may be a lack of blood to the lungs, which means the baby cannot get enough oxygen. The oxygen-poor blood is dark in colour so the baby's skin looks "blue" (see far left picture), and he or she may easily become breathless. If the baby's condition is very serious, he or she may have an emergency operation. Otherwise tests are carried out to find the exact cause and plan treatment.

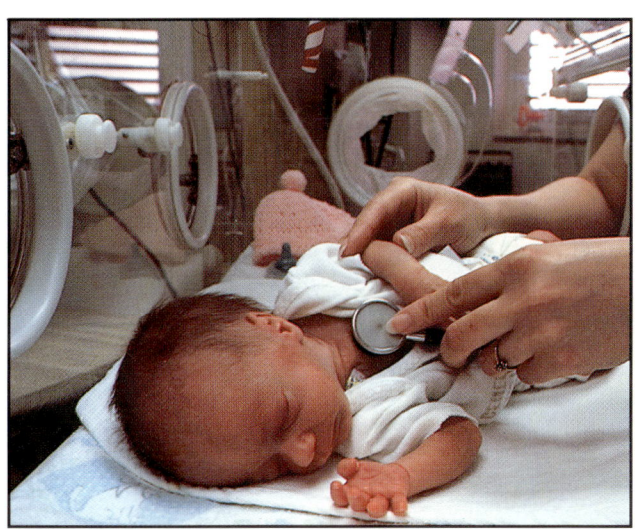

Checking the heartbeat of a baby in an incubator

Chest pain and angina

Like arteries elsewhere, the heart's own coronary arteries can be damaged by disease. Common problems are a furring-up inside by fatty tissue and hardening of the arterial wall. When this happens in a coronary artery, it is called coronary artery disease (CAD). If the body is at rest, CAD may not have any effect. But during activity, the heart pumps harder to supply extra blood to the muscles – which includes the heart muscle itself. The part of the heart muscle fed by the diseased artery does not receive enough oxygen or nutrients. It temporarily "starves" and responds by a warning pain, called angina. This is often felt as a "tightness" in the chest, sometimes spreading down the left arm. Angina can also be brought on by stress and may become serious in later stages.

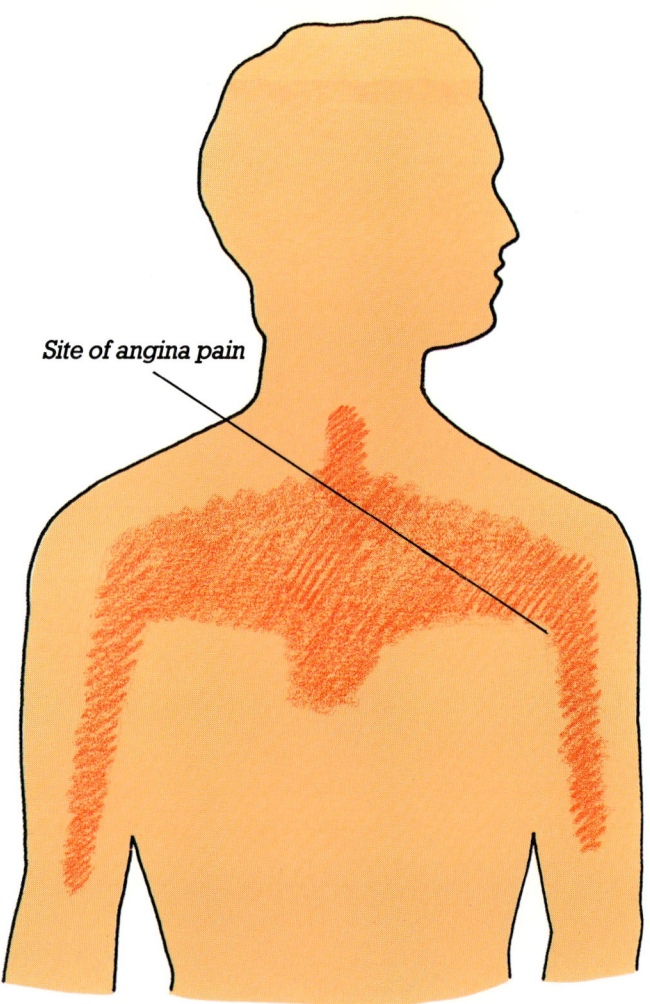

Site of angina pain

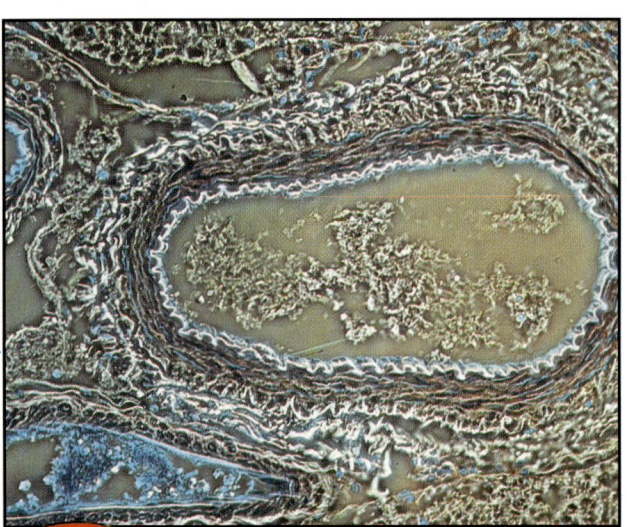

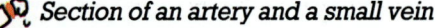
Section of an artery and a small vein

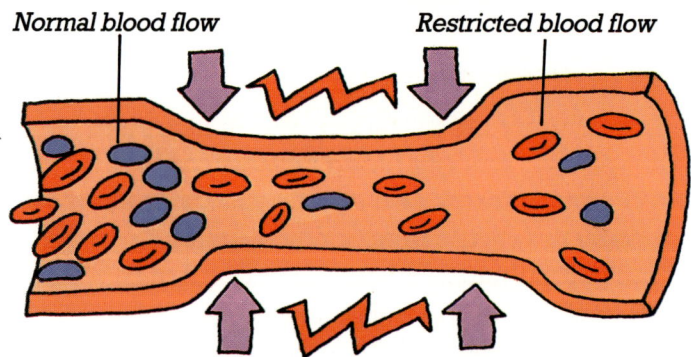

Normal blood flow   *Restricted blood flow*

Artery in spasm felt as pain

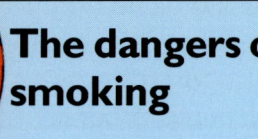

The dangers of smoking

People who smoke have a much greater risk of developing many health problems, from lung cancer to bronchitis and heart disease. The poisonous substances in tobacco smoke lower the oxygen-carrying ability of the blood, encourage the heart to beat faster and raise blood pressure. They may also lead to a greater risk of blood clots inside blood vessels.

So over the years, smoking gradually damages the heart, coronary arteries and blood vessels in other body parts. Overall, smokers have double the heart-attack risk of non-smokers. It is always worth giving up smoking. The chances of developing angina or a heart attack begin to decrease as soon as you stop. Giving up smoking and not smoking at all is the best "treatment" for any form of heart disease. It has a better record than either drugs or surgery. The key to success in giving up is strong motivation and being well informed of the dangers.

Furring-up of arteries

Arterial narrowing is caused by a two-part problem. One part is a build-up of fatty material, called atheroma, on the inside of the artery wall. The lumps of atheroma are known as plaques. The second part of the problem is hardening and stiffening of the wall of the artery, due to a build-up of rigid calcium-containing deposits. This means the wall can no longer expand to let through the high-pressure surge of blood with each heartbeat.

Both furring-up (atherosclerosis) and hardening (arteriosclerosis) reduce blood flow through the artery. If this happens in the coronary arteries that take blood to the heart, it can lead to angina (see left).

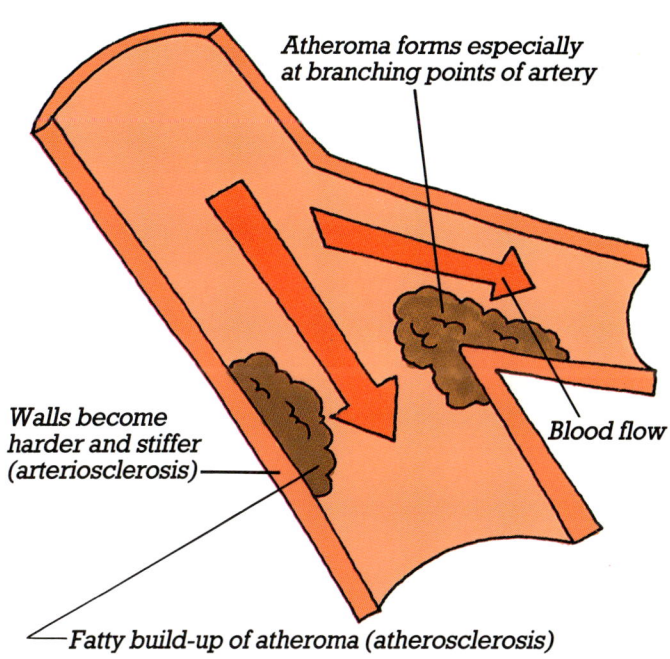

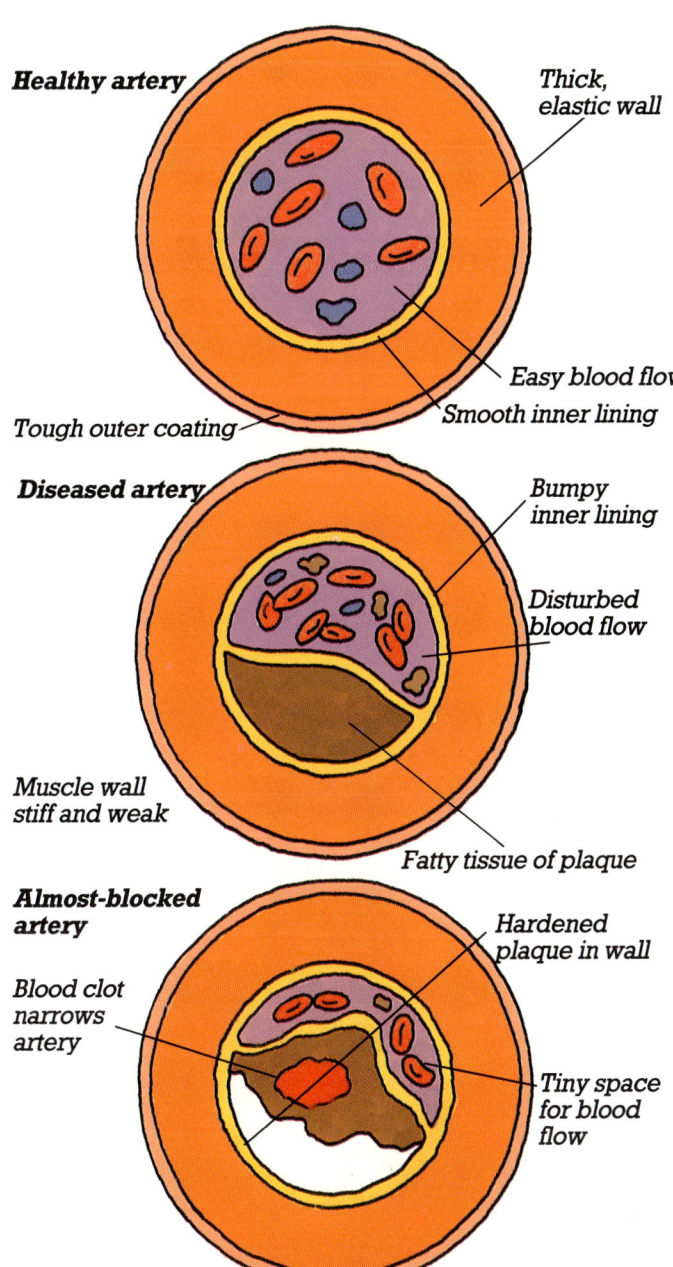

Heart attack

The most common form of "heart attack" has two main stages. They usually happen in a coronary artery already stiffened by arteriosclerosis and partly blocked by the lumpy plaques of atherosclerosis. The bumpy, narrowed nature of the coronary artery lining causes blood to flow through in a swirling, turbulent way. This makes the blood more likely to form solid lumps or clots (called thromboses). The clot, or coronary thrombosis, forms quickly in a narrowed portion of coronary artery and partly or completely blocks it (lowest diagram, left). Within a minute or two, the parts of heart muscle which were fed by the artery begin to go into spasm (uncontrolled tightening), as in angina. Then if the blood supply stays cut off, areas of the muscle begin to die, a process known as myocardial infarction.

In angina, the heart muscle recovers. In infarction, the dead muscle cannot recover. Depending on the size and position of the affected area, the heart's pumping ability weakens. Blood flows more slowly around the body, including to the coronary arteries, and so the situation can soon deteriorate. Some heart attacks are minor, and the victims make a good recovery. In other people, as the heart becomes weaker, its regular beating action may be disturbed. This complication is one of the most common causes of death as a consequence of a heart attack.

Thromboses and strokes

A thrombosis or blood clot in one of the arteries leading to the brain can have serious results. This problem (called cerebral thrombosis) is one form of stroke, in which part of the brain tissue is starved of oxygen and nutrients and dies. Another kind of stroke is cerebral haemorrhage, in which an artery to the brain breaks and its blood leaks out into the surrounding tissues. The effects of a stroke include pins-and-needles or numbness, weakness in moving a body part, or being unable to move the part at all (paralysis). The affected areas of the body depend on the site of the blockage. Since each side of the brain controls the opposite side of the body, a stroke in the right side of the brain affects the left side of the body.

The travelling embolus

Sometimes, a blood clot forms in a vessel and then breaks apart, so that fragments are swept away into the circulation. Alternatively the whole clot may come away from the vessel wall. Such a "foreign body" in the bloodstream is called an embolus.

Pulmonary embolism is a possible complication of deep-vein thrombosis. A clot in the leg vein comes away, travels up the wide main vein to the heart, then along the pulmonary artery to the lungs. As the artery divides into smaller branches, the clot becomes stuck and prevents blood reaching that part of the lung. If the embolus is large, it can cause breathlessness and chest pain.

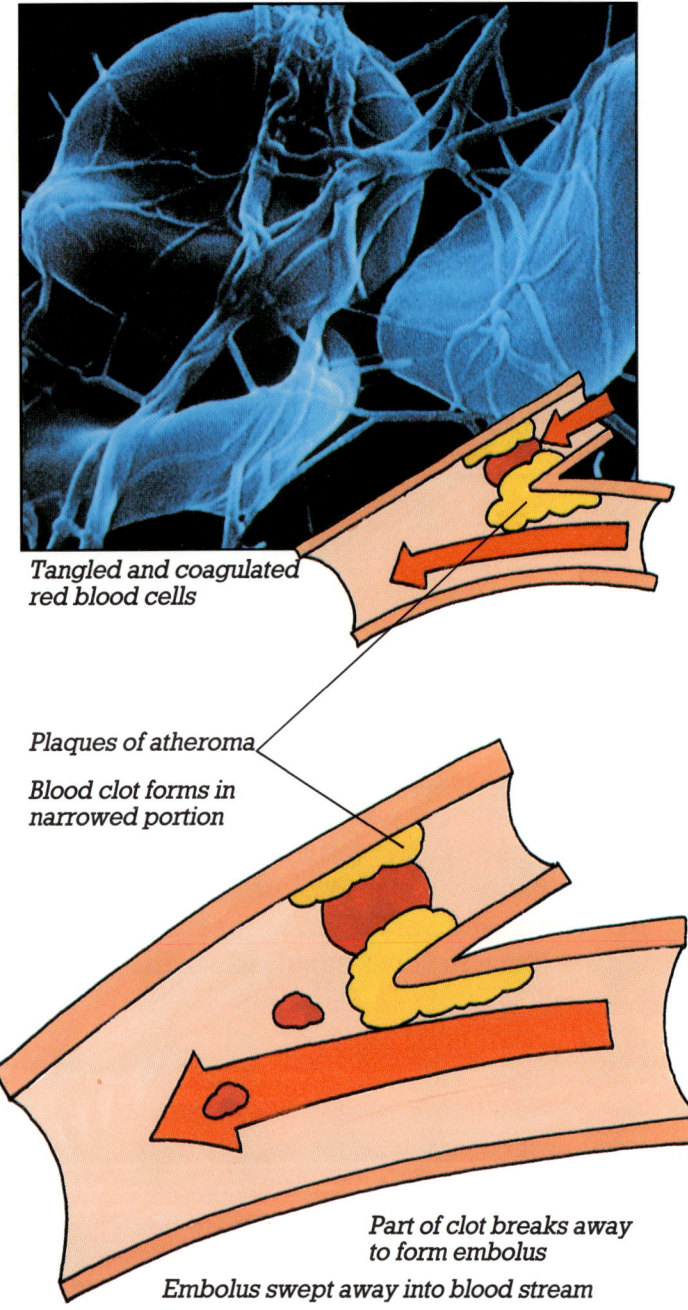

Tangled and coagulated red blood cells

Plaques of atheroma
Blood clot forms in narrowed portion

Part of clot breaks away to form embolus
Embolus swept away into blood stream

Blood clot (thrombosis)

Blood is likely to clot in any diseased blood vessel, not only a coronary artery. The damage done by the clot depends on where it forms. In a leg artery, it can prevent enough blood flowing to the leg muscles. This brings on cramp-like pain when walking or running. Alternatively, a clot can occur in one of the veins within the leg muscles (a deep-vein thrombosis). Blood cannot flow back up the vein to the heart, so the limb becomes swollen and painful. This becomes more likely if the leg muscles have not been used regularly, as used to happen when patients stayed in bed for a long time after surgery.

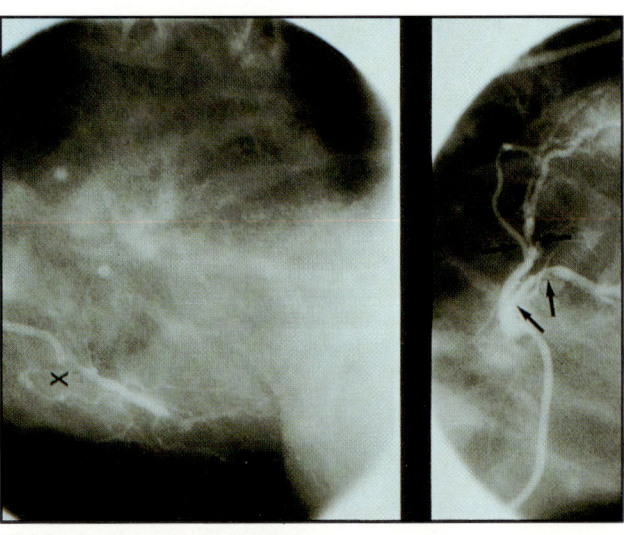

A blocked artery in the leg

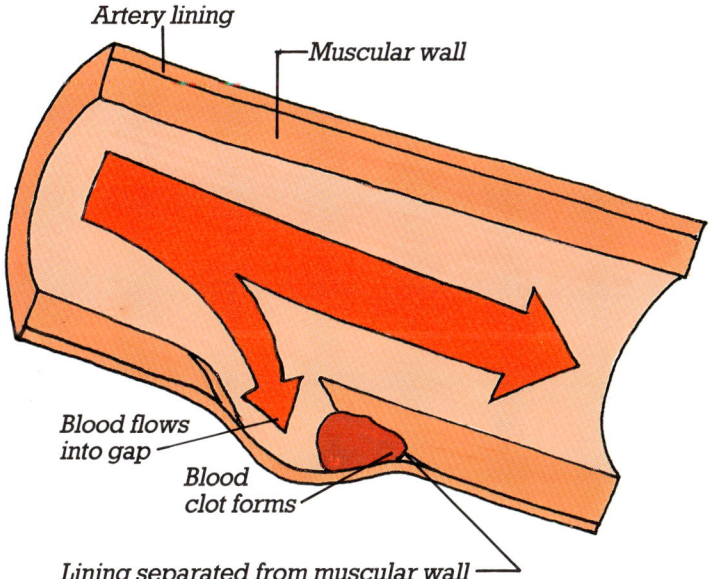

Aneurysms

An aneurysm is a balloon-like swelling in the wall of an artery. It may occur because the arterial wall is thin and weak, due to faulty development, or because of conditions such as atherosclerosis and high blood pressure, which weaken the wall. The pressurized blood flow gradually stretches the swelling, and it may eventually burst. Blood may also force its way between the artery lining and its muscular outer wall, with the risks of clotting and embolism. The signs of an aneurysm depend on where it forms and its size. In the main aorta, a burst can let so much blood escape from the circulation that it is fatal.

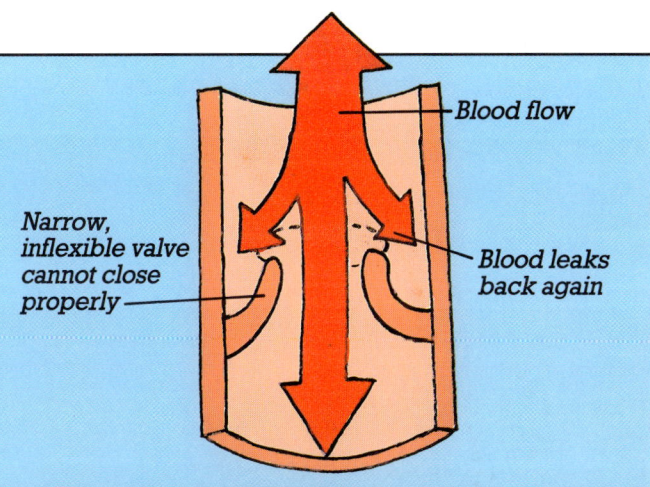

Valve problems

Healthy valves in the heart and blood vessels seal well to stop any backflow of blood. But a valve can become loose and floppy, so that blood leaks back through the wrong way (incompetence). Or a valve may become stiff and narrowed, so that blood cannot easily flow through it even in the correct direction. This is called stenosis. The usual "lub-dup" sound of the heartbeat is caused by the heart valves opening and closing. A faulty valve can change this sound slightly to produce a "heart murmur". In the past, rheumatic fever was a common cause of valve problems, often in the form of mitral stenosis. The mitral valve became rigid, causing breathlessness and eventual heart failure. Rheumatic fever tends to run in families and chiefly affects children between five and fifteen years of age. The symptoms are sore throats, fever and joint pain. It is a bacterial infection but can lead to damage to the heart valve.

An electrocardiogram or ECG is used to monitor heart performance

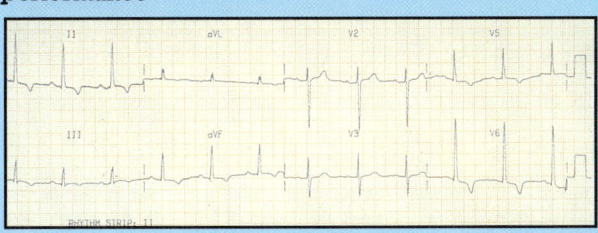

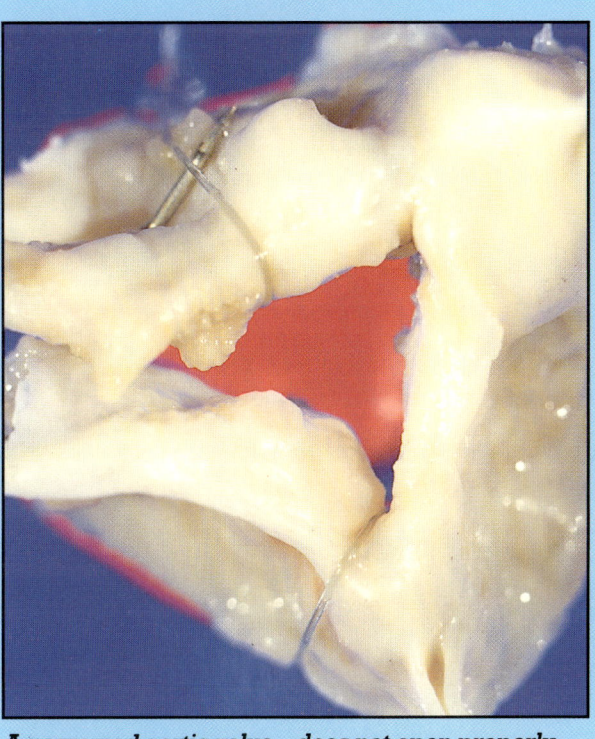

A narrowed aortic valve – does not open properly

Heart muscle disorders

The most common disorder that affects heart muscle is infarction (see page 11). However other disorders, known as cardiomyopathies, can develop in the heart muscle itself. In nutritional cardiomyopathy, the muscle is damaged by lack of the minerals and vitamins normally found in nutritious food. Alcoholics are at risk, partly because of the damaging effects of alcohol on heart muscle and also because of the poor diet they eat. In hypertrophic cardiomyopathy, the diseased heart muscle fibres cannot contract properly. The heart walls thicken to try and increase their pumping power (hypertrophy). But this can narrow the chambers inside, reducing blood flow. In myocarditis the heart tissues become swollen and sore, as a result of rare complications due to some other disease such as mumps. Doctors sometimes refer to a heart attack as myocardial infraction or MI for short. The term literally means destruction of the heart muscle. The symptoms can be similar to angina but they occur at rest and last much longer.

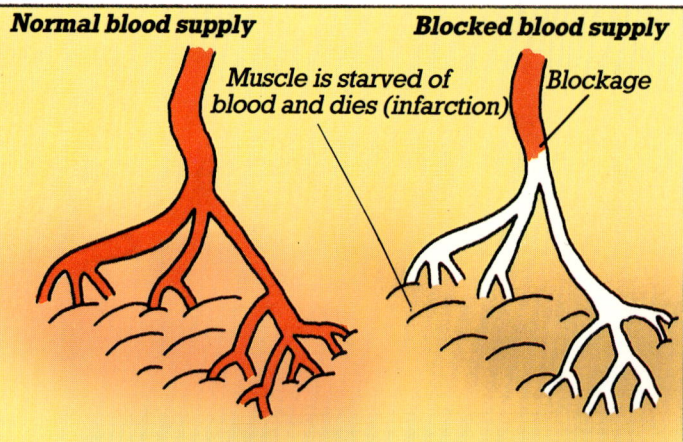

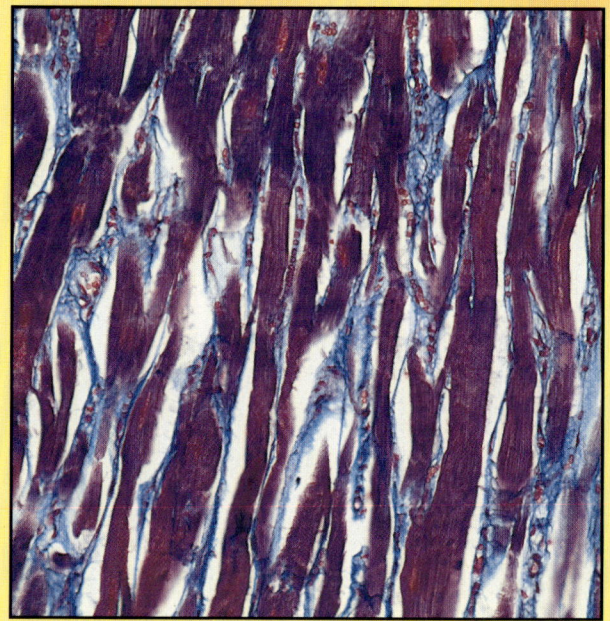

Micrograph of an inflamed heart muscle

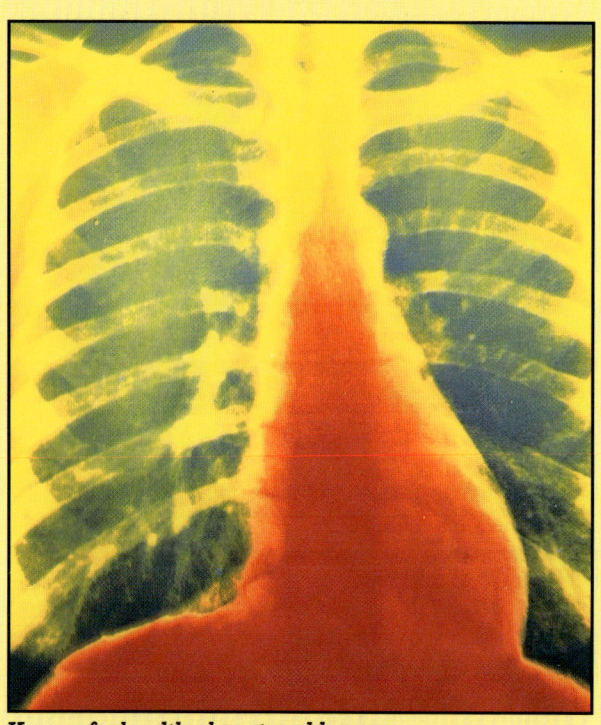

X-ray of a healthy heart and lungs

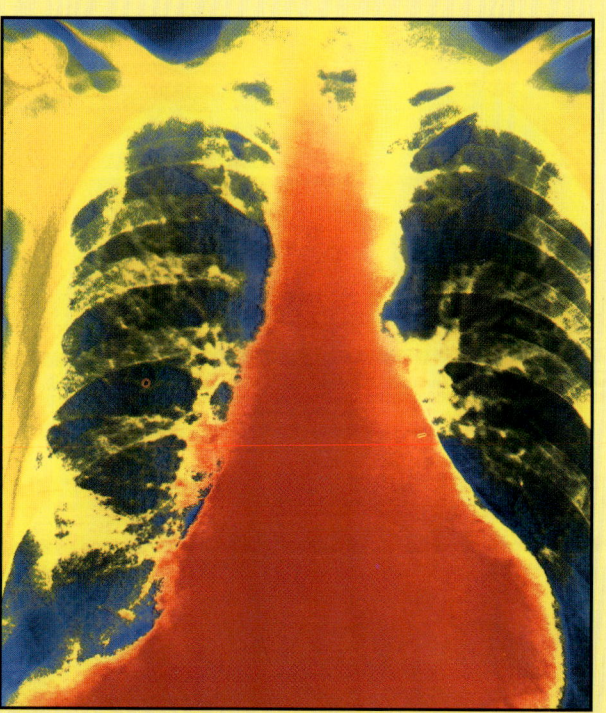

X-ray of an enlarged heart due to high blood pressure

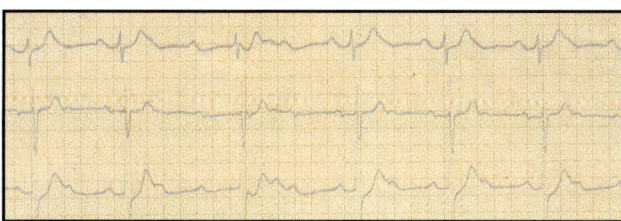
An ECG of bradycardia

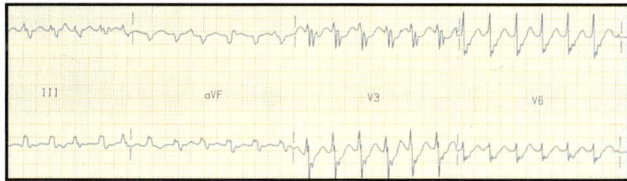
An ECG of tachycardia

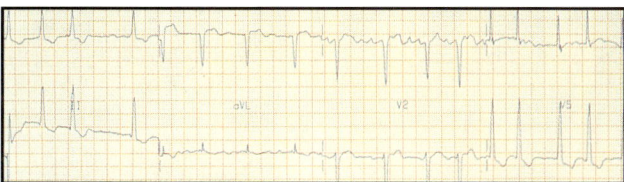
An ECG of aortic stenosis

Rhythm disorders

A healthy heart beats in a regular, co-ordinated way. In a rhythm disorder, the beating action becomes unco-ordinated, or the heart pumps too fast or too slowly for the body's needs. This may be due to a faulty natural pacemaker. This is a small area in the wall of the right atrium that "fires" bursts of tiny electrical signals which stimulate the muscle to contract. Or there may be a defect in the nerve-like fibres which conduct the signals. In bradycardia, the heart rate is too slow. In tachycardia, it is too fast. In fibrillation, the muscle fibres in the heart wall twitch or tremble without co-ordination. An ectopic beat is an "extra" heartbeat. Minor cases of these disorders are often felt as "heart flutters" or "palpitations". In severe cases the heart may stop completely.

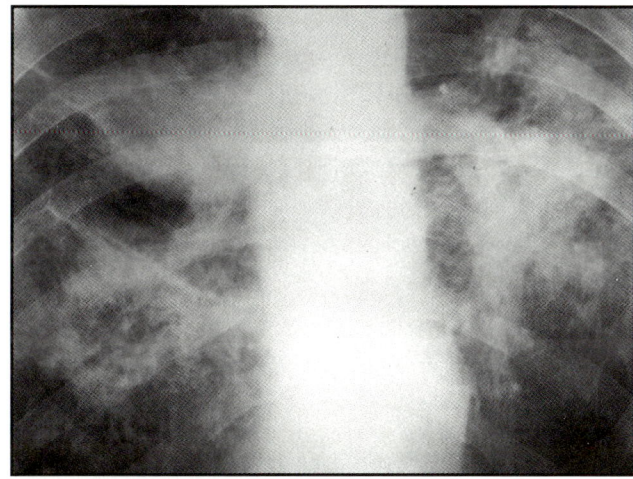
X-ray of congested lungs

Heart failure

Despite its name, heart failure is not an immediately life-threatening condition. It simply means that the heart cannot pump with enough force to continue circulating the blood round the body efficiently. This may happen, for example, if the valves are diseased. Heart failure usually comes on slowly and gradually. As a result, less blood is pumped into the body. If the left pump is failing, fluid collects in the lungs (congestion), bringing on breathlessness. For the right pump, fluid collects around the body, particularly in the legs. If both sides of the heart are affected, the condition is termed congestive heart failure.

Stress and the heart

In our fast-paced modern life, we often feel under stress. The body is designed to cope with a certain amount of stress. It reacts through its nervous and hormonal systems to make the heart beat faster and harder. This prepares the body for action, such as running away from danger. It worked well in our distant past, when the danger could have been a sabre-toothed tiger. Yet the sorts of "dangers" we face today, in the form of taking an examination or looking for a job, usually do not need much physical action. It is thought that too much mental stress, without letting the body react by "burning off" its excited state, may harm the heart. Stress can contribute to high blood pressure or coronary heart disease. Other factors often linked to a high-stress lifestyle, such as smoking, a poor diet and high blood pressure, also damage the heart. But a certain amount of stress is essential for our personal growth. It keeps us going and makes us creative. Too few challenges make our lives boring and frustrating.

TESTS AND TREATMENTS

Years ago, a serious heart disease was usually fatal. It was difficult for doctors to treat the heart, or even carry out tests to discover what was wrong, without interfering with its pumping action and so putting life at risk. Modern tests, drugs and surgery have revolutionized this. Doctors record the heart's electrical signals and beating sounds, scan its pumping action, and watch it beating "live" on a television screen. Once the disorder is diagnosed, there are dozens of highly specific heart drugs which can slow it down, speed it up or make it pump more regularly or more powerfully. If surgery is needed, a "heart-lung" machine can temporarily take over the pumping and oxygenating of the patient's blood. This allows surgeons to open up the heart, to repair a hole, unblock an artery or replace a diseased valve with an artificial implant. But the best treatment is our own active participation in maintaining good health, and so avoiding heart problems.

Heart examinations and tests

An experienced doctor can diagnose disorders simply by looking at a patient, feeling the pulse and listening to the heartbeat through a stethoscope. If the cause of a problem is not revealed, the doctor may then order specific tests. One is the electrocardiogram, or ECG. Up to 12 sensitive detectors are taped to the patient's skin, usually on the chest, head, wrists and ankles. The detectors pick up the tiny bursts of electrical signals from the heart, which "ripple" outwards through the body tissues to the skin. These signals are amplified by the ECG machine and displayed as a wavy trace on a chart or television screen. To the expert eye, the spikes and dips on the trace reveal problems such as rhythm disorders, dead patches of heart muscle or thickened walls (see page 14). Another test is the straightforward chest X-ray. This shows the size of the heart and if it is enlarged (hypertrophied). It may also show whether the lungs are congested, in cases of suspected heart failure.

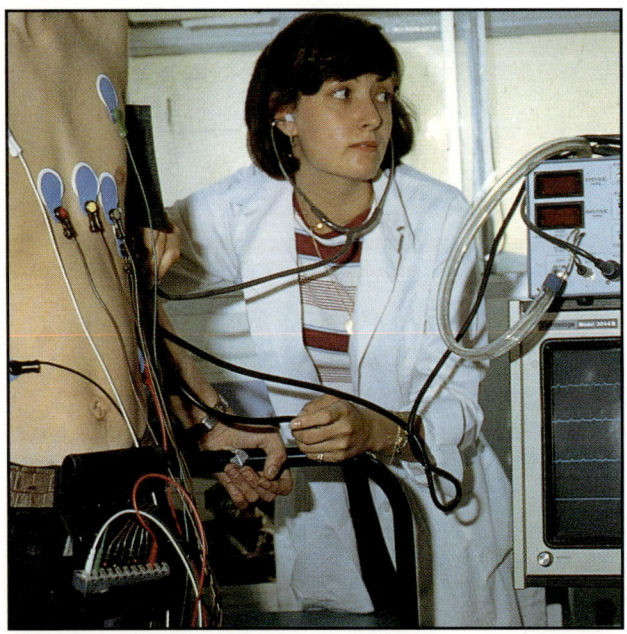

A person wired to an ECG machine

X-ray showing dead sections of heart muscle

An examination to show the state of a child's arteries

Scans and arteriography

An X-ray technique called computerized axial tomography (CAT) scans the body. A machine rotates around the body, taking pictures of successive "slices" using very low-dose X-rays. A computer combines the slices into one image. An ordinary X-ray shows bones well but gives little detail of soft organs such as the heart. In coronary arteriography, a long tube called a catheter is inserted into a leg vein and threaded up to the heart. A special harmless dye, which shows up on X-rays, is squirted through the tube. Then a series of X-ray pictures are taken as the dye travels through the coronary arteries.

Pulmonary arteries, outlined by the injection of a dye

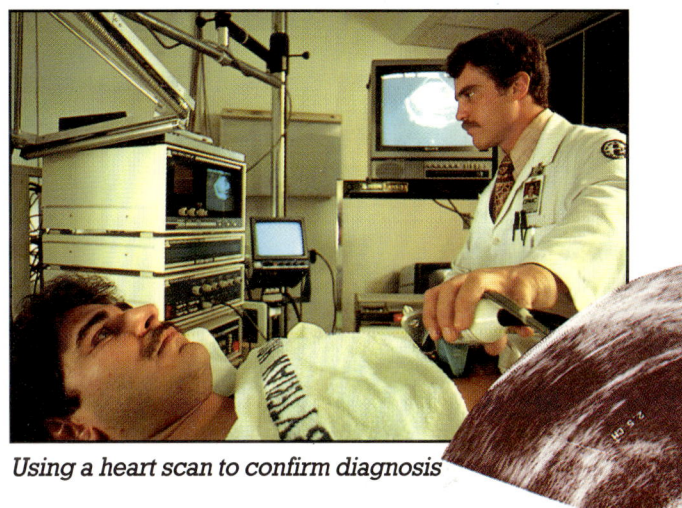

Using a heart scan to confirm diagnosis

Mending a hole in the heart

A hole in the heart, such as between the two atria, sometimes mends itself. If tests show that this is not happening, the opening can be closed using a plastic net "scaffolding". The patient is connected to a cardiopulmonary bypass ("heart-lung") machine. The surgeon can then cut open the heart and stitch the netting into the hole, grafting in other pieces of muscle as necessary to encourage healing. Scar tissue forms over the net, eventually mending the hole. In newborn babies that have a hole in the heart, the deoxygenated or "used" blood bypasses the lungs and continues to circulate around the body. This causes the baby's skin to appear blue in colour at birth, and so be referred to as a "blue" baby.

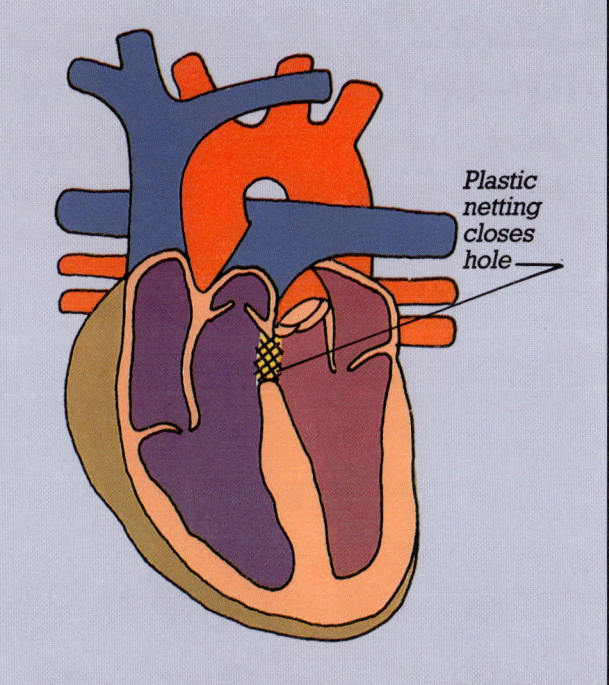

Plastic netting closes hole

17

Emergency procedures

In the United Kingdom, a person dies of coronary heart disease or its complications every three minutes. This terrible loss of life could be decreased if people took more care of their hearts (see page 28). It also helps to know emergency first aid (see page 30). And it is vital to find medical help as fast as possible. If expert treatment can be obtained within minutes, while a heart attack is still in progress, modern techniques can minimize the damage to the heart and other organs, and help the victim to recover well. Usually the patient is taken straight to hospital, if possible in a specially-equipped ambulance where oxygen and drugs can be given. He or she is taken to the Cardiac Care Unit (CCU, also called the Coronary Care Unit). It has ECG machines and other monitors, so that the specially trained staff can keep a minute-by-minute watch. The chest pain of the attack is reduced by injection of a strong pain-killing drug, and this also helps to ease the patient's worry and panic, and so reduce the strain on the heart. Blood samples are taken for laboratory tests. The heart may stop completely (cardiac arrest), usually because of ventricular fibrillation. An "electric-shock" can be used to jolt the heart back into action.

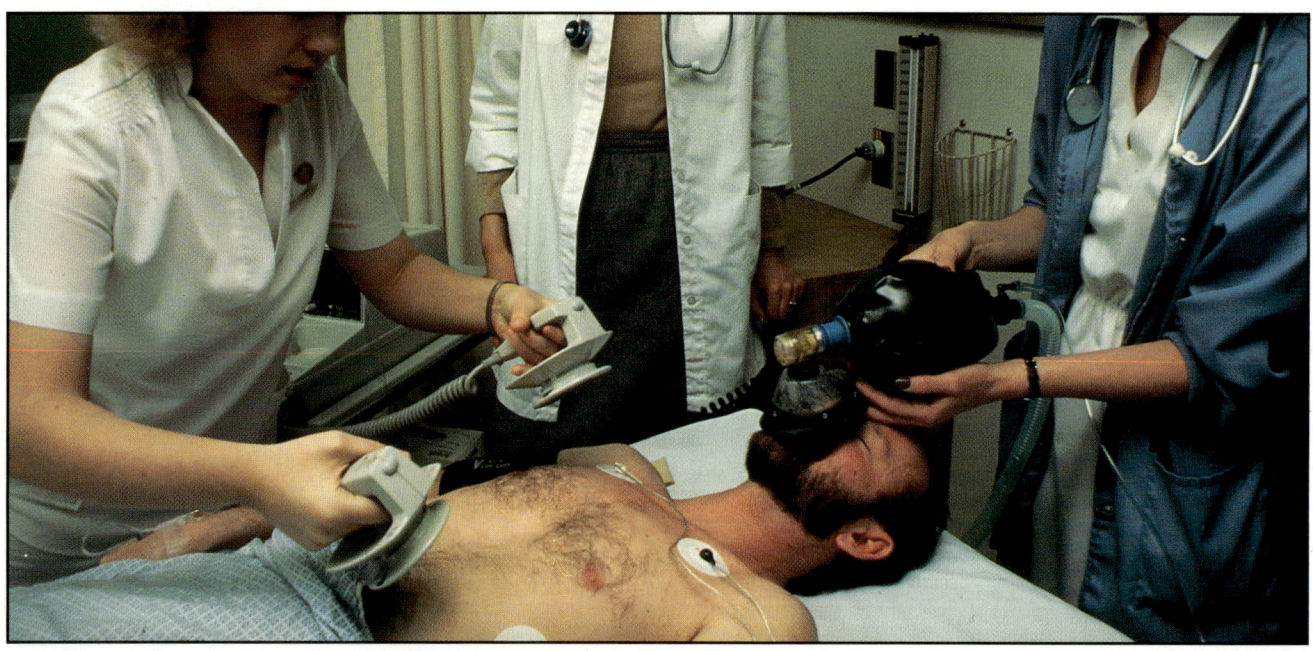

A cardiac arrest patient receiving emergency treatment

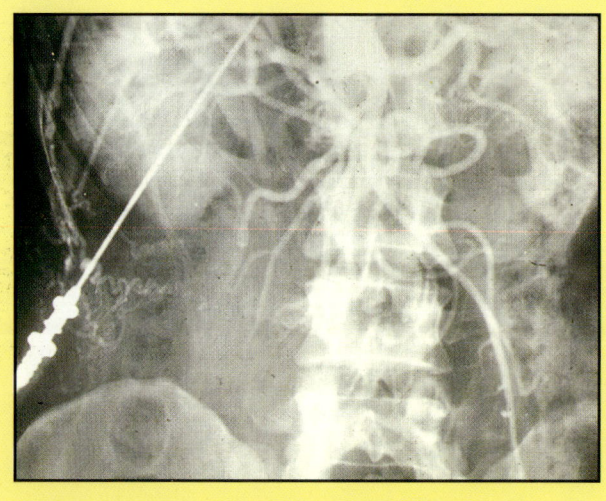

X-ray of an aortic vein with a blood clot

Dissolving the clot

In certain cases, newly-developed drugs called thrombolytics can dissolve a blood clot (thrombosis) that is blocking a coronary artery. But the drug must be given immediately, at least within the first few hours of the attack. If successful, the drug re-opens the artery and helps to reduce the amount of muscle death. Aspirin reduces the stickiness of blood platelets and the tendency for the blood to coagulate. If you are prescribed an anti-coagulant, you must not take any other drugs without consulting your doctor.

Angioplasty

The technique of coronary angioplasty was first carried out in 1977. It enables the heart surgeon to "ease away" a narrowing in a coronary artery, to reduce the pain from angina. A long tube, a catheter, is inserted into the heart. The catheter has a tiny sausage-shaped "balloon" at its tip. This is guided into the narrowed part of the artery with the help of an X-ray monitor. The balloon is then inflated to about 3 millimetres in diameter. It cracks and breaks up the lumps in the artery wall into smaller, less harmful pieces and "stretches" the artery wall. Angioplasty is suitable for about one-quarter of patients with angina.

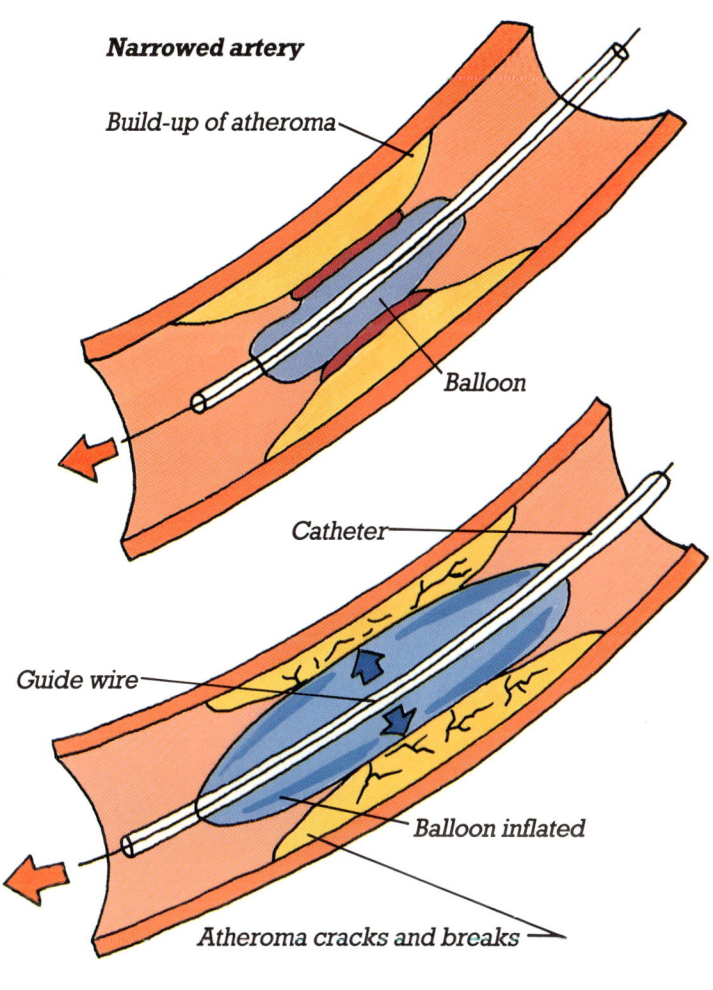

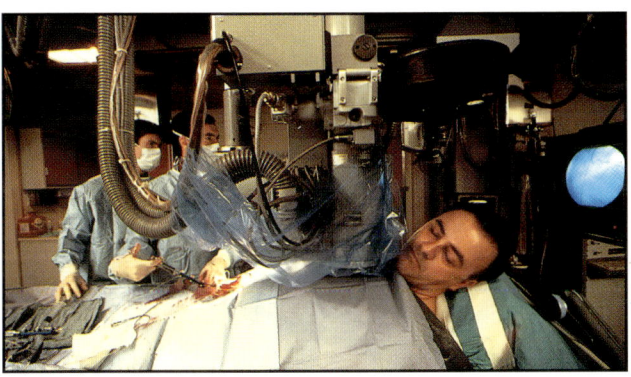

Using a balloon catheter to free an obstruction

Coronary artery bypass

Diseased and narrowed coronary arteries can be bypassed by less essential blood vessels transplanted from elsewhere in the body. A short length of leg vein is suitable, although a piece of the internal mammary artery from inside the chest often gives better long-term results. In this form of open-heart surgery, the transplanted vessels are stitched in place alongside the narrowed coronary arteries. Some patients have up to four bypasses made in the various coronary arteries. After the operation a patient will spend a few days in the intensive care unit of the hospital, during which time their heart will be carefully monitored. Most patients need to convalesce at home for between three and six weeks.

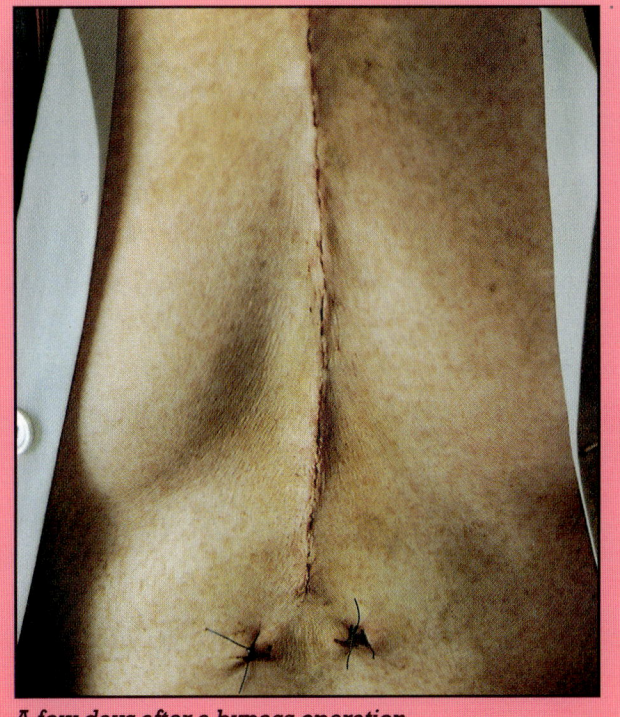

A few days after a bypass operation

Pacemakers

If the heart's own pacemaker fails to work, surgeons can implant an artificial one. There are various designs available. Most are the size of a slim matchbox and weigh around 50-100 grams. The metal case is waterproof and contains the electronic signal-generating circuit and a battery which lasts up to ten years. A lead from the pacemaker carries the signals to the heart. This can be inserted via a vein into the heart's ventricle (transvenous method). The case is implanted in the skin and muscle of the chest near the shoulder. Or, the wire can be attached to the heart's outside surface (epicardial method) with the case inserted into the front of the abdominal wall.

The fixed-rate type of pacemaker sends out a steady stream of pulses, no matter how the heart is pumping. The demand type fires only when the heart does not beat properly. The newest type of pacemaker changes its firing rate according to how the heart is performing. It can also store data about the heart's workings, and transmit it to a small programmer held close to the chest.

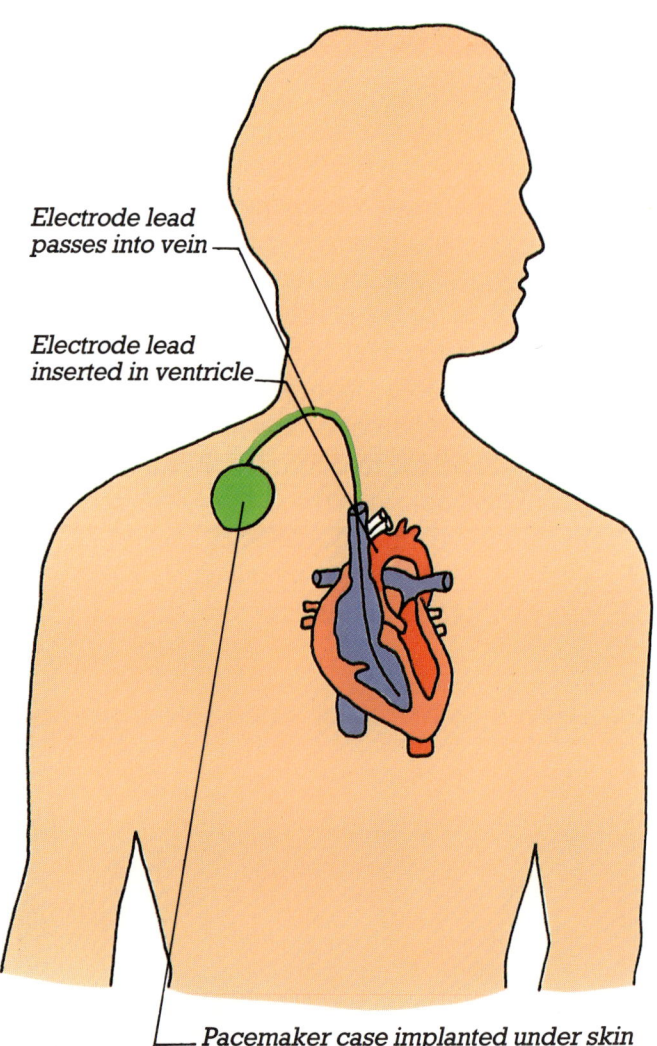

Electrode lead passes into vein

Electrode lead inserted in ventricle

Pacemaker case implanted under skin

In position

The transvenous method of implanting a pacemaker (above) is the most common. The operation usually takes less than one hour and it can be done under a local anaesthetic.

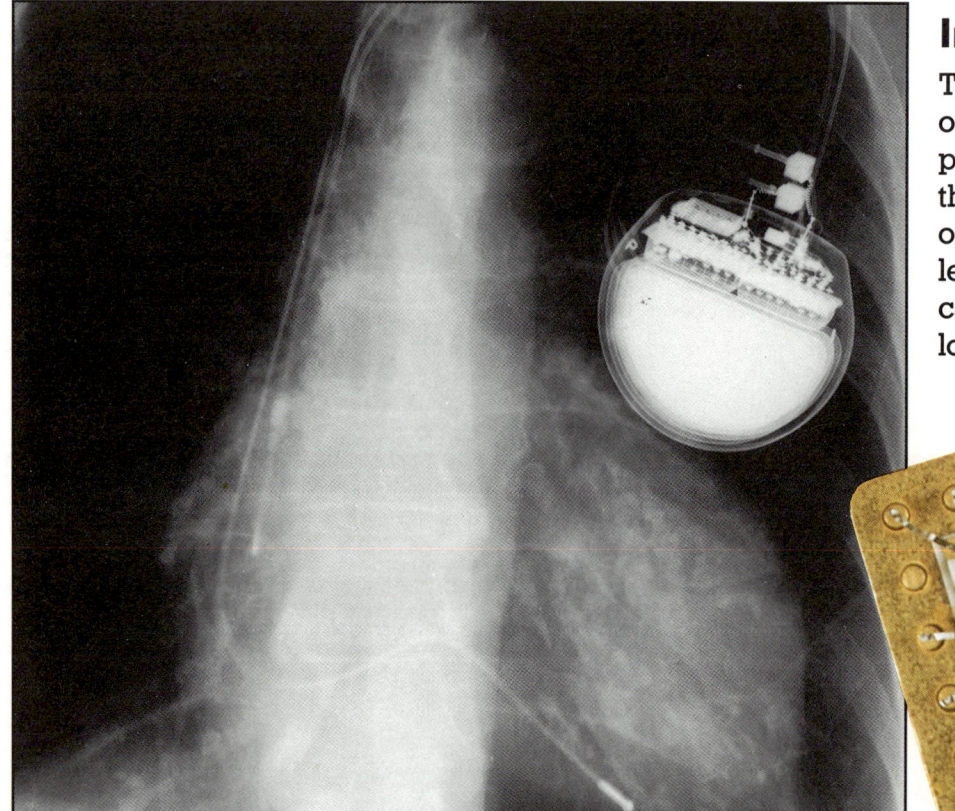

A pacemaker in position and a pacemaker's printed circuit board (right)

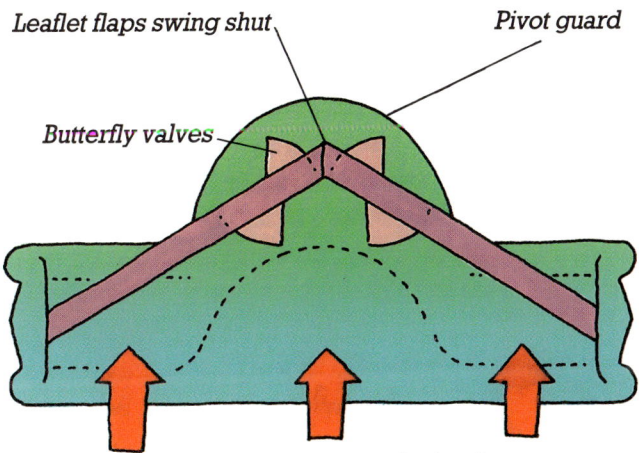

Back pressure of blood provides closing force

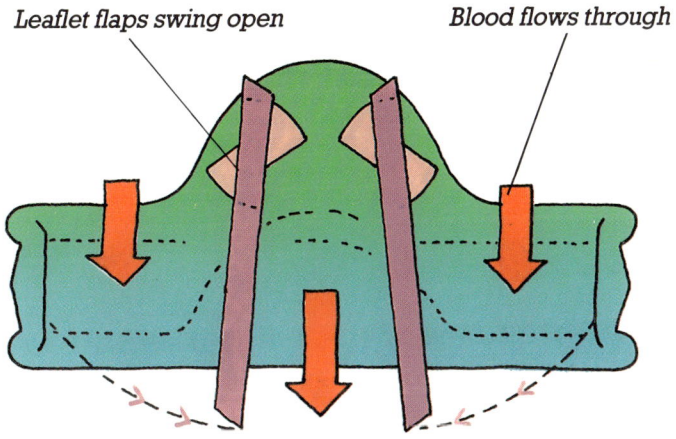

In a St Jude valve the pivot guard protects the mechanism from being clogged by an ingrowth of heart tissue. The blood flows through three roughly equal areas to avoid turbulence and damage to blood cells. The two leaflets close snugly to stop any backflow of blood.

New valves for old

In many cases, a diseased and leaky heart valve can be replaced by an artificial one. The cardiac surgeon opens the chest by cutting along the breastbone, and a heart-lung machine takes over so that the heart itself may be opened. If the faulty valve can possibly be repaired, this may be considered. But usually the valve is cut out and an artificial one sewn in its place.

There are two main kinds of artificial valve: tissue and prosthetic. Tissue valves are made of animal tissue, treated to stop the body reacting against it and "rejecting" it. There are several designs of prosthetic valve, mostly made of combinations of special metals and plastics. In the Starr-Edwards design, a ball bobs up inside a wire cage to let blood pass, but then falls down to seal against a lower ring to stop any backflow. In the butterfly type, two flaps (leaflets) swing open to allow blood through, then swing back to reseal. About 5,000 people have valve implants each year in the United Kingdom. Some have two or even three new valves. Those with prosthetic valves must take anticoagulant pills, which stop blood clotting on the "unnatural" materials. Taking anticoagulants means that the patient needs to have frequent blood tests. Patients must also take care with their teeth, since germs from the mouth can reach the bloodstream and infect an artificial valve.

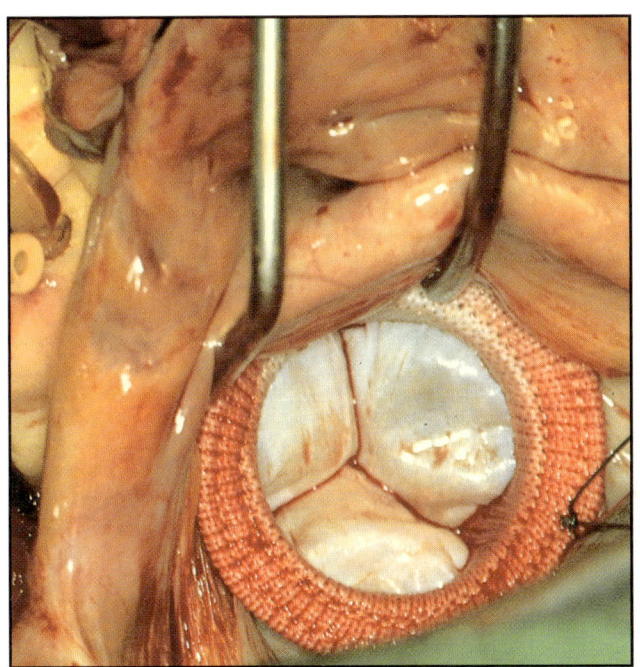

Open heart surgery showing replacement valve

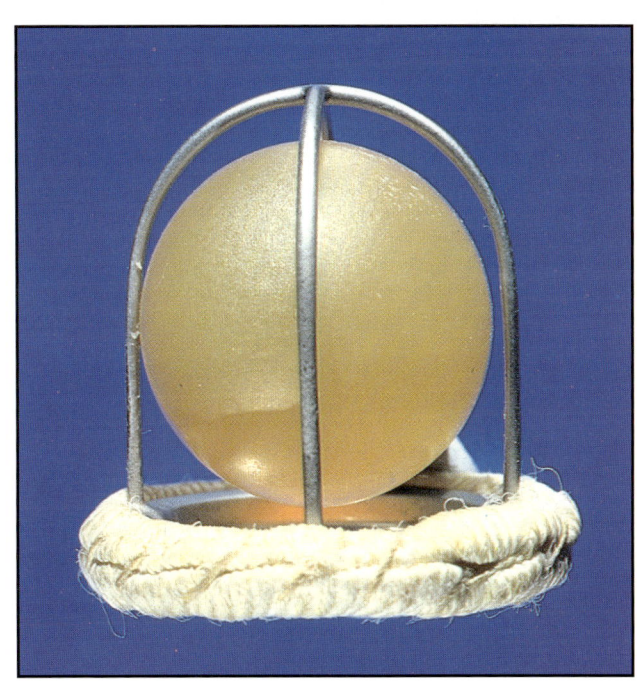

A ball type heart replacement valve

Heart transplants

The first heart transplant was carried out in 1967 in Cape Town, by South African surgeon Christian Barnard. The patient lived for another 18 days. Since then, techniques have advanced greatly. Many thousands of people around the world have somebody else's heart beating in their chest, and four out of five recipients are still alive two years after the operation. Transplants are considered for patients with cardiomyopathy, severe heart failure or severe coronary disease. It is essential to "match" the donor heart to the tissues of the recipient (patient), to lessen the problem of rejection. The donor heart must also be the right size, to cope with its new body. So patients may have to wait for some time before a suitable donor heart becomes available. Computer registers of donors and waiting recipients help to make the match quickly. The full operation takes about five hours, during which a heart-lung machine keeps the patient's blood circulating.

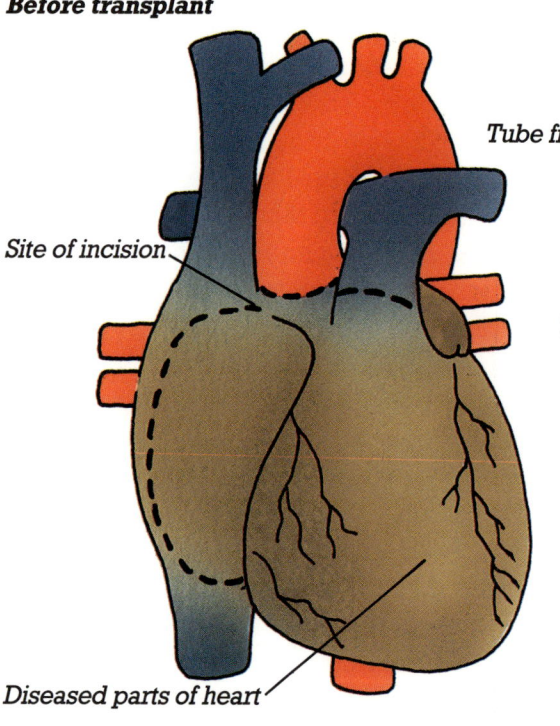

Before transplant

Site of incision

Diseased parts of heart

Part of a heart

In most heart transplants, only part of the heart is transplanted. The parts most commonly diseased are the ventricles, which are cut away along with the lower parts of the atria and the valves. The back walls of the two atria, with their connections to the main veins, are left in place. Tapes pulled tight stop blood flowing into the old heart. During surgery blood is diverted to the heart-lung machine.

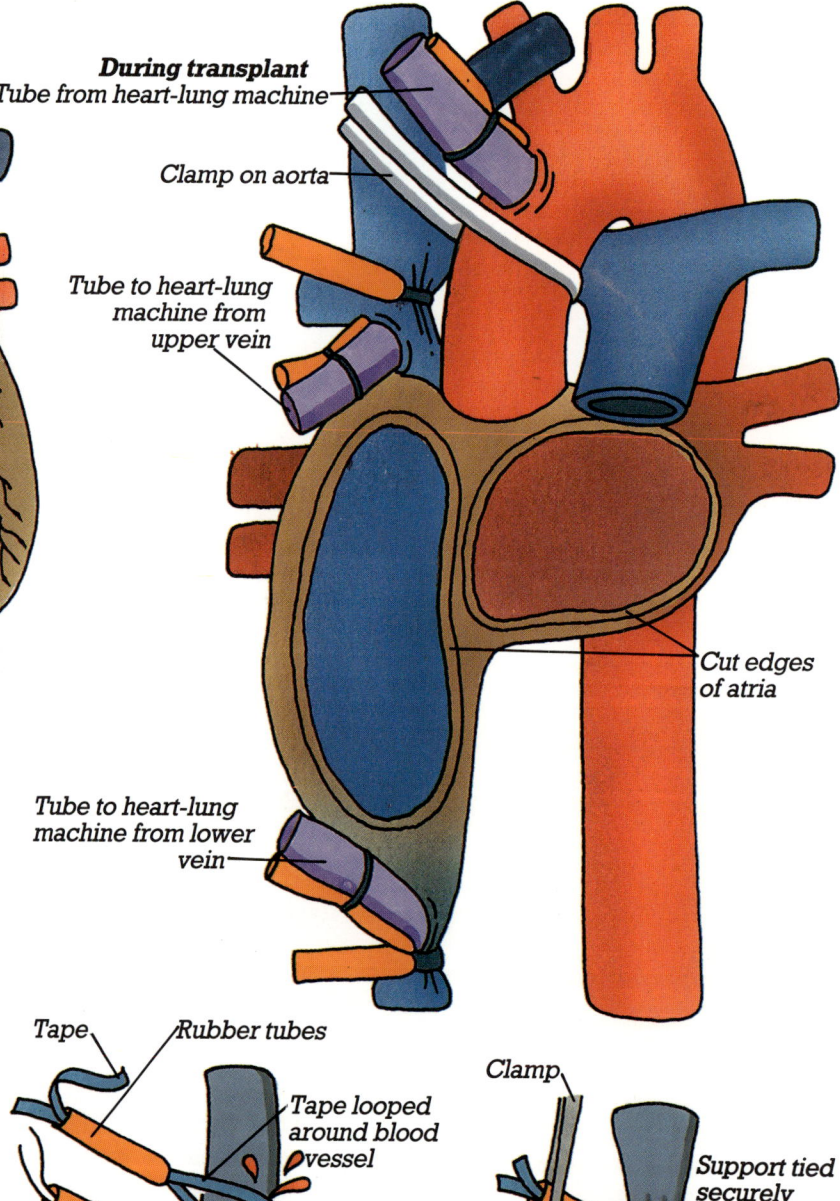

During transplant

During transplant
Tube from heart-lung machine

Clamp on aorta

Tube to heart-lung machine from upper vein

Cut edges of atria

Tube to heart-lung machine from lower vein

Tape — *Rubber tubes*

Tape looped around blood vessel

Clamp

Support tied securely to tube

Tape pulled tight

Tube to heart-lung machine

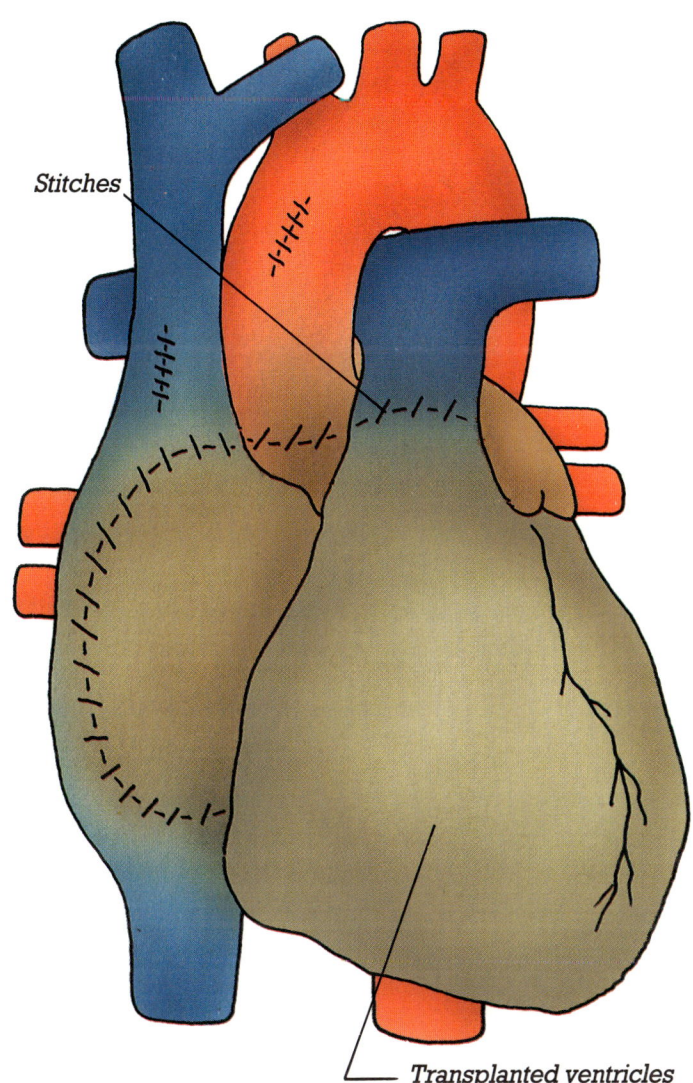
Stitches
Transplanted ventricles

After a transplant

After surgery, the patient takes a "cocktail" of drugs. This includes powerful immunosuppressives such as cyclosporin, which alleviate the rejection problem and allow the transplanted heart to settle down. However, immunosuppressive drugs also reduce the body's resistance to infection, so antibiotics are also given. During recovery, the patient is monitored in the Cardiac Care Unit. Samples of the heart tissue (biopsies) are taken weekly and analyzed to check for rejection or other problems. A heart transplant is the most extreme form of surgery carried out to help patients with heart disease. A heart transplant is only considered if the function of the heart muscle is so weakened that the pumping chambers are not able to pump out more than 30 per cent of the blood they receive. The main problem involved in heart (and other) transplants is the body's rejection of foreign tissue. Drugs that suppress the rejection process have been developed and, as a result, an increasing number of transplants are now successful. Artificial pacemakers can transform a short, inactive life to an active, normal life that can continue for many years, but more serious conditions may require surgery or transplants.

An artificial heart

The Jarvik 7 artificial heart is named after its American designer, bioengineer Robert Jarvik. It has two "ventricles" made of aluminium and plastic, and four mechanical valves. The device is powered by compressed air fed from a pump along a tube that passes through the chest wall into the heart region. This heart has been implanted into humans several times, and one patient lived for a further 112 days. However most experts agree further research is needed.

The advantages of using artificial organs are mainly conquering the problem of tissue rejection. Further research is being carried out into an external blood pump necessary to enable blood circulation.

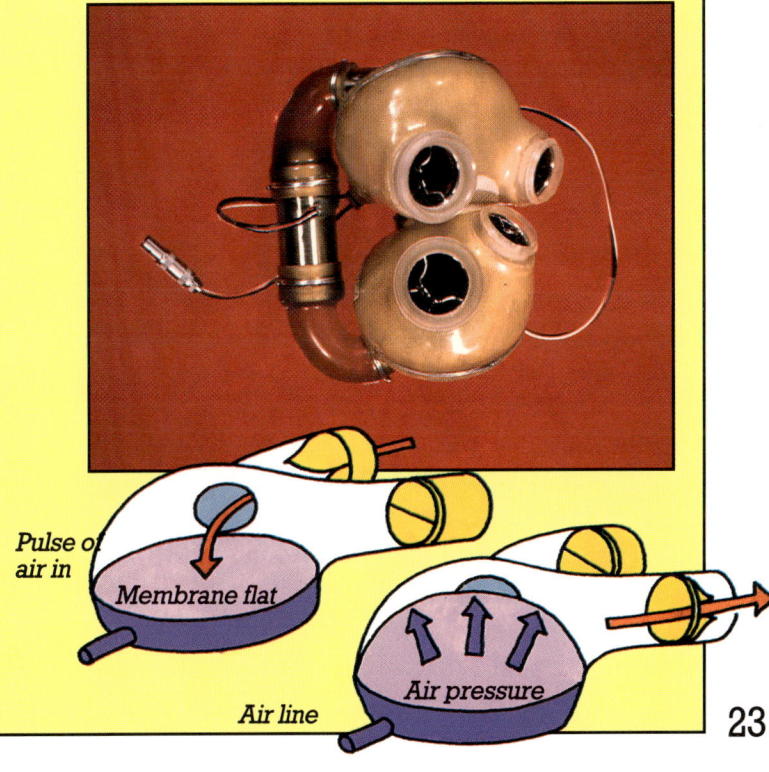
Pulse of air in — *Membrane flat* — *Air line* — *Air pressure*

LIVING WITH HEART DISORDERS

Heart disease affects different people in different ways. Some cope well. They learn everything they can about their condition, follow their doctors' advice thoroughly and plan their lives around diets, drugs and exercise programmes. But others see heart disease as a "death sentence". They lose interest and enthusiasm, and become fed up and inactive. This sort of reaction is common in the few weeks after a heart attack. Patients become aware that they will not live for ever.

But continued worrying slows down good recovery, and can turn into full-scale depression. With the help of family, friends, doctors and medical staff, negative feelings can be replaced by a more positive outlook. In many areas, cardiac rehabilitation classes ("coronary clubs") are held where patients can talk about their feelings and share experiences. Relaxation is vital for people with a heart disorder and we all need to develop a mechanism for coping with stress.

Drugs for heart conditions

There is a bewildering array of drugs for various heart conditions. Vasodilators encourage blood vessels to widen, for treating angina and heart failure. Diuretics stimulate fluid removal from the body (such as urine) and are also used for heart failure. Digoxin makes the heart pump more strongly and regularly, for heart failure and rhythm problems. Beta-blockers slow the heart rate, again for rhythm problems and angina. Anti-arrhythmics make the heart beat more regularly. Anticoagulants "thin" the blood and make clotting less likely. It is important for the patient to know the name of each drug, what it does, when it should be taken, and how often. Some drugs have side-effects, and patients should always mention any such problems to the doctor.

These medications are all long term, sometimes for the rest of a patient's life. Although it is important to continue to take the drugs according to prescribed doses, the patient must also realize that drugs are not the complete answer to a heart problem. It is the self-help measures that are also important.

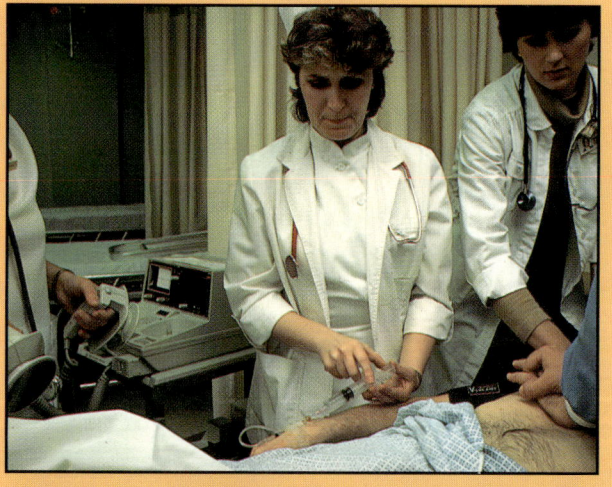
Inserting an anticoagulant drip

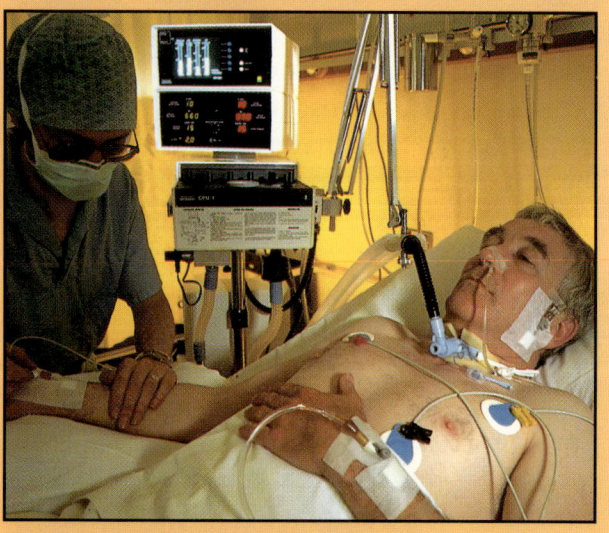
Nurse attending a man with a ventilator monitor

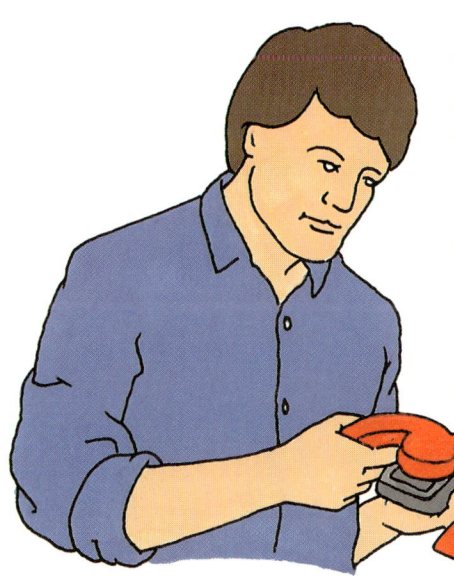

Telemetry

Modern microelectronics have come to the aid of heart patients. Microelectronics can be used in different ways to help. For example, advanced pacemakers can be programmed to work in a certain way, responding to a signal transmitted through the skin (see page 20). Signals received from the pacemaker about the heart's performance can also be received by some devices. The signals are converted into sound "bleeps" which can be sent along a telephone line to a receiver-analyzer at the heart clinic. In this way a doctor can keep a regular check on patients who live in remote areas and who cannot visit the clinic very often. He or she can also give advice over the telephone if the signals show a problem.

A cardiotrack receiver at a heart clinic

Exercise programmes

A graduated exercise programme is prepared for each heart patient, suited to his or her disorder and general physical fitness. After a heart attack, patients take only very gentle exercise, such as walking about the house with a rest in the afternoon. Exercises gradually build up to become longer and more vigorous, with aerobics and jogging. The programme should be followed closely. If a patient tries to do too much, the heart may suffer another attack. The important thing is to build up an exercise programme gradually. A person should not exercise to the point where they become short of breath, tired, dizzy or suffer from chest pains.

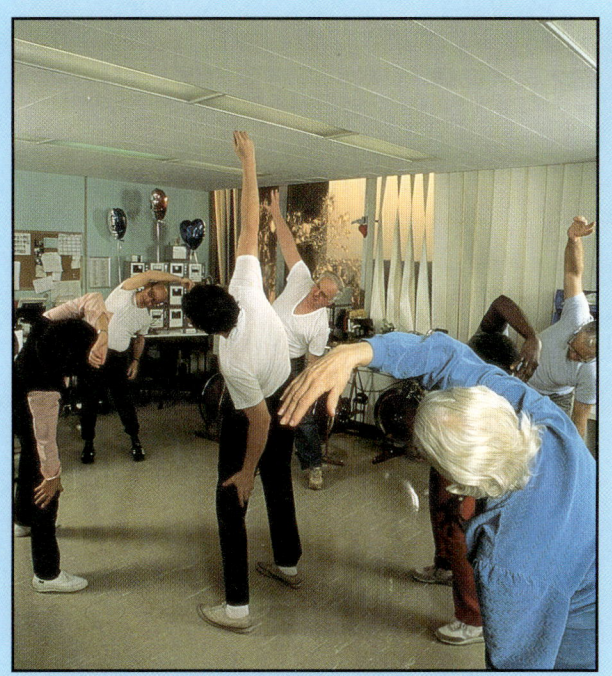

A cardiac rehabilitation class

25

Body weight and food

Being overweight increases the risk of many disorders, including heart attacks. After a heart attack, it is even more important to keep slim and eat less fatty foods. Patients should cut down on red meats, eggs and dairy produce, and eat more fresh vegetables, fruits and whole grains. This advice benefits the digestive system as well as the heart.

Changes in lifestyle, such as giving up smoking, eating healthier food, losing weight, taking regular exercise, can often bring the blood pressure back to normal, with a beneficial effect on the heart.

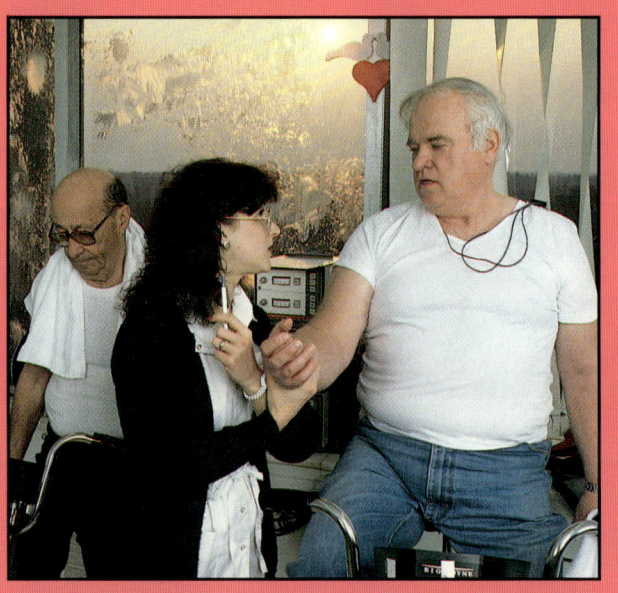

Blood pressure

Blood pressure which remains too high for a long period (hypertension) brings a greater risk of heart disease, stroke, and kidney and eye damage. But there may be no outward signs of high blood pressure until the damage to internal organs is well advanced. So adults are advised to have their blood pressure checked regularly by a doctor or nurse, especially when approaching middle age.

High blood pressure is very common and if uncontrolled can lead to serious complications – like the ones mentioned above. Everyone should therefore have their blood pressure checked every five years, or more frequently if it is on the high side.

Relaxation

The link between heart disease and stress is complicated, and research has not shown clearly if there is a cause-and-effect connection (see page 15). But for many reasons, which include preventing angina brought on by emotional stress, it is sensible to avoid too much high-pressure living. Put some time aside regularly to relax and "take your mind off things". Many people find that it helps to book yoga or relaxation classes in advance, because it is more difficult to put off the session and they feel committed to attend. Sports and physical exercise can also divert the mind from the everyday problems of school and work.

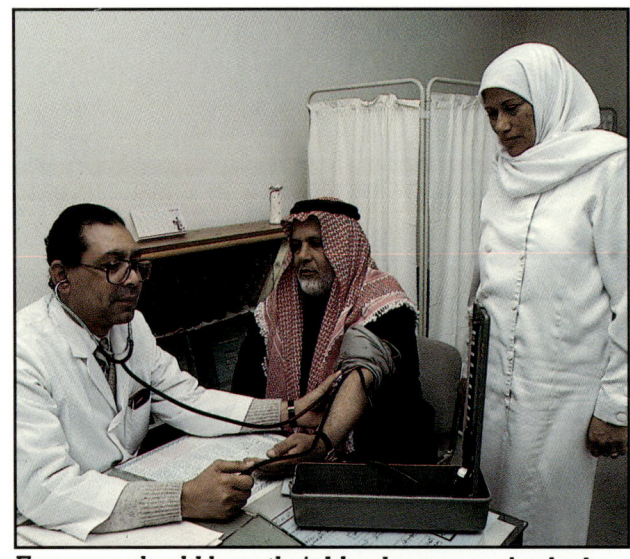

Everyone should have their blood pressure checked.

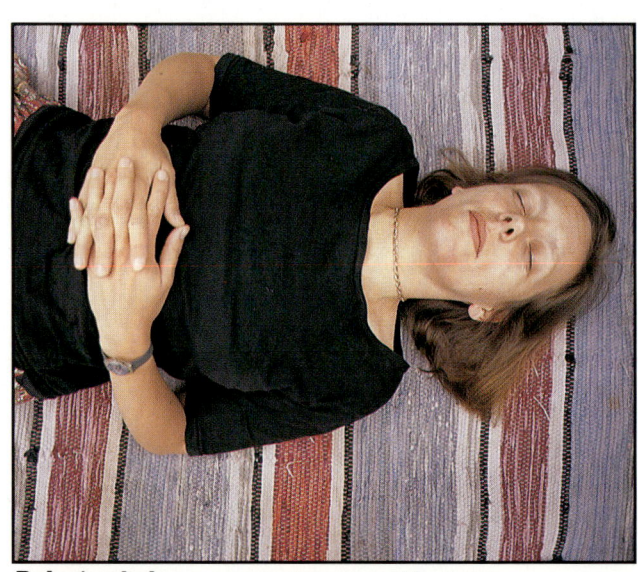

Relaxing helps a person to cope with stress.

LEADING A NORMAL LIFE

After a heart attack or some other serious heart condition, it is vital to try and return to normal life as quickly as possible – although following the doctors' advice. One risk is that family and friends treat you as a "cardiac cripple", not allowing you to take any exercise, and generally overprotecting you. Although well-meaning, this can delay your recovery and make you feel more ill than you really are! There are certain regulations to consider. For example, the car licence authorities and insurance companies must be told if you have had a heart attack.

Otherwise, most heart patients can go back to work after about three months. About four out of five heart attack patients resume their previous jobs. If the previous work was very strenuous, it might be worth changing jobs.

Pastimes and leisure activities can also be taken up again, although very risky or tiring activities might be best avoided. Some patients even say their health seems to improve after they have recovered from a heart attack. With new interests in health and exercise, they feel much fitter and more active than they did in the years that led to the problem.

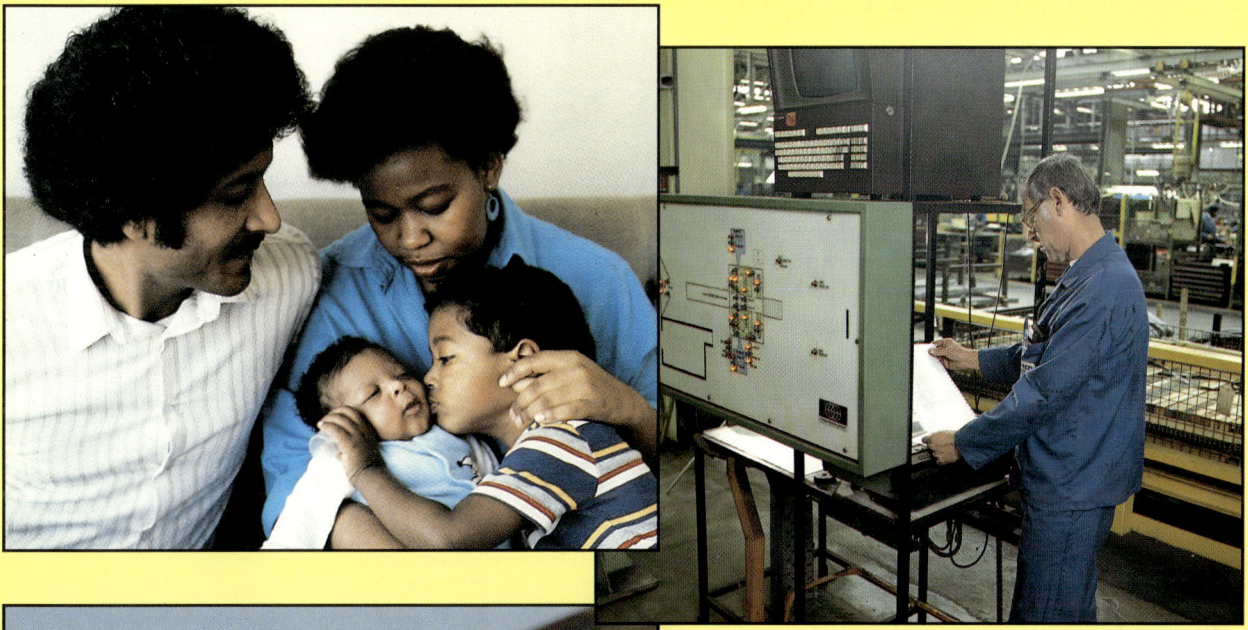

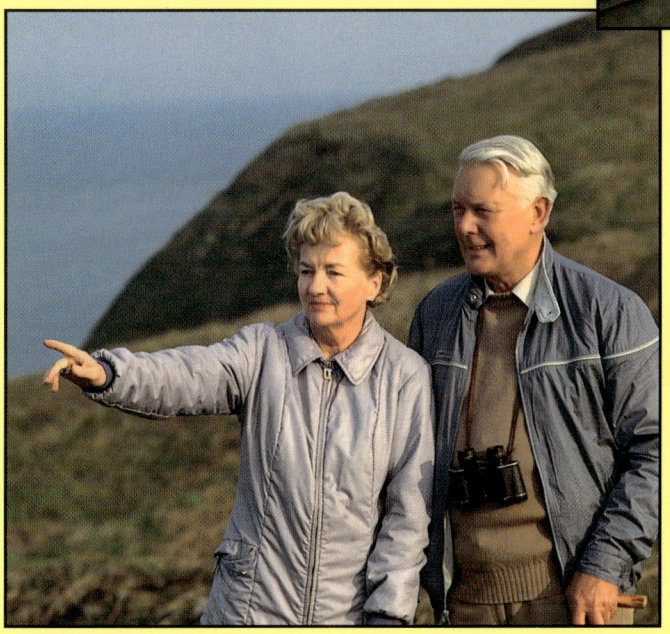

The state of mind affects the speed of recovery of people with a heart condition or who have had heart attacks. It is important to think about your lifestyle – ways of coping with stress, positive relaxation programmes, managing anger and having a healthy eating plan. There is much truth in the saying "you are what you eat" particularly as far as the heart is concerned. Taking regular exercise is also another factor towards staying healthy. Many find a new spiritual meaning to life – beyond that of mundane day-to-day living and materialistic goals which helps in the recovery process.

TAKING CARE OF YOUR HEART

Research has shown that furring-up of the arteries and heart disease can start during the teenage years or even earlier. So it is never too early to learn how to take care of your heart – and the rest of your body. Experts have drawn up guidelines aimed at keeping the heart and blood vessels healthy into old age. These include not smoking, keeping blood pressure normal, staying slim and avoiding obesity, avoiding too much fatty food, taking regular exercise, and (for women) keeping a check on the oral contraceptive pill. About 1 person in 500 is prone to coronary disease and heart attacks because he or she has a very high level of certain fats (lipids) in the blood. This problem tends to run in families and is termed FH (familial hypercholesterolaemia). It can be detected by blood tests, and treated by good health habits and perhaps lipid-lowering drugs. If you do introduce changes in your diet, do so gradually so your body can adjust.

Do not smoke

Smoking causes many serious diseases (see page 10). The best way to stop is never start. For those who want to stop, there are many methods. Stop in a group, with friends. Save the money you would have spent on cigarettes and buy a treat with it. Various "giving-up aids" include nicotine gum, dummy cigarettes and treatments such as acupuncture and hypnosis. The risk of having a heart attack rises with the number of cigarettes smoked but, in general, people who smoke are twice as likely to die from a heart attack as those who do not smoke. Smoking does put your health at risk.

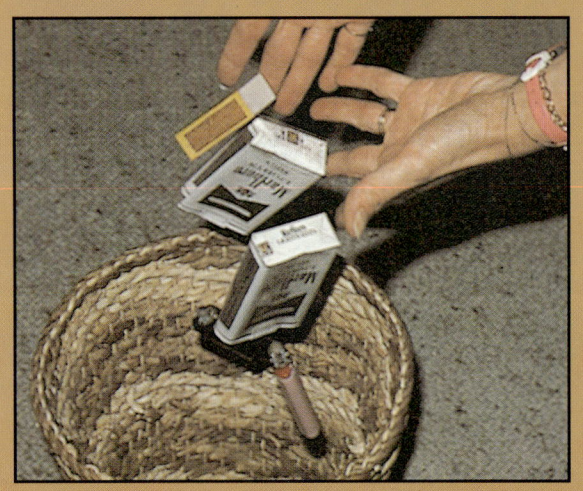

Drink in moderation

Too much alcohol can damage various parts of the body, including the heart, brain, liver and stomach. It can also make accidents more likely, especially when driving. Men are advised to drink no more than 21 units of alcohol each week, and women 14 units. (One unit is about half a pint of beer, one glass of wine or one measure of spirits.) Moderate drinking, even after a heart attack, is acceptable. If a person is used to drinking, a glass of wine with a meal, for example, may help convalescence by boosting morale. Giving up alcohol completely could cause additional stress.

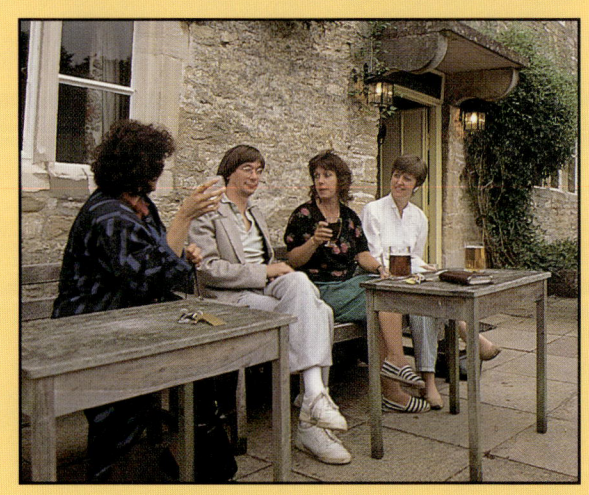

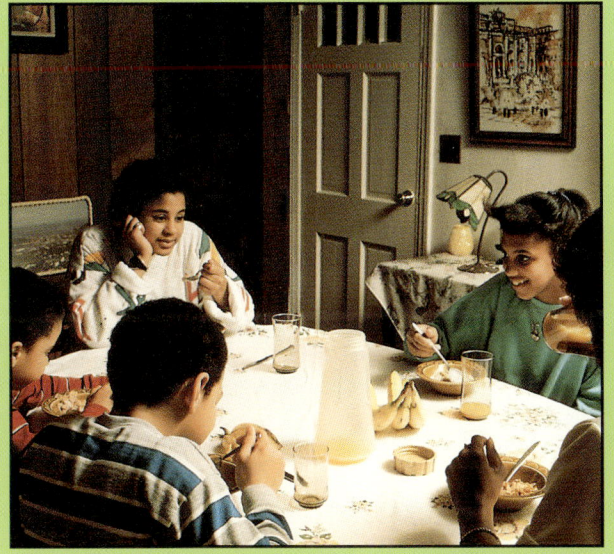

Eating the right food

We know that one of the main causes of illness is the way we live. Eating too much and choosing the wrong sort of food makes us less healthy. Help your heart by staying slim. Being overweight stresses the muscles and joints, and strains the heart by making it work excessively hard to pump blood around the extra fatty tissues. Check on a weight-for-height chart (ask your doctor) to see if you are in the "healthy" band. If not, go on a reducing diet. In any case, eat less fatty foods.

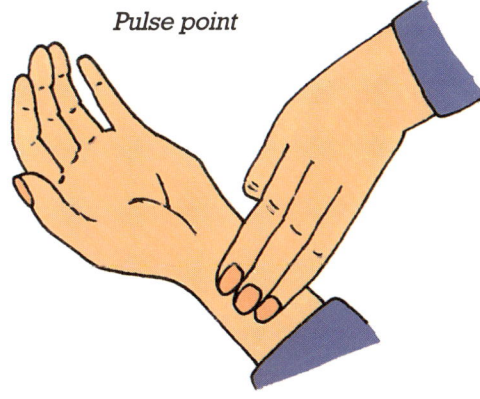

Pulse point

Pulse rate test

Feel for the pulse with your fingers on the inside of the wrist, below where it joins the thumb, as shown. The regular "pulse" is a pressure wave of blood spurting though the artery in the wrist, pushed by the heart. So pulse rate (number of wrist pulses in one minute) equals heart rate (number of beats in one minute). The average pulse rate at rest is around 70 in an adult. In very fit people, it may be as low as 50. During great excitement or hard exercise it can rise to 150 or more. A fit heart copes better with this strain, and so the pulse rate is unlikely to go so high. And the fit heart recovers more quickly, regaining its pulse rate after a few minutes. You can keep a track of an exercise programme by seeing how long your pulse takes to fall to 80 beats per minute after exercise. This time should become shorter as your heart becomes fitter.

Taking exercise

Physical activity is good for the body, to a degree. People with active jobs get most of the exercise they need at work. If not, doctors recommend at least three exercise sessions each week, of 20 to 30 minutes each session, and strenuous enough to make you pant and break into a sweat. If you have had hardly any activity for a long time, check with your doctor first, and then begin slowly and build up. It is best to try and build an exercise plan into the weekly routine, and take it in a form you enjoy, and that you can can afford if you have to pay. In this way you will be more likely to keep it up in the long term. Swimming, cycling and tennis are good all-round forms of exercise.

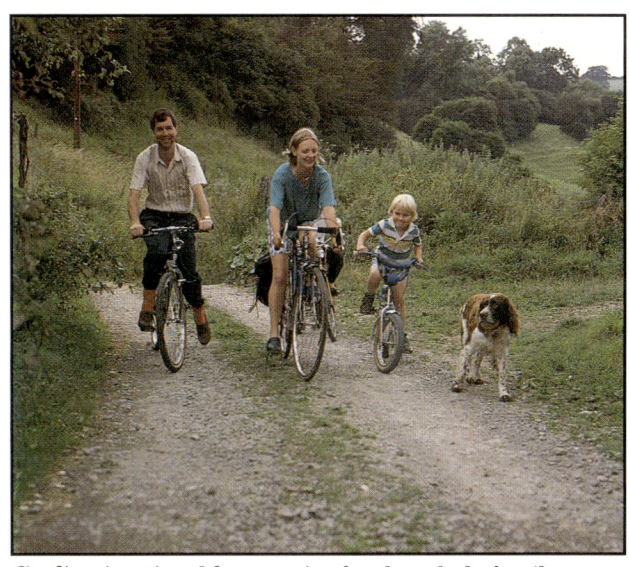

Cycling is enjoyable exercise for the whole family

First aid

Someone who is having a heart attack usually turns pale or grey, goes sweaty and breathless, and has pain in the chest, possibly spreading into the neck, shoulders and arms (particularly on the left side). If this happens, get medical help on the way at once. The quickest way is to dial 999 and ask for an ambulance. Say that you think it may be a heart attack. Loosen the clothing around the victim's neck and chest, and make the person warm and comfortable. If he or she collapses, place in the recovery position (as shown, right) to prevent choking.

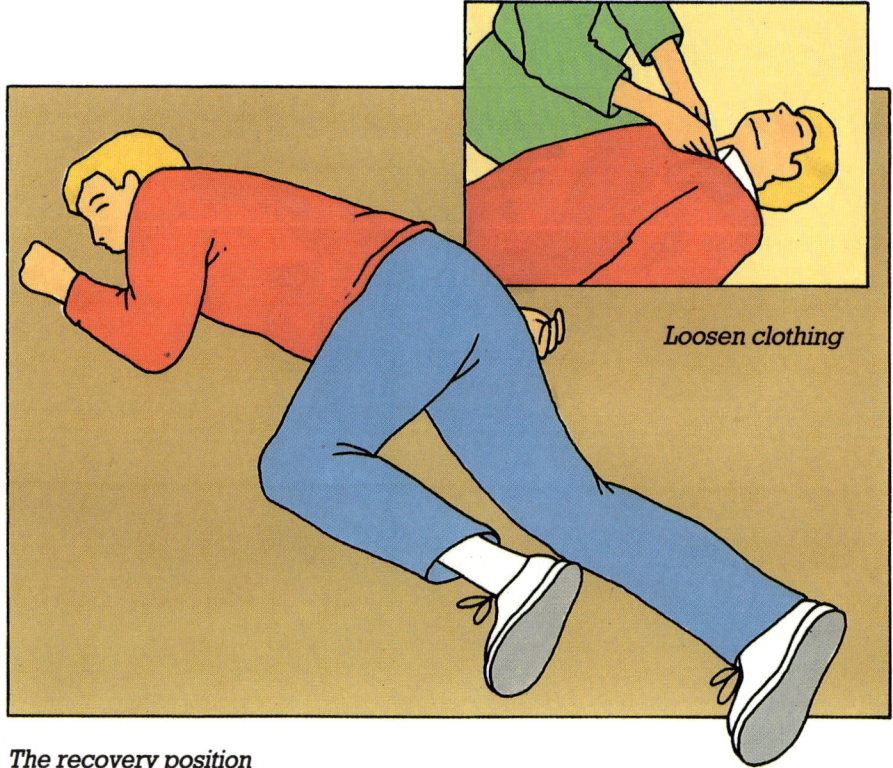

Loosen clothing

The recovery position for an unconscious victim

Donor Cards

Some body organs can give life to other people. If after your death you would like to donate an organ of your body to be used for transplantation and treating another person, you can carry a donor card with you. This gives your signed authorization to use an organ from your body – such as your heart – to save someone else, without delay.

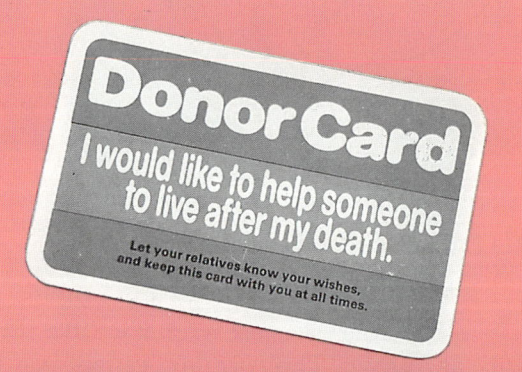

Useful information

British Heart Foundation
102 Gloucester Place
London W1H 4DH.
Tel: 01-935 0185.
Funds medical research into heart disease, campaigns for health education and publishes leaflets and booklets on circulatory disorders and all aspects of heart disease.

British Red Cross Society
9 Grosvenor Crescent
London SW1X 7EJ
Tel: 01-235 5454
Voluntary organization which runs training courses in first aid and what to do if someone has a heart attack. A first aid course is always a good asset to have in case of emergencies.

Action on Smoking and Health (ASH)
5-11 Mortimer Street
London W1N 7RH.
Tel: 01-637 9843.
Publicizes information and produces leaflets and statistics on the dangers of smoking and gives advice on how to give up.

Glossary

Aneurysm A localized dilation or bulging in an artery. Aneurysms can occur anywhere but they are most common and most troublesome when they are in a cerebral artery or in the aorta.

Angina Means a pain in the centre of the chest. It occurs when the heart is not receiving enough oxygen to meet demand.

Angioplasty A procedure, known in full as transluminal balloon coronary angioplasty, that is sometimes used in hospitals. It entails "squashing" the fatty deposits that block arteries, with small balloons.

Anticoagulants Drugs that thin the blood to prevent blood clots forming. The most commonly prescribed anticoagulant is called coumarin.

Aorta The largest artery in the body, measuring about 2.5 centimetres in diameter. It leads directly from the left ventricle of the heart and branches to form the rest of the system of arteries.

Beta blockers Drugs that block the effects of stress hormones (like adrenaline, for example) in the heart and blood vessels.

Blood pressure The pressure of the blood in the body's major arteries, measured in millimetres of mercury. Two measurements are taken: one of the highest pressure reached when the ventricles of the heart contract and one of the lowest pressure between contractions.

Bradycardia A slower than normal heart rate.

Congenital heart disease Heart malformations that are present from birth.

Coronary angiography A kind of X-ray which shows where a coronary artery is narrowed or blocked, and reveals more information about the general condition of the heart muscle and how well the heart is functioning.

Diuretic Drugs that lower the blood pressure by reducing the volume of fluid in the circulation.

Electrocardiogram The tracing of the electrical activity of the heart, as it beats and relaxes. It is usually monitored in a hospital.

Embolus A part of a blood clot that breaks off and is swept away by the blood stream to another part of the body.

Fibrillation When the heart is not beating but only quivering. Immediate electric stimulation is given to hopefully, jolt the heart back into beating normally again.

Hole in the heart Also known as ventricular septal defect (VSD). A hole in the wall between the two lower chambers of the heart.

Myocardial infarction Literally means destruction of the heart. It is also known as a heart attack, coronary thrombosis or cardiac infarction.

Pacemaker A group of specialized cells, situated in the wall of the right atrium of the heart, that transmit an electrical impulse to make the heart beat.

Stroke An interruption in the supply of blood to part of the brain. It can lead to impaired function in the areas of the body controlled by that part of the brain.

Tachycardia A faster than normal heart rate.

Thrombosis When a blood clot blocks an artery and disturbs the normal flow of blood.

Vasodilators Drugs that dilate the blood vessels and so reduce the resistance to the flow of blood in them.

INDEX

Aneurysm 13
Angina 10, 11, 13, 19, 26
Anti-arrhythmics 24
Anticoagulants 18, 21, 24
Aorta 4, 5, 8, 13
Aortic valve 4
Arterial duct 8
Arteries 4, 5, 6, 7,
    furring-up of 11
    hardening of 10, 11
Arteriosclerosis 11
Artificial heart 23
Aspirin 18
Atheroma 11
Atherosclerosis 11, 13
Atria 4, 7, 9, 15
Atrial septal defect (ASD) 9

Beta blockers 24
Blood clot 12, 18
Blood pressure 26
Blue baby 9, 17
Bradycardia 15

Cardiac arrest 15, 18
Cardiac Care Unit (CCU) 18
Cardiac defibrillator 18
Cardiac muscle 6
Cardiomyopathy 14
Capillaries 5
Cerebral haemorrhage 12
Cerebral thrombosis 12
Chest pain 10
Computerized axial tomography (CAT) 17
Congenital heart disorders 9
Congenital pulmonary stenosis (CPS) 9
Congestive heart failure 15
Coronary angioplasty 19
Coronary artery 11
Coronary artery bypass 19

Coronary artery disease (CAD) 8, 10

Diet 26, 28, 29
Digoxin 24
Diuretic 24

Ectopic beat 15
Electrocardiogram (ECG) 7, 15, 16
Embolus 12, 13
Exercise 25, 28, 29

Familial hypercholesterolaemia (FH) 28
Fibrillation 15
First aid 30

Heart
    artificial valves 21
Heart attack 11, 28
    recovery 27
Heartbeat 7
Heart failure 15
Heart-lung machine 17, 22
Heart murmur 13
Heart muscle 6
Heart transplants 22
High blood pressure 13
Hole in the heart 9, 17
Hypertrophy 14

Infarction 14

Lungs 5

Mitral valve 4, 7, 13
Mumps 14
Myocardial infarction 14
Myocarditis 14

Open heart surgery 19

Oval window 8

Pacemaker 4, 15, 20
Patent ductus arteriosus (PDA) 9
Pulmonary artery 4, 5, 8
Pulmonary embolism 12
Pulmonary valve 4
Pulse 29

Relaxation 26
Rheumatic fever 13
Rhythm disorders 14

Septum 4
Smoking 10, 15, 28
Stenosis 13
Stress 15
Stroke 12

Tachycardia 15
Telemetry 25
Thrombosis 11, 12, 18
Tissue rejection 23
Tricuspid valve 4, 7

Valves 6
    problems 13
Vasodilators 24
Veins 4, 6, 7
Vena cava 4
Ventricles 4, 7, 9
Ventricular septal defect (VSD) 9

X-ray 16, 17

Yoga 26

Photographic Credits:
Cover: Tom Stoddart; pages 5, 6, 8 both, 9, 10, 12t, 13 both, 14 all, 15t and 2 middle, 16 both, 17 all, 18t, 19t, 21 both, 23 both, 24 both, 25 both and 26t: Science Photo Library; page 12b: National Medical Slide Bank; pages 15b, 18b and 20l: Biophoto Associates; page 12b: F. Killerby; pages 20r and 27tr: J. Allan Cash Library: page 26bl: Robert Harding; pages 26br, 28r and 29b: Roger Vlitos; page 26tl: Anthea Sieveking/Network; page 27b: Image/Biophoto Associates; page 28l: Vanessa Bailey; page 29t: Zefa.